Soils in Waste Treatment and Utilization

Volume I
Land Treatment

Authors

Wallace H. Fuller
Arthur W. Warrick

Department of Soils, Water and Engineering
University of Arizona
Tucson, Arizona

CRC Press, Inc.
Boca Raton, Florida

Library of Congress Cataloging in Publication Data

Fuller, Wallace Hamilton.
  Soils in waste treatment and utilization.

  Bibliography: p.
  Includes indexes.
  Contents: v. 1. Land treatment -- v. 2. Pollutant
containment, monitoring, and closure.
  1. Sewage disposal in the ground. 2. Land treatment
of wastewater. 3. Soil pollution. 4. Soils.
I. Warrick, Arthur W. II. Title.
TD760.F85   1985   628.4'456   84-7047
ISBN 0-8493-5151-0 (v. 1)
ISBN 0-8493-5152-9 (v. 2)

    Direct all inquiries to CRC Press, Inc., 2000 Corporate Blvd., N.W., Boca Raton, Florida, 33431.

© 1985 by CRC Press, Inc.
International Standard Book Number 0-8493-5151-0 (Volume I)
International Standard Book Number 0-8493-5152-9 (Volume II)

Library of Congress Card Number 84-7047
Printed in the United States

PREFACE

Soils reflect their environment and, therefore, differ just as plants and animals do. They have easily recognizable characteristics which are used for classification. Their individual properties are acquired from the many forces acting upon them. These forces may be grouped into soil forming factors of, climate, biological systems, topography, parent material (mineral matter), and time. Thus, soils from one distinct climatic and geologic region differ from those of other regions. This book relates the different soil properties to their capacity to control movement of waste pollutants through them such as might be envisioned in a huge chromatographic column. It is these differences that makes it so essential to understand the reactions and interactions between soils and wastes as an essential function in soil management for waste control.

In these books, we bring together and correlate recent information known to have a prominent influence on the rate of movement of pollutants from wastes and their leachates in the soil. The emphasis is on those characteristics most amenable to modification and their management, such that secure and rational choices of disposal can be made. Identification of limits of the state of the art are carefully defined.

This book presents five distinct but related subject matter sections. The first topic relates to soil as a natural system to prepare the reader with a basic knowledge of soil properties as they must become involved in waste management at the disposal facility. The soil, its depth-profile, and certain characteristics are identified and described.

Historically, soils have demonstrated their function as waste utilization (resource) systems. In fact, soil organic matter originates from natural plant and animal residues that annually reach the soil surface and become incorporated. The farmer and home gardener have long extended waste utilization through application to the land of animal manure, green manure, composts, and other residues by use of simple as well as sophisticated techniques. Certain wastes, therefore, have been considered resources to be husbanded for improvement of soil productivity. In the second topic, we look at opportunities to manage industrial wastes as resources. Discussion of wastes as a resource is separated into two chapters based on use on cultivated and noncultivated land.

The third topic deals with those wastes that, at the present time, must be disposed of without immediate obvious benefit. The topics include such disposal options as land treatment, landfills, trenches, encapsulations, and other burials. Soil as lining material for landfills, impoundments, and ponds is included to describe the function of soil as an attenuation, absorption, and filter barrier for pollutants. Soil may be used as an effective medium for management of gases, odors, and aerosols as well as for solids and liquids.

Soils can retain pollutants either by ponding the liquid at the surface through infinitely slow permeability or by attenuation of pollutants leaching from hazardous waste while allowing acceptable leakage of potable quality water to reach groundwater. Furthermore, soils can biodegrade organic-containing wastes to harmless constituents. The fourth topic therefore, presents descriptions of research methods used for predicting both biodegradation and mobility of pollutants through soils as affected by specific soil and leachate properties. Progress on screening protocol for predicting waste treatability, models for predicting metal pollutant movement through soil, and effects of hydrogen ion concentration of acid wastes on pollutant retention and soil failure are discussed. Again, unfavorable as well as favorable situations are identified.

The final subject relates to site selection for waste disposal, monitoring, site-facility closure and continued surveillance. Predisposal, disposal, and postdisposal monitoring are considered.

Because soil management for pollution waste control is a new and unusually broad subject to encompass under the cover of one book, the level of presentation and consequent comprehension is not established at a constant level as we would like. Available literature and data are scattered, fragmentary, and often conflicting. Consequently, for some topics the emphasis is as a review, in others more as a statement of fact, and in still others almost as a journal paper or research report. Despite these circumstances, we hope to reach the main audience of concerned people who are responsible for planning, designing, managing and operating wastes and wastewater disposal facilities. In addition, educators and researchers probably can benefit.

There is no doubt that newer and more accurate information is developing in great abundance each year which will soon antiquate this book and call for a revision. However, this work is presented as a framework for a realistic base of departure on which to hang new guidelines and new ventures in a rapidly advancing field of soils as waste treatment/resource utilization systems.

The Authors

THE AUTHORS

Wallace H. Fuller, Ph.D., is a professor and biochemist in the Department of Soils, Water and Engineering at The University of Arizona, Tucson, Arizona, where he teaches and is leader of the waste research program. He received the B.S. and M.S. degrees in Soil Science, Washington State University and Ph.D. in Soil Bacteriology and Biochemistry, Iowa State University. His early experience as soil surveyor in Wisconsin, Missouri, and Iowa and Research Associate at Washington State University as agrochemist, and Iowa State University as head of War Hemp Industry investigations provides a practical background for the long research tenure in land treatment of unwanted residues as resources, as well as waste requiring final and secure disposal. Early training (1945 to 1948) in atomic energy and radionuclide research in biophysical chemistry with the USDA, ARS, Agricultural Research Center, Beltsville, Maryland, equips him well for developing critical research and publications in waste control.

During Dr. Fuller's tenure as Head of Agricultural Chemistry and Soil Department at The University of Arizona (1956 to 1972), he continued research activities and published over 175 scientific journal papers, authored two books on arid land soils (*Soils of the Desert Southwest* and *Management of Southwest Desert Soils*), and wrote many chapters in books on waste management. Another book, *Scientific Management of Hazardous Wastes,* from Cambridge University Press (1983) is co-authored by Dr. Fuller with two British authors. He shared honors as Special Associate with Water Resource Associates, Inc., whose Green-belt Flood Control Project was selected by the National Society of Professional Engineers as one of the top ten outstanding engineering achievements of 1974.

Dr. Fuller's experience in waste management includes 14 years on the Arizona State Water Quality Control Council; 16 years as State Chairman of the Advisory Committee to the Arizona State Chemist Office on agricultural chemicals, feeds, pesticides, and minerals; Consultant and Advisor to U.S. Department of Public Health (HEW) and U.S. EPA National Project Review Council (1960 to 1972); and Consultant to scientific enterprises in South America, Panama, and Mexico governments. He has been awarded many research grants, contracts, and cooperative projects from industry, the Atomic Energy Commission, the U.S. Department of Agriculture, U.S. Environmental Protection Agency, and private enterprises as principal investigator. He also is a noted national and international consultant on water quality, hazardous waste management, environmental quality control, and arid-land soils.

A. W. Warrick is Professor of Soil Physics at The University of Arizona in Tucson. He received a B.S. degree in Mathematics and a M.S. and Ph.D. (1967) in Soil Physics at Iowa State University. Research and teaching topics include flow through porous media alonng with associated transfer of solutes and materials. Emphasis is on theoretical description of these processes and modeling. Sampling and the quantification of spatial variability through geostatistics and other techniques are of particular current interest. He has been involved with several projects dealing with waste disposal; including septic tanks, radioactive wastes, solvents, and gaseous wastes.

Dr. Warrick has served as chairman of the S-1 Division of the Soil Science Society of America as well as an associate editor of the SSSA Journal. He has authored over 75 scientific journal papers or book chapters.

ACKNOWLEDGMENT

We are deeply indebted to many individuals and organizations for the use of material. We especially thank those who provided soils from throughout the United States to make much of the research discussed here possible:

> Drs. S. W. Buol, North Carolina State University; L. R. Follmer, University of Illinois; R. E. Green, University of Hawaii; D. M. Hendricks, University of Arizona; M. R. Roulier, U.S. E.P.A.; E. P. Whiteside, Michigan State University; and A. L. Zachary, Purdue University.

We are especially grateful to the U.S. Environmental Protection Agency and project officers: M. H. Roulier of MERL in Cincinnati, Ohio, who contributed time and encouragement for the heavy metal attenuation and model development and L. G. Swaby, ORD U.S. EPA, Washington, D.C., who as project officer, contributed to the successful completion of the research on hydrogen ion effects in soils presented in Chapter 6, Volume II. Special appreciation goes to the University of Arizona staff who were involved in the waste control research, as related to groundwater quality and food chain protection, included in this book, foremost of whom are: N. E. Korte; E. E. Niebla; B. A. Alesii; J. F. Artiola; J. Skopp; Ronda Bitterli; and P. J. Sheets. Aziz Amoozegar-Fard deserves special recognition and thanks for his long devotion to soil-waste interaction research and mathematical model computations.

Foremost in support of these volumes is the patience and understanding of Winifred Fuller who deserves more than can be put into words.

We appreciate the opportunity the University of Arizona College of Agriculture and Department of Soils, Water and Engineering provided for the completion of this research program and manuscript preparation. Significant encouragement was provided by our department head, W. R. Gardner.

Long hours of work in the preparation of the manuscript were generously supplied by Mary Schreiner in text review; by Susan Angelon, who devoted unlimited time and talent to the art preparation, graphs, and photographs; by Olivia Ayala, typist; by Sheri Musil with the final manuscript review; and by Lisa Littler, Coordinating Editor.

Wallace H. Fuller
Arthur W. Warrick

TABLE OF CONTENTS

Volume I

Land Treatment

SOIL AS A NATURAL SYSTEM

SOIL AS A WASTE UTILIZATION SYSTEM

Chapter 2
Waste Utilization on Cultivated Land

SOIL AS A WASTE TREATMENT SYSTEM

Chapter 4
Waste Treatment on Land Primarily for Disposal...............................195

Volume II

**Pollutant Containment, Monitoring,
and Closure**

SOILS INVOLVED WITH WASTE BURIALS

PREDICTING POLLUTION TRANSFORMATIONS AND MOBILITY

Chapter 5
Models for Predicting Pollutant Movement through Soil 123

A cloak of loose, soft material, held to the earth's hard surface by gravity, is all that lies between life and lifelessness. Crumbling rock, grit and grime, and decaying residue — abrading by wind and water — weather into soil — Mother Earth. This loose hide lives — yields, yet does not yield to the forces of climate, having formed through the ages from meteorological, geological, and biological action on rock. The soil not only was born out of fire, flood, and ice but it lives and continues to renew life. At first, animal and plant residues decayed into simpler constituents, renewing nutrient elements available for new life in a perpetual cycle, but now, wastes accumulating from the population bulge of human beings add serious proportions to the burden of cycling and recycling.

Wallace H. Fuller
Soils of the Desert Southwest
University of Arizona Press, 1975

Soil as a Natural System

Chapter 1

THE SOIL, PROFILE AND CHARACTERISTICS

I. THE SOIL

A. Description

Most people can best recall what a soil profile looks like from road cuts along highways. The fresh cuts slice through soft mantles of earth of varying depths and downward into stony parent materials, or, if the cut is deep or the soil shallow, into bedrock that required blasting to reach roadgrade. On close observation, certain physical features stand out. The soft topsoil layers vary in depth, color, clay, silt, sand, gravel, and stones, stratifications and structure, moisture, and vegetation. Soils also feel different underfoot: sandy or sticky, hard or soft, stony or smooth. Garden soils look rich and mellow and may be spaded with ease in contrast with shallow soils on steep hillslopes of wilderness areas. Often soil characteristics differ within small distances or small areas such as an acre. These differences extend not only to physical properties but to differences in soil fertility and plant (crop) productivity. Such differences reflect the numerous combinations of soil-forming factors (climate, vegetation, parent material, topography, and time) imposed on the geological material. These differences also determine the effectiveness of the soil as a waste treatment and waste utilization system. Almost all waste ends up on the land. Air and streams are only transport systems back to the land, and ocean dumping is now restricted (Figure 1). Therefore, waste disposal programs must be based on a thorough knowledge of the soil characteristics and soil management options as specifically related to all phases of operations from site selection to final closure. The management of the soil by man brings into focus a new dimension of soil reactions and can alter the characteristics of specific soil properties such as texture, kind and content of minerals, structure, acidity, alkalinity, aeration, and permeability. The purpose of this book is first to bring together and correlate the best available information and most recent knowledge of soil characteristics that appear to be prominent in influencing the rate of movement of pollutants from wastes that reach the land and their leachates and second to emphasize those characteristics most amenable to modification and their management, such that secure and rational choices of disposal can be made.

B. Three-Phase Soil System

The nature of the component parts of soil determines its capacity to stabilize the movement of pollutants. The broadest grouping of these components is a three-phase system of solid, liquid, and gas.

The solid phase occupies roughly 50% of most soils, and pore space the other half. Also, roughly half of the pore space in a good top soil will contain water and the other half gas (air). A saturated soil, therefore, would have all the pore spaces filled with water. The proportion of water to gaseous phase, however, is every changing. The three phases, solid, liquid, and gas, share common boundaries. The soild phase boundaries are fixed compared with the mobile liquid and gaseous phases. The three phases meet and interact at the particle surfaces and interfaces between the liquid and gaseous phases. Thus, the extent of the surface area per unit volume of soil must be given serious consideration as an important factor in waste disposal operations as it is reflected in soil texture. The three phases are recognized separately for convenience of discussion, whereas in nature they are not divisible but are in intimate association and highly sensitive to changes in one another. For example, ions and molecules interchange readily among themselves and reactions between them seldom go to completion but sustain various degrees of moderate stability on a geochemical time scale.

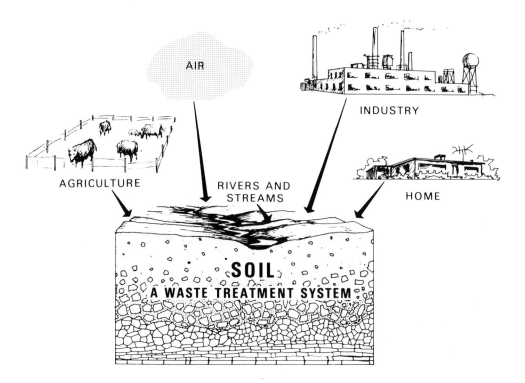

FIGURE 1.   The soil, a waste treatment/utilization (resource) system.

This allows for some reactions to go readily to completion, but not for others. Thus, the characteristics of the three phases must be understood both collectively and individually in any pollutant abatement program.

1. The Solid Phase

Soils form from the constant destruction of rocks as a result of weathering. The crumbling mountain rock that yields the perpetual slide to the valley, the erasing of the gravestone chisel markings, and eroding away of ancient stone structures all testify to rock destruction by weathering. The resulting fine particles, or unconsolidated rock material, form into soils. Although the parent rock (e.g., granite vs. limestone) influences the characteristics of the soil, other factors in soil formation make it possible for different kinds of soil to develop on the same rock source. The Russian scientist, V. V. Dokuchaiev (1846 to 1903) was the first to document this phenomenon and C. F. Marbut (1863 to 1935), the first Chief of the USDA Soil Survey, followed the lead by demonstrating the dominating effects of climate in soil formation. More recently, Hans Jenny[1] showed that multiple factors act in soil formation, such as parent material, climate, topography, biotic (organisms), age or time, and additional factors.

a. Soil Organic Matter

Natural soils contain organic and inorganic materials in the solid phase. The organic matter concentrates on the soil surface and in the upper or A layer as a result of decaying of plant and animal life. The concentration decreases abruptly with depth such that the B layer below is almost wholly mineral matter in most soils.

b. Total Element Analyses

The total elemental composition of soils changes from one site to another depending largely on the composition of the parent material. The extent of this relationship is evident from data in Table 1   which presents the content of various elements in the lithosphere (solid crust of the earth to a depth of 10 miles) in contrast to those in soils. Not all of these elements are in soluble forms, therefore, the soil water solution may contain them in such small amounts they are detectable with difficulty by usual analytical equipment. On the other hand, strong acids, bases, and hightly concentrated reducing solutions found in many waste streams can make these elements soluble and, therefore, sufficiently mobile to be a threat to the pollution of underground water sources.[2] The data in the last column of Table 1 provide a means of estimating maximum concentrations of each element that may appear in a soluble state in the natural soil solution based on a soil water content of 10% of the dry weight. The data can be adjusted for other soils of different moisture contents and elemental compositions. These data are also useful to predict maximum solubilities of an element in the soil and to predict minerals that cannot persist in soils on a longtime basis. In this way they become a valuable tool for planning long-term waste disposal on land.

c. The Lithosphere

The composition of the lithosphere also provides some insight as to soil composition (Figure   2). Oxygen and silicon represent about 75% of the earth's crust. Aluminum content is given by Clark[3] to be 7.85% and iron 4.5%. It is no accident then that Al, Si, and O_2 unite during weathering processes to form secondary minerals of aluminosilicates with some Fe substitutions. These secondary clay minerals along with primary silicate minerals represent the major part of the mineral content of soils (Figure 3).

d. Primary Silicate Minerals

Six prominent groups of primary silicate minerals, as compiled by Jenny[1] provide additional information on the solid phase of soils. Rock minerals have a skeleton of O^{2-} and OH^- anions arranged as tetrahedra and octahedra with small, multivalent cations in their interstices:

Olivine (olive green) is composed of isolated $(SiO_4)^{4-}$ tetrahedra bonded mainly to Mg^{2+} and Fe^{2+} and formulated chemically either as Mg_2SiO_4 or Fe_2SiO_4 or mixed $(Mg,Fe)_2SiO_4$.

Augite (black, prismatic) contains individual chains of linked Si-O tetrahedra having the units $(SiO_3)^{2-}$. The strands are held together by Ca^{2+}, Mg^{2+}, and Fe^{2+} (formula: $Ca(Mg,Fe)Si_2O_6$).

Hornblende (black, prismatic-fibrous) is made up of double chains of Si-O tetrahedra with units of $(Si_4O_{11})^{6-}$, joined to Ca^{2+}, Mg^{2+}, Fe^{2+}, OH^-, with the formula $Ca_2(Mg,Fe)_s(OH)_2Si_8O_{22}$.

These three "ferromagnesian" minerals serve as hosts for minor elements, and they are often associated with the mineral apatite which contains phosphorus.

Quartz (colorless, with greasy luster) represents a three-dimensional framework of SiO_4 tetrahedra, all corners being shared except, of course, those protruding at the outer surfaces. The formula is SiO_2 (silicon dioxide) and the mineral is highly resistant to weathering. Randomized SiO_2 frameworks are attributes of the noncrystalline opals, agates, flints, cherts, and quartzites.

Feldspars (whitish) are the most common of all rock minerals. Like quartz they have a framework of linked tetrahedra, but in one fourth of them Al^{3+} substitutes for Si^{4+}. To overcome the resulting deficiency of positive charges additional cations, mostly K^+, Na^+, and Ca^{2+}, are incorporated in the open spaces. Common feldspars are orthoclase ($K Al Si_3 O_8$), albite ($Na Al Si_3 O_8$), and anorthite ($Ca Al_2 Si_2 P_8$). The last two may appear as mixed crystals called plagioclase. The formulas are idealized because the crystals are never pure. Orthoclase is the most resistant to weathering.

Table 1
THE CONTENT OF VARIOUS ELEMENTS IN THE LITHOSPHERE
AND IN SOILS

Element	Atomic weight (g)	Content in lithosphere (ppm)	Common range for soils (ppm)	ppm	Selected average for soils molar conc. at 10% moisture log M
Ag	107.87	0.07	0.01—5	0.05	−5.33
Al	26.98	81,000	10,000—300,000	71,000	1.42
As	74.92	5	1—50	5	−3.18
B	10.81	10	2—100	10	−2.03
Ba	137.34	430	100—3,000	430	−1.50
Be	9.01	2.8	0.1—40	6	−2.18
Br	79.91	2.5	1—10	5	−3.20
C	12.01	950		20,000	1.22
Ca	40.08	36,000	7,000—500,000	13,700	0.53
Cd	112.40	0.2	0.01—0.70	0.06	−5.27
Cl	35.45	500	20—900	100	−1.55
Co	58.93	40	1—40	8	−2.87
Cr	52.00	200	1—1,000	100	−1.72
Cs	132.91	3.2	0.3—25	6	−3.35
Cu	63.54	70	2—100	30	−2.33
F	19.00	625	10—4,000	200	−0.98
Fe	55.85	51,000	7,000—550,000	38,000	0.83
Ga	69.72	15	5—70	14	−2.70
Ge	72.59	7	1—50	1	−3.86
Hg	200.59	0.1	0.01—0.3	0.03	−5.83
I	126.90	0.3	0.1—40	5	−3.40
K	39.10	26,000	400—30,000	8,300	0.33
La	138.91	18	1— 5,000	30	−2.67
Li	6.94	65	5—200	20	−1.54
Mg	24.31	21,000	600—6,000	5,000	0.31
Mn	54.94	900	20—3,000	600	−0.96
Mo	95.94	2.3	0.2—5	2	−3.68
N	14.01	—	200—4,000	1,400	0.00
Na	22.99	28,000	750—7,500	6,300	0.44
Ni	58.71	100	5—500	40	−2.17
O	16.00	465,000		490,000	2.49
P	30.97	1,200	200—5,000	600	−0.71
Pb	207.19	16	2—200	10	−3.32
Rb	85.47	280	50—500	10	−2.93
S	32.06	600	30—10,000	700	−0.66
Sc	44.96	5	5—50	7	−2.81
Se	78.96	0.09	0.1—2	0.3	−4.42
Si	28.09	276,000	230,000—350,000	320,000	−2.06
Sn	118.69	40	2—200	10	−3.07
Sr	87.62	150	50—1,000	200	−1.64
Ti	47.90	6,000	1,000—10,000	4,000	−0.08
V	50.94	150	20—500	100	−1.71
Y	88.91	—	25—250	50	−2.25
Zn	65.37	80	10—300	50	−2.12
Zr	91.22	220	60—2,000	300	−1.48

Lindsay, W. H., *Chemical Equilibrium in Solids,* John Wiley & Sons, New York, 1979, 24, 449. With permission.

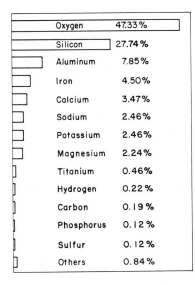

Oxygen	47.33%	
Silicon	27.74%	
Aluminum	7.85%	
Iron	4.50%	
Calcium	3.47%	
Sodium	2.46%	
Potassium	2.46%	
Magnesium	2.24%	
Titanium	0.46%	
Hydrogen	0.22%	
Carbon	0.19%	
Phosphorus	0.12%	
Sulfur	0.12%	
Others	0.84%	

FIGURE 2.   Composition of the solid crust of the earth to a depth of 16 kilometers. (From Clark, F. W., The data of geochemistry, *U.S. Geol. Survey, Bull.*, 770, 1924, 15.)

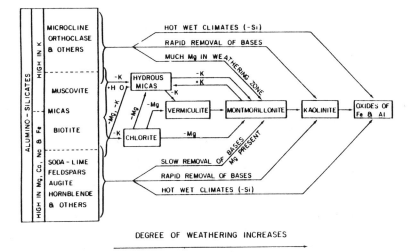

FIGURE 3.   General conditions for the formation of the various silicate clays and oxides of iron and aluminum. (Reprinted with permission of Macmillan Publishing Co., Inc., from Brady, N. C., *Nature and Properties of Soils*, 8th ed., chap. 1. Copyright© 1974 by Macmillan Publishing Co., Inc.)

Micas are flaky minerals with layer lattices that are akin to those of pyrophyllite, $Si_4P_{10}Al_2(OH)_2$, the symbol in italics pertaining to cations in the octahedral sheet. The many types of mica found in nature result from partial substitutions of ions of similar size and dissimilar charge, Si^{4+} and Al^{3+} in tetrahedra and Mg^{2+}, Fe^{2+}, Fe^{3+}, and Al^{3+} in octahedra.

When all tetrahedra are filled with Si^{4+} and all octahedra with Mg^{2+}, the mineral is talc, with the formula $Si_4O_{10}Mg_3(OH)_2$. In the common micas one fourth of tetrahedral Si^{4+} is replaced by the slightly larger Al^{3+} which distorts somewhat the ideal hexagonal pattern and leaves a '' + '' charge deficiency as in some of the clays. An electrical imbalance is alleviated by K^+ ions between the layers, one for each Al^{3+} and partially in hexagonal

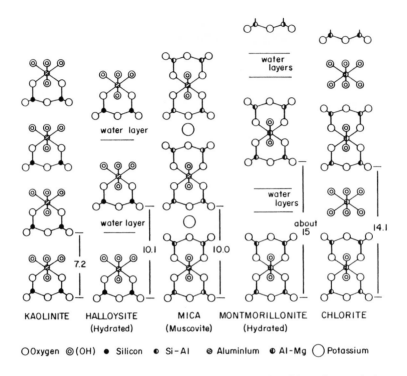

FIGURE 4.   Diagrammatic examples of idealized succession of layers in some lattice-type aluminosilicates. (From Brindley, G. W., *X-Ray Identification and Structures of Clay Minerals,* Mineralogical Society, London, 1961, 32.)

cavities of the tetrahedral sheets, which are slightly forced apart. The K^+ holds adjacent layers tightly together, as shown in Figure 4. Along a stack of layers (C-axis) the period of structure repetition of C-spacing is 10 Å. For muscovite mica the resulting formula is $K(AlSi_3O_{10})Al_2(OH)_2$. In biotite mica the octahedral layer corresponds to that of talc with Fe^{2+} substituting partially for Mg^{2+} , producing the formula $K(AlSi_3O_{10})(Mg,Fe)_3(OH)_2$.

The feldspar and micas are primary aluminosilicates, yet all six are primary silicate minerals. The idealized formulas given above omit the biologically important micronutrient accessories Zn, Mn, S, Co, F, and others.

Another important group of minerals from which soils are derived are the oxides (Table 2).   For example, when oxygen units with silicon, iron, aluminum, and manganese the metal oxides form as colloidal-size particles. In the presence of water they assume various hydrated forms. The hydrous oxides react with many pollutants and strongly affect attenuation of heavy metals in aqueous waste streams.

Despite the uniformity of the deeper earth's crust, the elements are not uniformly distributed over the surface. The variability in plant nutrients exemplifies these differences, for example, phosphorus is concentrated in deposits of mining quality at one location and has deficiencies for plant growth at another. Historic geological events resulted in many variations in the original earth, which yielded the igneous, sedimentary, and metamorphic rocks from which soils were formed; and these in turn also vary in structure, composition, and rate of disintegration. Since they contain minerals (Figure 3) that do not weather at the same rate, such soils can be expected to differ somewhat according to the proportionalities of their original mineral components (Table 3).

e. Aluminosilicates

The clay in soils contains mostly secondary minerals called aluminosilicates. Since aluminosilicates form from primary minerals they are more stable as the free energy of the

Table 2
SOME EXAMPLES OF HYDROUS OXIDE
CLAYS IN SOILS[a]

Aluminum Compounds (colorless, white)	Iron Compounds (red, yellowish, and brown)
Amorphous hydroxides $Al(OH)_3$ or $^{1/2}(Al_2O_3 \cdot 3H_2O)$	Amorphous hydroxides $Fe(OH)_3$ or $^{1/2}(Fe_2O_3 \cdot 3H_2O)$
Crystalline hydroxides Gibbsite and bayerite $Al(OH)_3$ or $^{1/2}(Al_2O_3 \cdot 3H_2O)$	Paracrystalline hydroxide Ferrihydrite
Crystalline oxyhydroxides Boehmite and diaspore $AlOOH$ or $^{1/2}(Al_2O_3 \cdot 1H_2O)$	Crystalline oxyhydroxides Goethite and lepidocrocite $FeOOH$ or $^{1/2}(Fe_2O_3 \cdot 1H_2O)$
Crystalline oxide Corundum Al_2O_3	Crystalline oxide Hematite and maghemite Fe_2O_3 Magnetite $FeO \cdot Fe_2O_3$ or Fe_3O_4

[a] Compiled from various sources (Dixon and Weed, 1977).[4]

Table 3
REPRESENTATIVE MINERALS FOUND IN SOILS

Oxides/Hydroxides	
Si—oxides	Quartz, tridymite
Fe—oxides/hydroxides	Goethite, hematite, limonite
Al—oxides/hydroxides	Gibbsite, boehmite, diaspore
Silicates	
Neosilicates	Olivine (Mg), garnet (Ca, Mg, Mn^{2+}, Ti, Cr), tourmaline (Na, Ca, Li, Mg, BO_3), Zircon (Zr)
Inosilicates	Augite (Ca, Mg), hornblende (Na, Ca, Mg, Ti), feldspars
Phyllosilicates	Talc (Mg), biotite (K, Mg, F), muscovite (K, F), clay minerals: illite (K), Kaolinite, montmorillonite, vermiculite (Mg)
Tectosilicates	Albite (Na), anorthite (Ca), orthoclase (K), zeolites (Ca, Na, K, Ba)
Carbonates	Calcite ($CaCO_3$), dolomite ($MgCa(CO_3)_2$)
Halides	Halite (NaCl), sylvite (KCl), carnallite ($KMgCl_3 \cdot 6H_2O$), ($CaCl_2nH_2O$)
Nitrates	Soda—nitre ($NaNO_3$), nitre ($KNO_3)_2$, Calcium nitrate ($Ca(NO_3)_2$)
Phosphates	Apatite ($Ca5(F, Cl, OH)(PO_4)_3$), vivianite ($Fe_3(PO_4)_2 \cdot 8H_2O$)
Sulfates	Gypsum ($CaSO_4 \cdot 2H_2O$)

products of a chemical reaction must always be less than the free energy of the reactants. Aluminosilicates in the soil vary dramatically in structure and composition because of elemental substitution in the structure (Table 4), and may occur either as unsubstituted or substituted forms.[4,5]

Three types of aluminosilicates, smectites (montmorillonite), hydromicas (illite), and kaolinite, dominate the soil 2 μm-clay (Figures 4 and 5). Because of their small particle size they are not found in significant amounts in the silt or sand fraction (Figure 6).

f. Representative Referee Soils

The final soils resulting from all of the factors in soil formation vary to the ultimate extent that no two are identical. The surveyor, therefore, delineates soils on a basis of a range of like characteristics. When the range of characteristics becomes so broad, the soil would

Table 4
SELECTED ALUMINOSILICATES FOUND IN SOIL MATERIAL

Class	Name	Formula
Al—Si	Sillimanite	Al_2SiO_5
	Kyanite	Al_2SiO_5
	Andalusite	Al_2SiO_5
	Halloysite	$Al_2Si_2O_5(OH)_4$
	Dickite	$Al_2Si_2O_5(OH)_4$
	Kaolinite	$Al_2Si_2O_5(OH))_4$
Na Al—Si	Nepheline	Na $AlSiO_4$
	Albite	Na $AlSi_3O_8$
	Beidellite	$Na_{0.33} Al_{2.33}Si_{3.67}O_{10}(OH)_2$
K Al—Si	Microcline	K $AlSi_3O_8$
	Muscovite	K $Al_2(AlSi_3O_{10})(OH)_2$
Ca Al—Si	Pyroxene	Ca $AlSiO_6$
	Anorthite	Ca $Al_2Si_2O_8$
	Leonhardite	$Ca_2Al_4Si_8O_{24} \cdot 7H_2O$
Mg Al—Si	Chlorite	$Mg_5Al_2S_3O_{10}(OH)_8$
	Fluorphlogopite	$KMg_3AlSi_3O_{10}F_2$
Substituted Al—Si	Vermiculite	$(Mg_{2.71} Fe(II)_{0.02}Fe(III)_{0.46}Ca_{0.06}K_{0.1})$ $Si_{2.91}{}^-Al_{1.14}O_{10}(OH)_2$ $O_{10}(OH)_2$
	Hydrous mica (montmorillonite)	$K_{0.6}Mg_{0.25}Al_{2.3}Si_{3.5}O_{10}(OH)_2$ $M^+{}_{0.}{}^+(Si_{3.81}Al_{1.71}Fe(III)_{0.22}Mg_{0.29})$
	Smacttite	

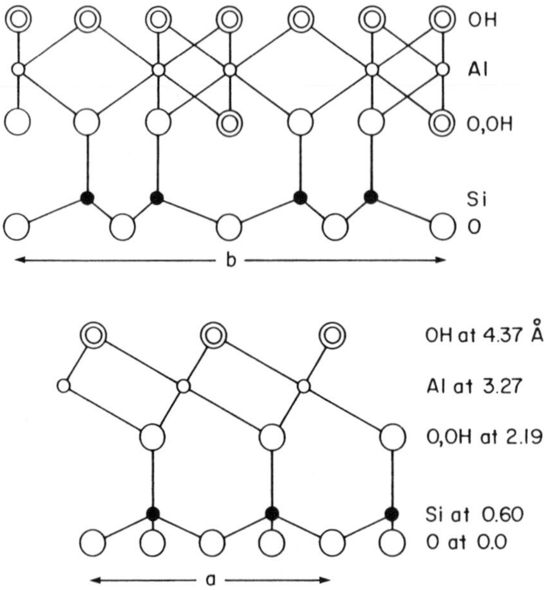

FIGURE 5.   Idealized examples of Kaolin as illustrated along a axis and along b axis. The measurement of layer separation above oxygen is in Å units. (From Brindley, G. W., *X-Ray Identification and Structures of Clay Minerals,* Mineralogical Society, London, 1961, chap. 11.)

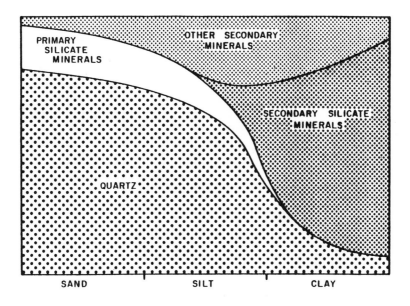

FIGURE 6. General relationship between particle size and kind of minerals present. (Reprinted with permission of Macmillan Publishing Co., Inc., from Brady, N. C., *Nature and Properties of Soils*, 8th ed., chapt. 1. Copyright© 1974 by Macmillan Publishing Co., Inc.)

naturally fall into another land use; crop productivity class, or management class, another soil series is established. Examples of variability among soils and their clay mineralogy is demonstrated in Table 5 for a group of soils used in evaluating the effectiveness for attenuating pollutants from various aqueous waste streams.[6]

2. The Liquid Phase

Water represents the natural liquid phase of soils. Agricultural people call it soil solution and refer to it as moisture when the pore spaces are unsaturated. Soil receives water directly from rainfall and indirectly by irrigation, watershed runoff, and stream flooding. Soil waters contain dissolved substances and further dissolve substances from the solids. In humid climates where rainfall is abundant, the dissolved solids (TDS) content of the soil solution is relatively low as compared with that of arid climates (Table 6). Although considerable attention has just been given to the solid phase of the soils, in practice the liquid phase (soil moisture) of soils has received more attention than any other factor in agricultural production.

The soil water that bathes the roots, soil particles, and wets the capillary fringes of the underground water is the transport system for all pollutants with few exceptions. In a few instances, the pollutant established as a gas or water must give way to organic solvents. Regardless of these exceptions, the water content of the soil still determines pollutant movement.

The greater the abundance of water in a soil, the greater is the opportunity for pollutants to move. Chemical and biological activity becomes correspondingly reduced as the soil moisture approaches the wilting point for plants (15 bars). The activity reduces further until it ceases almost completely as the soil reaches an air-dry condition. Thus, the soil moisture necessary to promote biodegradation at a maximum level must be maintained.

Chemical, physical, and biological reaction rates depend on the nature of the liquid phase in soils. Aerobic and anaerobic conditions resulting from saturation levels of the soil greatly determine degradation products of microbial activity. Chemical reactions may be altered to a large extent by variations in redox conditions associated with wetting and drying, and shrinking and swelling as they influence the physical properties of the soil. Consequently,

Table 5
CLAY (<2μm) AND MINERAL COMPOSITION

Soil Series	Location	Clay %	Smectite (Montmorillonite)	Beidellite	Vermiculite	Chlorite	Mica	Kaolinite	Other (amount)
Davidson	N. Carolina	52	0	0	0	1	0	5	Gibbsite (1)
Molokai	Hawaii	52	0	0	0	0	2	4	Gibbsite (1)
Nicholson	Kentucky	49	0	0	5	0	1	1	Quartz (2)
Fanno	Arizona	46	3	0	0	0	2	1	Quartz (2)
Mohave (Ca)	Arizona	40	3	0	0	0	4	2	Quartz (2)
Chalmers	Indiana	31	4	0	2	2	0	2	Quartz (2)
Ava	Illinois	31	2	0	3	0	2	3	Quartz (2)
Anthony	Arizona	15	4	0	1	1	3	2	Quartz (2)
Mohave	California	11	2	0	0	0	4	3	Quartz (2)
Kalkaska	Michigan	5	0	0	0	3[b]	1	2	Quartz (2)
Wagram	N. Carolina	4	0[a]	0	0	3[b]	1	4	Quartz (3)

[a] The amount of each mineral present is represented as: predominant 5, large 4, moderate 3, small 2, trace 1, and not detected 0.

[b] Chloritic intergrade = mixed layer.

Table 6
CATION EXCHANGE CAPACITY, EXCHANGEABLE CATIONS, pH, AND SOLUBLE ION CONCENTRATION IN SATURATED PASTE OF SOILS

Soil Series	Order and Great Soil Group	Soil Paste pH	Sat. Soil Extract Elec. Cond. (μmhos/cm)	TDS[a] (salt, ppm)	Cation Exchange Capacity	Exchangeable Cations meq/100 g soil			
						Na	K	Ca	Mg
			Humid Climate						
Wagram sl	*Ultisol*, R. Y. Podzolic	4.2	225	1440	2	0.20	0.40	0.60	0.60
Ava si c l	*Alfisol*, G. B. Podzolic	4.5	157	1000	19	0.43	0.23	1.33	2.63
Kalkaska s l	*Spodosol*, Podzol	4.7	237	1520	6	0.10	0.06	0.80	0.10
Canelo c l	*Alfisol*, G. B. Podzolic	5.4	240	1550	6	0.10	0.15	0.80	0.19
Davidson c l	*Ultisol*, R. B. Lateritic	6.2	169	1070	9	0.31	0.16	3.40	1.58
			Subhumid Climate						
Chalmers si c l	*Mollisol*, Prairie	6.6	288	1840	22	0.20	0.10	20.60	8.60
Nicholson si c	*Alfisol*, G. B. Podzolic	6.7	176	1260	37	0.70	0.19	21.80	8.82
			Semiarid to Arid Climate						
Fanno c	*Alfisol*, R. Brown	7.0	392	2510	33	0.35	0.87	17.10	7.82
Mohave s l	*Aridisol*, Red Desert	7.3	615	3940	10	0.40	0.50	8.90	1.80
Mohave (Ca) c l[b]	*Aridisol*, Red Desert	7.8	470	3000	12	0.41	0.88	6.20	4.30
Anthony s l	*Entisol*, Alluvial	7.8	328	2100	10	0.15	0.38	3.85	1.23

[a] TDS — total dissolved solids.
[b] Mohave (Ca) means free lime (CaCO₃) is present in the soil as a natural constituent.

Fuller, W. H., Investigation of Landfill Leachate Pollutant Attenuation by Soils, U.S. Environmental Protection Agency, Cincinnati, 1978, 239.

the way wastes react to the environment is often a response to moisture content in the soil. Because of its importance to pollutant migration rates, soil water will be treated in an independent section in this chapter.

3. The Gaseous Phase

The air of the soil and atmosphere are not appreciably different in composition except where large amounts of organic materials have been turned under or where air exchange is limited by water occupying nearly all of the available pore spaces (waterlogging). Air flow in and out of the soil is essential for plant survival and the rapid decomposition of plant and animal residues and organic industrial wastes. Since soil gases occupy the same soil pore spaces as water, one phase fluctuates opposite to the other. This relationship may be expressed[7] as:

$$f_a = f - \theta_v \qquad (1)$$

where

f_a = volume fraction of air
f = the total porosity (fractional volumes of soil not occupied by solids)
θ_v = the volume fraction of water (volumetric water content)

The volume fraction of air (f_a) is an index of soil aeration. The field-capacity water (or just "field-capacity") is a term well-known to soil scientists as "the amount of water retained by a soil after the initial rapid change of internal drainage".[7] A field-air capacity can be defined corresponding to field-water capacity. Field-air capacity is not the same for all soils but varies with a number of soil characteristics such as texture, structure, organic matter, compaction and the profile itself. High water contents result in a deviation in pore space gas quality similar to the atmosphere to gas containing more CO_2 and less O_2. Because the rate of diffusion of gases in water is slower than in air, oxygen is more likely to be limiting for maximum rate of biological and chemical reactions when the pore spaces are dominated by water. The composition of the soil air becomes proportionately higher in CO_2 and lower in O_2 as the soil reaches the stage of waterlogging or as mechanical compaction restricts gas interchange with the atmosphere. Various gaseous forms of nitrogen (e.g., N_2, NH_3, NO, NO_2), sulfur (e.g., H_2S, SO_2, SO_3), hydrocarbons (primarily methane, CH_4), and hydrogen (H_2) appear in the soil when oxygen is limited by these conditions. Elevated amounts of these gases, above what is usually found in aerated soils, occur also as a result of incorporation of large quantities of biodegradable plant (crop) residues or other readily decomposable wastes in the soil. A combination of restricted aeration by water and/or compaction with high applications of biodegradable organic wastes also can radically alter the soil-air composition. The composition of soil air is much more highly variable than the atmosphere. Such factors as temperature, moisture content, time of the year, pH, depth below the surface, and rate of exchange of gases, influence composition. On the other hand, internal soil temperature and humidity fluctuations are small compared with those in the atmosphere. Soil air approaches 100% relative humidity most of the time except at the surface.

Two different mechanisms, convection and diffusion cause the exchange of gases between the soil and the atmosphere.[7] Briefly, convection results from a gradient of total gas pressure and the entire mass of air streams from a zone of high pressure to low. In diffusion, the moving force is a gradient of partial pressure of concentration of any of the gases present in the soil or atmosphere air.[7] Further information on the gaseous phases of soils may be found in reviews.[7-10]

C. The Soil Profile

The soil profile begins at the land surface and includes all of the material down into the parent material. The *Glossary of Soil Science Terms*[11] defines soil profile as "a vertical

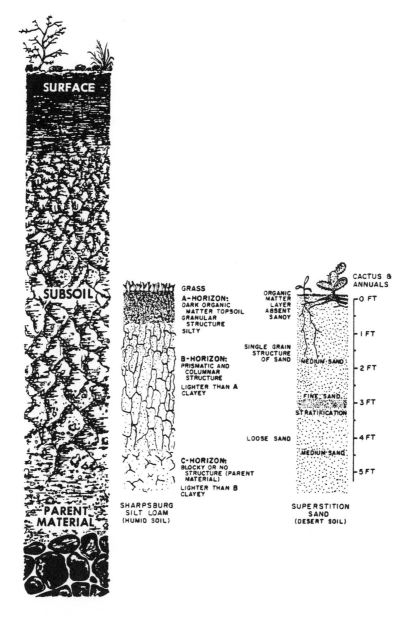

FIGURE 7. Typical soil profiles showing horizons that differentiate during weathering under two different climates. (From Fuller, W. H., *Soils of the Desert Southwest,* The University of Arizona Press, Tucson, 1975, 38.)

section of the soil through all its horizons and extending into the parent material'' (Figure 7). The key to an understanding of the profile and the way the different soils fit into the waste disposal operations on land is to study the soil horizon characteristics that make up the soil body as a whole. Examples of three soil profiles are reproduced in Figures 8, 9, and 10. The definition of horizon is precise:[12]

"Soil Horizon — a layer of soil or soil material approximately parallel to the land surface and differing from adjacent genetically related layers in physical, chemical, and biological properties or characteristics such as color, structure, texture, consistency, kinds and numbers of organisms present, degree of acidity or alkalinity, etc. The following table lists the designations and properties of the major soil horizons. Very few, if any, soils have all of these horizons well developed but every soil has some of them.''

Horizon[12] designation	Description
0	''Organic horizons of mineral soils. Horizons: (i) formed or forming in the upper part of mineral soils above the mineral part; (ii) dominated by fresh or partly decomposed organic material; and (iii) containing >30% organic matter if the mineral fraction is >50% clay, or >20% organic matter if the mineral fraction has no clay. Intermediate clay content requires proportional organic-matter content.
01	Organic horizons in which essentially the original form of most vegetative matter is visible to the naked eye. The 0l corresponds to the L (litter) and some F (fermentation) layers in forest soils designations, and the horizon formerly called Aoo.
02	Organic horizons in which the original form of most plant or animal matter cannot be recognized with the naked eye. The 02 corresponds to the H (humus) and some F (fermentation) layers in forest soils designations, and to the horizon formerly called Ao.
A	Mineral horizons consisting of: (i) horizons of organic matter accumulation formed at or adjacent to the surface; (ii) horizons that have lost clay, iron, or aluminum with resultant concentration of quartz or other resistant minerals of sand or silt size; or (iii) horizons dominated by (i) or (ii) above but transitional to an underlying B or C.
B	Horizons in which the dominant feature or features is one or more of the following: (i) an illuvial concentration of silicate clay, iron, aluminum, or humus, alone or in combination; (ii) a residual concentration of sesquioxides or silicate clays, alone or mixed, that has formed by means other than solution and removal of carbonates or more soluble salts; (iii) coatings of sesquioxides adequate to give conspicuously darker, stronger, or redder colors than overlying and underlying horizons in the same sequum, but without apparent illuviation of iron and not genetically related to B horizons that meet requirements of (i) or (ii) in the same sequum; or (iv) an alteration of material from its original condition in sequums lacking conditions defined in (i), (ii), and (iii) that obliterates original rock structure, that forms silicate clays, liberates oxides, or both, and that forms granular, blocky, or prismatic structure if textures are such that volume changes accompany changes in moisture.
C	A mineral horizon or layer, excluding bedrock, that is either like or unlike the material from which the solum is presumed to have formed, relatively little affected by pedogenic processes, and lacking properties diagnostic of A or B but including materials modified by: (i) weathering outside the zone of major biological activity; (ii) reversible cementation, development of brittleness, development of high bulk density, and other properties characteristic of fragipans; (iii) gleying; (iv) accumulation of calcium or magnesium carbonate or more soluble salts; (v) cementation by accumulations such as calcium or magnesium carbonate or more soluble salts; or (vi) cementation by alkali-soluble siliceous material or by iron and silica
R	Underlying consolidated bedrock, such as granite, sandstone, or limestone. If presumed to be like the parent rock from which the adjacent overlying layer or horizon was formed, the symbol R is used alone. If presumed to be unlike the overlying material, the R is preceded by a Roman numeral denoting lithologic discontinuity.''[12]

The soil profile varies in depth depending on its age and the intensity of weathering factors associated with the land at that particular location. Thus, the soil profile includes all generic horizons (A, B, and C) that are present and ranges in thickness from a few centimeters to several meters. A less academic and more practical use approach relates the soil to the unconsolidated material that generally overlies coarse stony debris or consolidated bedrock. Such material, indeed, often looks like soil, having been subdivided by the abrasion of glacial, wind and water action. In addition, the soil scientist continues to discover buried profiles developed in another historic era and often formed under climatic conditions quite

FIGURE 1.8.   Mohave loam profile from Arizona is an Aridisol with lime accumulation in the B horizon. (From Fuller, W. H., *Soils of the Desert Southwest,* The University of Arizona Press, 1975, 32.)

unlike those of today. Geologists may refer to both a soil profile and a weathering profile.[13,14] Quantification of the exact limits as to what constitutes a soil has yet to emerge. Therefore, this text will consider "soil" in its broadest interpretation as the regolith (i.e., all the weathered material above the bedrock) in the discussion of waste treatment, since all unconsolidated weathered materials participate in waste modification.

D. The Soil Classification

During the weathering of parent material into a more restrictive concept of soil than the whole regolith, all factors combine to create a heterogeneously layered profile even though the original geological material may have been highly homogeneous. The more mature the soil, the greater the differentiation between the horizons (A, B, and C). Soils, therefore,

FIGURE   9.   A buried soil profile formed in ancient history during a humid climate
cycle, New Mexico. (From Fuller, W. H., *Soils of the Desert Southwest,* The University
of Arizona Press, Tucson, 1975, 77.)

are characteristically heterogeneous rather than homogeneous with depth, and as such, make
predictions of pollutant migration rate difficult. Mechanical mixing of the soil to achieve
homogeneity, improves the opportunity, not only to predict migration rates, but for retention
of pollutant movement. The scientists that classify and map soils use the numerous variations
in soil horizons as revealed by the soil profile's individuality for the identification and
classification into pedons.[12]

Since only a rare few of the soil properties can be identified from the surface, the nature
of the soil is studied to its full depth of horizons or profile. A pedon is defined as "a three-
dimensional body of soil that has lateral dimensions large enough to include representative
variations in the shape and relation of horizons and in composition of the soil."[12] The area
of pedon ranges from 1 to 10 m², depending on the variability of the soil. The pedon is the
natural basic unit of a soil and covers "a real extent by lateral displacement."[15]

FIGURE 10. Vinton loamy fine sand is a deep Entisol (Alluvial soil) that is highly stratified in sand, silt, and clay. (From Fuller, W. H., *Soils of the Desert Southwest,* The University of Arizona Press, Tucson, 1975, 32.)

Soil classification techniques are designed to group like-soils into like-categories[11] (Figure 11). The soil maps serve as references for general predictions of pollutant migration rates through soil and as an aid in determining ways in which soils may be modified to minimize movement of pollutants from waste disposals. Therefore, wherever like soils occur, we expect migration of a certain waste to react similarly when in the same soil category even though they may not actually be at the same location. On the other hand, since climate and vegetation along with other soil-forming factors dictate the kind of soil that forms, the soils forming vary from location to location. A soil developed near Cleveland, Ohio, for example, cannot be expected to attenuate a pollutant the same as a soil developed in the desert near Las Vegas, Nevada (Figure 11). Furthermore, consider two soils located within similar rainfall and parent material conditions. The one in a cold winter climate like Michigan will have podzolic (Spodosol) characteristics and the one from the warm climate of Hawaii will have lateritic (Oxisol) characteristics. Indeed, the two soils do not retain heavy metals from municipal solid waste leachate equally well.[6,16] On the other hand, two soils having similar texture and climatic conditions, one from Illinois and the other from Arizona that belong to the same soil order (Alfisol) retain heavy metals from municipal solid waste (MSW) leachate

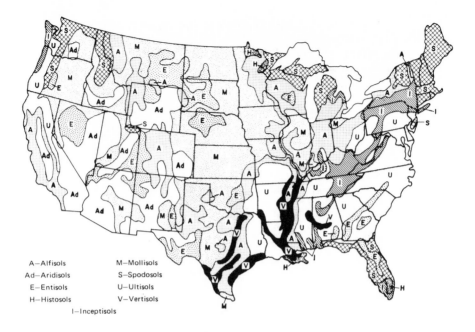

A—Alfisols   M—Mollisols
Ad—Aridisols   S—Spodosols
E—Entisols   U—Ultisols
H—Histosols   V—Vertisols
        I—Inceptisols

FIGURE 11.   National locations of the 9 soil orders as listed. (From USDA, Soil Survey Staff, *Soil Series in the United States, Their Taxonomic Classification,* USDA, Soil Conservation Service, Washington, D.C., 1968, 433.)

to about the same extent.[6] The USDA Soil Conservation Service soil survey and classification. scheme is an important tool useful from the first in deciding on the selection of a site, the kind of waste suitable for that particular soil, to the final determination of closure procedure.

The USDA Soil Conservation Service [17] soil surveyors recognize 10 soil orders (Table 7), the broadest category in the classification system. They suggest that the lowest categories the family and series serve purposes that are pragmatic; the series names are geographical and the technical family names are descriptive. The factors used to distinguish families of mineral soils within a subgroup are listed to illustrate how family differences can relate to pollutant attenuation characteristics. Indeed, they have been shown[6,18,19] to significantly affect movement of toxic and heavy metals and organic solvent pollutants. The justification for their presentation here becomes evident upon reviewing them:

- Particle-size classes (sand, silt, clay)
- Mineral classes (primary and secondary clay minerals)
- Calcareous and reactions (pH) classes (soil acidity or alkalinity)
- Soil temperature classes
- Soil depth classes
- Soil slope classes
- Soil consistency classes
- Classes of coatings (on sands)
- Classes of cracks

The polypedon finally permits us to arrive at the specific soil unit that is mapped in the field and from which a functional soil survey map emerges. The polypedon is a unit just large enough for classifcation of "A soil". A classified soil "consists of contiguous similar pedons that are bounded on all sides by 'non-soil' or by pedons of unlike character".[12] The polypedon may surround others but the limits are reached when there is no more soil or

Table 7ᵃ
THE TEN SOIL ORDERS, THEIR APPROXIMATE GREAT SOIL GROUPS, AND SIMPLIFIED DISTINGUISHING CHARACTERISTICS

Order	Approximate equivalents	Description
Entisol	Azonal soils with some low humic gley soils	Soils without pedogenic horizons. Very little development with properties mostly inherited from parent material.
Inceptisol	Ando, solbrun acid, some brown forest, low humic gley and humic gley soils	Soils usually moist, with pedogenic horizons of alteration of parent materials but not of accumulations.
Aridisol	Desert, reddish desert, sierozem, solonchak, some brown, and reddish brown soils, and associate solonetz	Soils with pedogenic horizons, low in organic matter, and dry more than 6 months of the year in all horizons.
Mollisol	Chestnut, chernozem, brunizem (prairie), rend-zinas, some brown forest, and associated so-lonetz and humic gley soils	Soils with nearly black, organic—rich surface horizons and high base supply.
Spodosol	Podzols, brown podzolic soils, and ground-water podzols	Soils with accumulations of amorphous materials in subsurface horizons. Soils with illu-vial B_2 horizon having free Fe and Al oxides, clay, humus.
Alfisol	Grey-brown podzolic, grey wood soils, non calcic brown soils, degraded chernozem, and associated planosols	Grey to brown surface horizons, medium to high base saturation and subsurfaces of clay assimilation, usually wet but may be dry during warm season.
Ultisols	Red—yellow podzolic soils, reddish-brown la-tertitic soils of the U.S., and associated plan-osols and half-bog soils	Soils that are usually moist, with horizon of clay accumulation and a low base supply
Oxisols	Laterite soils, latosols	Soils represented by warm, wet climates such as tropics, highly weathered, with B horizon mostly of sesquioxides or 1:1 clay.
Histosols	Bog soils	Organic soils
Vertisols	Grumusols	Soils with high content of swelling clays and wide deep cracks at some season

ᵃ   Modification from Reference 12.

where the pedons have properties that significantly differ. The soil mapping unit is defined in terms of taxonomic units and may contain more than one kind of soil. The different soils in the mapping unit are similar from a practical basis of use. The mapped unit may contain separate generic soils as judged by academia.

The soil survey map should become an essential point of departure for site selection and for all types of waste disposals. The limitations of the soil classification system as well as its advantages should be understood. Some wastes may be so corrosive as to alter the natural soil characteristics so drastically that the original distinguishing criteria for classification disappears. The soil then becomes an absorbing material devoid of most of the identifying characteristics that were useful for classification. Strong acid-waste streams, for example, can change the soil pH, exhaust the carbonates, solubilize native metals, and alter permeability, obliterating distinguishing horizon features. The soil texture or particle size distribution resists alteration by strong wastes and solvents longer than most soil properties. For soil-disposal-management planning, soil texture or particle size distribution is one of the most important physical properties about which to have available information.

II. BIOLOGICAL CHARACTERISTICS

A. The Microflora

The soil is teeming with a great variety of microorganisms and small animals which leads

Table 8
APPROXIMATE NUMBERS OF
ORGANISMS COMMONLY
FOUND IN THE SOIL

Organism[a]	Estimated Numbers[b] per gram of soil
Bateria	3,000,000 to 500,000,000
Actinomycetes	1,000,000 to 20,000,000
Fungi	5,000 to 900,000
Yeasts	1,000 to 100,000
Algae	1,000 to 500,000
Protozoa	1,000 to 500,000
Nematodes	50 to 200

[a] There also are large numbers of slime molds, virus, phages, and insects, worms, anthropods, and mycoplasma.

[b] Numbers are based on plate counts.

to the truism often found in textbooks — that the soil is a "living system". The most abundant numbers may be found associated with energy sources used as substrates for growth and reproduction. Natural food, energy, and nutrients are well represented by the soil organic matter and the continual deposits of animal and plant residues. The most prominent soil microorganisms occur as bacteria, actinomycetes, fungi, algae, protozoa, and nematodes (Table 8).[20]   The live weight of these microorganisms is estimated to range from 0.5 to over 4 t/ha in the surface 15 cm.[21]

The biological characterization of natural soils is centered on the soil microorganisms since they exert a profound impact on waste constituent mobility. Because their activity is not readily discernable, they are often overlooked. Vast quantities of organic residues (e.g., 10 to 20 t/ha) from agricultural crops decompose each year with the release of carbon, nitrogen, hydrogen, plant nutrients, and small amounts of chance heavy metals. Such cycling of nutrients is essential for new generations of living organisms. Wastes and toxic substances from plants and applied pesticides are degraded with the carbon and energy sources utilized for growth and reproduction. The ultimate end products under aerobic conditions are essentially carbon dioxide, water, and inorganic metals. Also, insoluble inorganic nutrients are solubilized and soluble compounds of nitrogen and sulfur are oxidized or reduced depending on the oxygen level (redox) of the soil. Not to be omitted is the energy derived by the oxidation of metals, e.g., iron and manganese, and methylation of metals such as mercury, selenium, and tellurium (Table 9).[20]

Microorganisms are present in all soils. In fact, even in the presence of high amounts of toxic substances complete destruction of the microflora is rarely possible. Sterilization of field soils with chloropicrin and fumigation with toluene, 1 to 2 dibromoethane, formaldehyde, kerosene or alcohol, for example, results in selectively killing a few organisms, temporarily inhibiting the activity of others, and not affecting still others. It is well known that soil "sterilization" is, at best, temporary for most of the indigenous soil microorganisms. The point to be made here is that the soil microflora in the natural habitat is highly versatile and resistant to destruction even by the most drastic treatment and will recover fully in time.

Another important characteristic of the soil microflora is that they are a well established ecological unit resistant to invasion or settlement of new microorganisms not indigenous to the habitat. Inoculation of the soil with new and different types of organisms has been shown not to be successful except where a host plant (such as a legume or infecting plant) is available.

Table 9
SOME MICROBIAL TRANSFORMATIONS OF INORGANIC SUBSTANCES

Element	Microorganism	Physiological activity[a]
As	*F. ferroxidans*	As_2S_3 oxidized to AsO_3^{3-}; AsO_4^{3-}; $SO_4^{2-}(?)$[b]
	Heterotrophic bacteria	AsO_3^{3-} oxidized to AsO_4^{3-}
	Achromobacter	
	Pseudomonas	
	Xanthomonas	
	M. lactilyticus	AsO_4^{3-} reduced to AsO_3^{3-}
Cd	*Desulfovibrio*	$CdCO_3 + SO_4^{2-} + 8H^+ + 8e^- = CdS + 4H_2O + CO_3^{2-}$
Cu	*T. ferroxidans*	$Cu_2S + 4H_2O = 2Cu^{2+} + 6H^+ + H_2SO_4 + 10e^-$
	F. ferroxidans	$CuS + 4H_2O = Cu^{2+} + 6H^+ + H_2SO_4 + 8e^-$
	Desulfovibrio, C. nigrificans	Cu^{2+} and SO_4^{2-} reduced to CuS; $Cu_{10}S_9$; Cu_2S
	M. latilyticus	$Cu(OH)_2 + H^+ \, e^- = CuOH + H_2O$
Fe	*T. ferroxidans* ⎫	
	Ferrobacillus spp. ⎬	$Fe^{2+} = Fe^{3+} + e^-$
	Gallionella ⎭	
	Leptothrix ochracea ⎫	
	Sphaerotilus ⎬	Adsorption, precipitation
	Protozoa, algae	
	M. lactilyticus ⎫	
	B. circulans ⎬	$Fe^{3+} + e^- = Fe^{2+}$
	B. polymyxa ⎭	
	Desulfovibrio, C. nigrificans	$Fe^{3+} + SO_4^{2-} + 8H^+ + 9e^- = FeS + 4H_2O$
Ni	*T. ferroxidans*	$NiS + 4H_2O = Ni^{2+} + 8H^+ + SO_4^{2-} + 8e^-$
	Desulfovibrio	$NiCO_3 + SO_4^{2-} + 8H^+ + 8e^- = NiS + 4H_2O + CO_3^{2-}$
		$Ni(OH)_2 + SO_4^{2-} + 10H^+ + 8e^- = NiS + 6H_2O$
S	*Thiobactericeae* ⎫	
	Thiorhodaceae	
	Chlorobacteriaceae	
	Beggiatoaceae ⎬	$H_2S = S^o + 2H^+ + 2e^-$
	S. natans	$H_2S + H_2O = SO_4^{2-} + 10H^+ + 8e^-$
	Achromatium	
	Leucothrix[c] ⎭	
	Bacteria, actinomycetes, fungi	Polysulfides reduced to thiosulfate and sulfide
	All microorganisms	$S^o + 2e^- + 2H^+ = H_2S$
Se	*M. selenicus*	$H_2Se + 4H_2O = SeO_4^{2-} + 10H^+ + 8e^-$
	M. lactilyticus ⎫	$Se^o + 2e^- + H^+ = SeH^-$
	C. pasteurianum	
	D. desulfuricans	
	Neurospora ⎬	$HSeO_3^- + 4e^- + 5H^+ = Se^o + 3H_2O$
	C. albicans	
	Baker's yeast	
V	*M. lactilyticus* ⎫	
	D. desulfuricans ⎬	$H_2V_{O4} + 2e^- + 2H^+ = VO(OH) + H_2O$
	C. pasteurianum ⎭	
Zn	*T. ferroxidans*	$ZnS + 4H_2O = Zn^{2+} + SO_4^{2-} + 8H^+ + 8e^-$
	Desulfovibrio	$^1/_5 \, |2ZnCO_3\cdot3Zn(OH)_2| + SO_4^{2-} + 9\,^1/_5\,H^+ + 8e^- = 5^1/_5\,H_2O + \,^2/_5\,CO_3^{2-} + ZnS$
Metal chelate	Heterotrophic microorganisms	Oxidation of chelating agent with precipitation of metal moiety

[a]   Oxidative or reductive half-reactions listed most nearly describe the particular microbial activity cited.
[b]   Proof of sulfate production lacking.

Modified from Microbial formation and degradation of minerals, Silverman, M. P. and Ehrlich, H. L., in *Advances in Applied Microbiology*, Vol. 6, pp. 153 to 206, 1964.

The single most important means of altering the natural soil microflora is to alter the energy or food source. The microflora adjusts to the kind of material and circumstance with which it is confronted, whether gas, liquid, solid, organic or inorganic substance. Microorganisms will attack almost any material at some given concentration (Table 10). They synthesize as well as biodegrade. The transformations may be beneficial or detrimental to good waste disposal operations and management. The soil microorganisms in waste management processes either perform entirely or mediate:

- Degradation of carbonaceous wastes
- Transformation of cyanide to mineral nitrogen compounds and dentrification to inert N_2 gas
- Initiation of metal ion oxidation-reduction
- Production of CO_2 with subsequent formation of weakly ionized carbonic acid
- Production of simple organic acids, methane, hydrogen, and carbon monoxide
- Production of small or large molecular species upon which trace contaminants may be absorbed
- Production of complex organic compounds which may react with waste contaminants
- Production of small-sized organic debris which can infiltrate small pore spaces and move downward in the soil profile

B. Biodegradation

Biodegradation is the most effective and economical means of eliminating the capacity of most organic materials and compounds to pollute the environment. As a tool in pollution control, we have just begun to understand how to use biodegradation processes efficiently, yet it has been available for time immortal. Microorganisms demonstrate great capabilities to produce unusual products and perform unusual functions in commercial industry for mankind; yet in the field of pollution abatement, biodegradation as a waste control management practice has been developed little beyond its original crude state.

1. Important Biological Processes

The three most effective biological processes influencing pollution control in waste disposal are

- Oxidation/reduction
- Mineralization/immobilization
- Synthesis of organic constituents/complexation reactions

These represent broad categories, useful primarily to establish a basis for discussion. Within each category will be cycles, and cycles within cycles. In order that biodegradation should take place, a favorable environment must be maintained. Although there is no one set of environmental factors ideal for all organisms, there is an environment most generally considered to be optimum for organic matter decomposition in the soil. The most suitable temperature is between 25 and 35°C (i.e., in the mesophilic range). Soil moisture should not exceed the level that impedes aeration or oxygen diffusion and should be above that of the wilting point for plants as the lower limit (Figure 12). Other factors controlling biodegradation include pO_2, pH, energy source, electron acceptor, and essential nutrients.

2. Temperature

Although microbial decomposition of organic materials occurs most rapidly in the mesophillic (25 to 35°C) range, higher and lower temperature levels favor some other organisms, i.e., psychrophiles (0 to 20°C) and thermophiles (45 to 65°C). Thus, biodegradation will

Table 10
INORGANIC COMPOUNDS AND IONS AS FINAL ELECTRON ACCEPTORS AND
ENERGY YIELD FOR THE SOIL ORGANISMS SHOWN

Organism	Reaction	kcal/mol
Nitrosomonas	$NH_4^+ + 1\frac{1}{2} O_2 \cdots \longrightarrow NO_2^- + 2 H^+ + H_2O\cdot$	$\Delta G = -66$
Nitrobacter	$NO_2^- + \frac{1}{2} O_2 \longrightarrow NO_3^-$	$\Delta G = -17.5$
Thiobacillus thiooxidans	$S^\circ + 1\frac{1}{2} O_2 + H_2O \longrightarrow H_2SO_4$	$\Delta G = -118$
Hydrogenomonas	$H_2 + \frac{1}{2} O_2 \longrightarrow H_2O$	$\Delta G = -57$
Desulfovibrio	$SO_4^= + 4 H_2 \longrightarrow S^= + 4 H_2O$	$\Delta G = -83 \ (-21)$
Methanobacterium	$4 H_2 + CO_2 \longrightarrow CH_4 + 2 H_2O$	$\Delta G = -31$
Ferrobacillus and Gallionella	$Fe^{++} \longrightarrow Fe^{+++} + e^-$	$\Delta G = -11$
Carboxyldomonas	$CO + \frac{1}{2} O_2 \longrightarrow 2CO_2$	$\Delta G = -66$
Thiobacillus thioparus	$H_2S + \frac{1}{2} O_2 \longrightarrow H_2O + S^\circ$	$\Delta G = -41.5$
Actinomycetes	$KCN + 2H_2O + \frac{1}{2} O_2 \longrightarrow KOH + NH_3 + CO_2$	$\Delta G = $ not calculatable

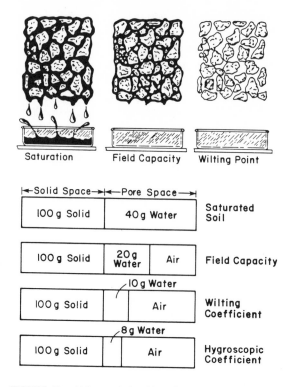

FIGURE 12.   Volume relationships of solids, water, and air in a
well granulated silt loam at different moisture levels. (Reprinted
with permission of Macmillan Publishing Co., Inc., from Brady,
N. C., *Nature and Properties of Soils*, 8th ed., 190. Copyright©
1974 by Macmillan Publishing Co., Inc.)

occur over a wide range of temperatures, although it takes place more slowly near the
extremes of 0 and 65 C.

3. Moisture

Soil moisture optimum for maximum biodigestion usually occurs at the upper limit of
soil moisture at a level just where the amount of water occupying the soil channel walls and
pore spaces does not interfere with soil oxygen exchange with the atmosphere. Usually this
is just slightly below the "normal field capacity" water (Figures 12 and 13). Moisture
retention curves for soils of the prospective site provide a useful tool and guide for decisions
on the optimum moisture range.[22] The relationship between thickness of water film and
water tension with which the water is held at the liquid-air interface is illustrated. Microorganisms must have a water film available in which to work (Figure 13). The value for
field capacity is often chosen as 0.1 to 0.33 bars, but this is at best a crude approximation
as the true value depends on the profile, history, and factors other than texture alone. The
wilting point is reasonably well represented as water retained at 15 bars. The soil, if not
overburdened with readily decomposable organic residues, is considered aerobic and oxidation occurs via aerobic and facultative aerobic microorganisms.

When soil water fully occupies the pore spaces, anaerobic conditions determine the biodegradation processes. When the anaerobic and facultative anaerobes dominate, decomposition
slows down, and intermediate degradation products accumulate. Waterlogging characterizes
the soil, and gases such as methane, hydrogen, nitrogen, carbon monoxide, and carbon
dioxide evolve. Soil and waste management can influence which process persists.

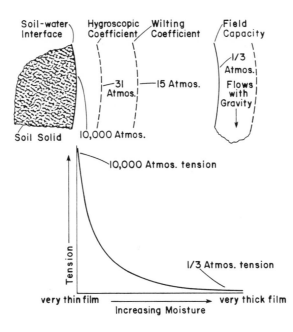

FIGURE 13.   Thickness of film of water and the tension with which water is held at the liquid-air interface. (Reprinted with permission of Macmillan Publishing Co., Inc., from Brady, N. C., *Nature and Properties of Soils*, 8th ed., 193. Copyright© 1974 by Macmillan Publishing Co., Inc.)

4. Oxygen Status

Soil aeration is necessary to maintain aerobic conditions. Manipulation of soil and waste disposal determines the oxygen level, and oxidation takes place in the soil as explained above. In the presence of adequate oxygen, complete or nearly complete oxidation of carbonaceous wastes is accomplished.

5. Acidity and Alkalinity (pH)

Biodegradation of wastes can occur over a wide range of pH values. As with the other factors, decomposition at the extreme ranges is slower than at the mean. The optimum range for most soil microorganisms is between pH values 5.5 and 8.5. Many organisms are not greatly affected by a unit change from their specific optimum. The nature of the waste and loading rate as it influences oxygen availability can influence soil pH levels either favorably or unfavorably. Often the greatest influence of pH levels on the microflora is the modification in availability of nutrients.

6. Carbon/Nitrogen Ratio

One of the most critical factors in maximizing biodegradation of carbonaceous wastes is the C/N ratio. Mature plant residues such as straw have a C/N ratio near 80:1 which is too wide for the maximum rate of decomposition. The nitrogen content is too low. The ratio of 30:1 is considered the approximate threshold for nitrogen demand beyond which additional N will not hasten the rate of decomposition. Generally, if a readily decomposable carbonaceous material has a concentration of 1.5% N, nitrogen will not be limiting. Since the ease of biodegradation varies with the kind of carbon compound or nature of the carbonaceous material, the N demand is not fixed. The nitrogen (N) factor was developed to take this into account. Nitrogen factor is defined as "the weight of nitrogen (N) immobilized in the decomposition of a given weight of material at a maximum rate". For example, if municipal solid waste (mostly paper) has a 0.6% N content and the expected demand for N for maximum decomposition is 1.5% N, the difference is 0.9% N. Therefore, the requirement of 0.9%, is multiplied by the dry weight figure for the solid waste and added to the waste to insure maximum biodegradation. To explain this further, if 2 t (or 4000 lb) of MSW were spread on an acre of land, 4000 × 0.009 = 36 lb additional N are needed. If urea fertilizer of 45% N were used, it would take 80 lb of urea fertilizer per acre to satisfy the nitrogen need. The equation is; 4000 (lb) × 0.009 (N) × 2.22 (urea N) = 80 lb urea/a. Every carbonaceous material has a nitrogen factor but does not always have a N requirement. If the C/N ratio, for example, is narrow and the material has more than about 1.5% N, as legume hay, green grass clippings, and most green plant residues, the materials do not require additions of N. In contrast, newpaper is very low in nitrogen as are oily wastes and, therefore, require added N. The N-factor, also, varies with different organic constituents and carbonaceous materials because they differ in ease (rate) of decomposition. Lignin and oily wastes decompose slowly, therefore, the nitrogen demand is low at any given time. Glucose, starch, and soluble plant materials decompose rapidly, thus the nitrogen demand (N-factor) is much higher than for lignin and oil.

C. Oxidation/Reduction

Oxidation/reduction processes influence pollutant behavior in soils as much as any biological or chemical factor. Microorganisms initiate acute reducing conditions when aeration is poor and an abundant supply of "available" oxidizable substances, such as nitrogenous wastes, green plant residues, and certain readily biodegradable, non- or low-toxic organic materials are present. Under anaerobic or reducing conditions, nitrates (NO_3^{-1}), sulfate (SO_4^{-2}), arsenate (AsO_4^{-2}), chromate (CrO_4^{-6}), ferric iron (Fe^{+3}), and others along with organic compounds are the electron acceptors. In contrast under aerobic conditions oxygen usually is the electron acceptor. The hydrous oxides of iron, manganese, and aluminum also play an important role in the oxidation/reduction soil processes. Organic matter is a cardinal component of redox reactions in soil.

The usual agricultural forest or range soils is aerobic. However, if soil becomes waterlogged, anaerobic conditions develop, acids and carboxyl radicals accumulate, and metallic elements are reduced. The pH values generally decrease as the hydrogen activity (acids) builds up. The associated evolution of CO_2 from degradation of carbonaceous materials also contributes to hydrogen ion activity and/or lowering of pH values. At lower soil solution pH values, the opportunity for trace contaminants to move through soils increases. Under reducing conditions, i.e., usually the absence of free molecular oxygen, chemical or biological oxygen-demanding reactions consume oxygen at a greater rate than O_2 enters into the system. Microorganisms find a substitute for O_2 in metabolic processes. These substitutes may contain combined oxygen as nitrates (NO_3^-) or sulfates (SO_4^{2-}) or may involve electron transfer without the involvement of oxygen as with Ferric (Fe^{3+}) or manganic (Mn_4^+) substances.

Oxidizing conditions favor attenuation as opposed to reducing conditions. Cyanide is an exception. When it is oxidized to NO_3^-, it is highly mobile. The reduced form NH_3^- is less mobile. Precipitates develop in anoxic landfill leachates upon exposure to air (oxic conditions). Many of the trace contaminants become a part of this insoluble residue.

Reducing (anoxic) conditions favor accelerated migration of heavy metals as compared with oxic conditions. For example, trace contaminants As, Be, Cr, Cu, Fe, Ni, Se, V, and Zn, are much more mobile under anaerobic than aerobic soil conditions, all other factors being the same. Water logging such as below sanitary landfills favors accelerated mobility of most trace constituents unless, of course, reduced substances such as HS^- are produced than unite with metals to form insoluble precipitates.

When wastes are disposed of in ponds, lagoons, or deep fills, management of redox status is not always practical. However, in land spreading or spray irrigation operations, oxidizing (aerobic) conditions can be promoted by allowing the soil-waste system to dry between waste applications. The determination of the tendency for soils towards oxidation or reduction conditions should play a part in establishing management practices for waste and wastewater disposal sites.

Oxidizing conditions are usually associated with higher pH values than reducing conditions. Thus, trace elements and especially heavy metals are expected to be less subject to movement under high pH conditions. However, there are situations in which the two are not related. In some instances highly anaerobic and reducing conditions may be associated with the latter stages of decay or decomposition when very stable organic complexes form, and alkali earths, and other basic ions accumulate.

The reduction processes in leachates, aqueous waste streams, and soils are energy-requiring systems. Decomposable compounds providing this necessary energy may be either organic or inorganic, but organic matter usually is present for the most active reduction. Organic products results from the decomposition of plant and animal material are much more effective in making heavy metals more mobile under anoxic than oxic conditions. The influence of reducing conditions on the mobility of trace and heavy metals is well documented by experimental evidence. For example, the change of Mn^{4+}, Fe^{3+}, and Cu^{2+} to Mn^{2+}, Fe^{2+}, and Cu^+ takes place when decomposable organic materials are added to soils as follows:

$$MnO_2 + 4H^+ + 2e \rightarrow Mn^{2+} + 2H_2O$$

$$Fe_2O_3 + 6H^+ + 2e \rightarrow 2Fe^{2+} + 3H_2O$$

$$CuO + 2H^+ + e \rightarrow Cu^+ + H_2O$$

Where drainage is poor in soils associated with disposal lagoons, ponds, lakes, landfills, wastewater spreadings, and industrial aqueous waste stream discharges, redox potentials decrease and trace and heavy metals become more soluble and mobile.[23-30]

The reduction of iron in soils is biologically dominated. The reaction is most prominent at low pH and anoxic conditions when readily available organic matter is present in abundance. Iron also can be used as an energy source by iron-oxidizing bacteria. Organic matter is not needed for oxidation to proceed, although some microorganisms use both.

The presence of sulfate-reducing bacteria and sulfate also favors reduction of iron. The sulfide produced, reduces ferric iron to ferrous sulfide. Hydrogen sulfide (H_2S) production can be considered as an unusually effective means of inactivating heavy metals. Most of the heavy metals form highly insoluble sulfides. Iron sulfide formation in some natural wet soils has resulted in the compaction of subsoil layers by clogging the pores. Oxidation/reduction of iron in most soils with regularly fluctuating water tables, may result in the accumulation of indurated iron pans composed of oxide and hydrous oxides of iron. The suggested establishment of anaerobic soil zones for control of nitrates, through the denitrification mechanisms, can create such an alternate anaerobic/aerobic layering pattern. Iron pans in old rice paddies exemplify these formations well. The ubiquitousness of manganese influences iron mobility as well as other heavy metals, and like iron, maganese hydroxides clog pores inhibiting the downward flow of water.[1] Iron and manganese precipitate in a mixed system more than they do singly in pure systems.[31,32] Some oxidation-reduction reactions of certain inorganic substances are presented in Table 10. Metals can be oxidized or reduced by at least one type of soil microorganism, depending on the availability or lack of free oxygen and other substances in the habitat such as readily available electron donors (Table 10).

D. Mineralization and Immobilization

When microbial decomposition results in the release of an element from an organic form to an inorganic state, the process is called mineralization. Immobilization is the reverse: an

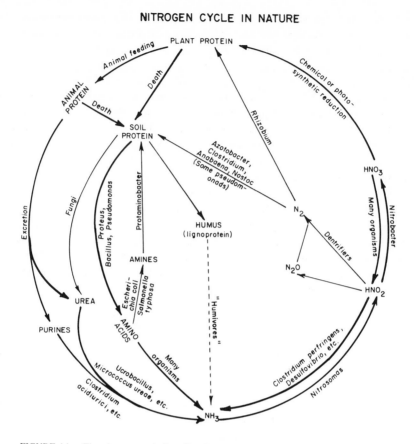

FIGURE 14.   The nitrogen cycle in soils. (Reprinted with permission of Macmillan Publishing Co., Inc., from Thimann, K. V., *The Life of Bacteria*, 2nd ed., 10. Copyright© 1955, 1963 by Macmillan Publishing Co., Inc.)

element is converted from an inorganic state to the organic form in microbial or plant tissues, rendering the element not readily available to other organisms or plants (Glossary of Soil Science Terms, 1978).[11] The incorporation of elements (including heavy metals) into biological tissues, results in their being "fixed" with mobility controlled by cell or cell tissues. On the other hand, for the elements that are relatively immobile in soils as inorganic complexes, incorporation of pollutants into cells may be thought of as a mechanism for them to migrate as minute particles and cell debris when the tissues die and decay. Phosphorus movement in organic form is an example.[33,34] Another example is the absorption of ^{45}Ca and ^{89}Sr by algae[35] and relative absorption of ^{45}Ca and ^{90}Sr by fungi.[36] Also, ^{45}Ca and ^{89}Sr movement in soils, and uptake by barley plants as affected by $Ca(Ac)_2$ and $Sr(Ac)_2$ treatment of the soil[37] and movement of algal- and fungal- bound radiostrontium as chelated complexes in calcareous soil.[38]

The process of immobilization through microbial incorporation is important in the nitrogen cycle (Figure 14).   If this cycle did not occur, the dependent animal kingdom would soon starve. Whereas carbon lost as CO_2 is liberated to the atmosphere, nitrogen is held more tightly and accumulates as the C/N ratio narrows finally to about 10/1 in most humus. Cyanide, for example, is oxidized in the soil into ammonium and finally to nitrate. During the transformation processes, the N is immobilized, thus preventing it from being lost to leaching. The carbon is oxidized to harmless CO_2. As nitrogen again becomes mineralized slowly, plants have an opportunity to take it up and into their protein pool. Upon death of the cell tissues which contain protein (animal, plant, or microorganism), the N returns to the soil-organic matter pool where it slowly mineralizes into a form usable by plants again, is lost to subsurface depths by leaching, or is denitrified to N_2 gas and escapes into the atmosphere (Figure 14).

The examples of other element cycles which illustrate biological influence on mobility of elements are carbon (Figure 15), phosphorus (Figure 16), and sulfur (Figure 17). Such cycles could be drawn for any element including all the potentially hazardous heavy metals with existing data and a little educated speculation.

Since the trace and heavy metals which concern us most enter in organic combination as living cells and tissues of microorganisms of soil, plants, and animals, the biological mechanisms of mineralization and immobilization act significantly in attenuation of the potentially hazardous pollutants.

1. Complexation with Organic Constituents

Organic complexes in soils which accumulate as a result of microbial synthesis as well as degradation of organic residues have a relatively high capacity to combine strongly with trace and heavy elements. Hodgson,[39] for example, delineates three organic fractions which can be identified in relation to immobility of metals:

- Lignin-like compounds of high molecular weight and similar organics, mostly cyclic in nature
- Organic acids and bases, short chained in nature (short-chained humic acids included)
- Soluble constituents which become insoluble when combining with heavy metals

The importance of organic substances in attenuation and immobilization of potentially hazardous radioactive pollutants has been well documented.[33,34,38,40]

Several mechanisms are responsible for the formation of complexes between organic matter and metals.[41] These are; ion exchange, surface absorption, chelation, and complex coagulation and peptization reactions. The chelation ligands, reactive organic groups, and radicals that bind elements chemically in the polymeric components of soil organic matter include carboxyl, hydroxyl, and amino groups to list a few. Specific ligands of microbial origin are present not only in soil organic matter[42] but in organic matter of municipal solid waste leachate[43] and in sewage sludge.[44,45] Decomposing wastes from certain industrial operations — canning, wood products, tanning, and meatpacking — are known to produce microbial slimes, gums, cell debris, fulvic and humic acids, and humus. Polymers of lignin, polysaccharides (uronides) tannins, polyphenols, proteins, and quinones have high molecular weights.

Low molecular weight substances (aliphatic acids, amino acids, and organic phosphates, certain phenolic, and volatile acid complexes) are all well known as biodegradation products of organic wastes.[16,46]

Fulvic and humic acids are important ligands that can bring about translocation and movement of Fe and Al in soils.[47] These acids, first identified in natural soils as a product of biodegradation of plant and animal residues, now are identified as products of many organic wastes.[48-53] The aromatic "nucleus" of fulvic acid, for example, with its hydroxyl, carboxyl, and carbonyl groups makes it a likely compound for influencing the mobility and attenuation of metals in soils. The organic constituents of municipal solid waste leachates have been shown to be highly influential in affecting the rate of movement of heavy metals through soils.[6,51,54,55] No doubt a number of chelating substances in addition to fulvic and humic compounds appear in leachate from organic solid wastes.[49,50] That microbial activities on wastes in soils are responsible for the production of a great variety of organic substances like complex pollutants, and represent an important factor in the retention and mobility of potentially hazardous substances through soils, appears to be well established.

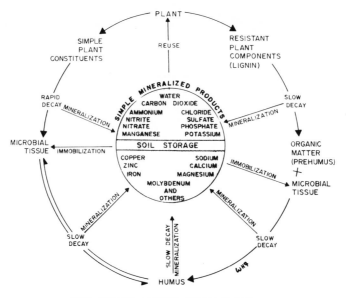

THE ORGANIC MATTER CYCLE IN SOIL

FIGURE 15.   The carbon and organic matter cycle in soils showing biodegradation with release of carbon dioxide, mineral elements, and humus formation. (From Fuller, W. H., *Soils of the Desert Southwest,* The University of Arizona Press, Tucson, 1975, 95.)

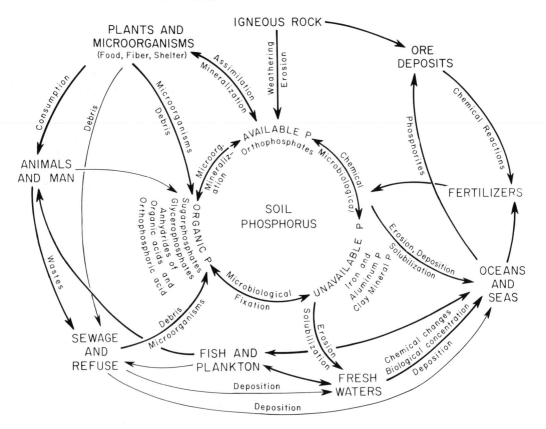

FIGURE 16.   The phosphorus cycle in soils showing the universal distribution of phosphorus. (From Fuller, W. H., *Soils of the Desert Southwest,* The University of Arizona Press, Tucson, 1975, 94.)

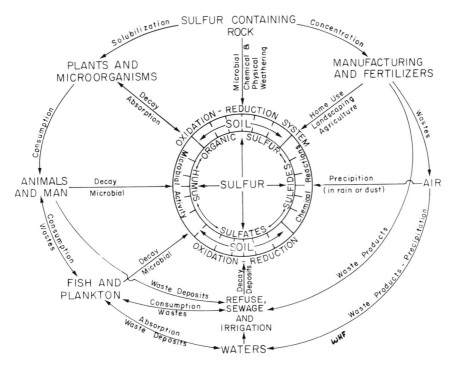

FIGURE 17.   The sulfur cycle in soils showing oxidation of sulfur to sulfuric acid and sulfate formation and reduction to sulfide. (From Fuller, W. H., *Soils of the Desert Southwest,* The University of Arizona Press, Tucson, 1975, 96.)

Organic complexes in municipal landfill leachates immobilize many of the trace metal contaminants by organic precipitation in the presence of air.[56] Some of the organic precipitates are irreversible unless the pH drops to 3.0 or below. The organic constituents in waste streams and leachates of low pH values (3 or below) may be expected to complex with trace contaminants poorly. The hydrogen ion concentration tends to keep heavy metals and trace elements in solution. At pH levels near neutral and slightly alkaline, a host of unidentified and identifiable biological products present both in soils and aqueous organic "waste streams" are capable of immobilizing metals. Flaig[57] and Allison[58] discuss some of these compounds in their reviews concerning the retention of metals by organic constituents in soil. Suggested chemical reactions between some of the better known organic complexes and the selected potentially hazardous waste constituents will be discussed further in the Section III.

2. Cometabolism

Classically, microorganisms in the soil completely biodegrade an organic molecule in water or soil in innocuous inorganic constituents through mineralization. During the process the organism converts some of the carbon to cell constituents in the accompanying energy release. The assimilation results in an increase in numbers and biomass of microorganisms. The simplified transformation is

$$\text{Organic molecule} \xrightarrow[\text{in water or soil}]{\text{biodegradation}} CO_2 + H_2O +$$

$$\text{inorganic minerals} + \text{ENERGY} \rightarrow \text{microbial cells} + (\text{biomass increase})$$

In the above, the organic substrate is mineralized in a growth-linked process. The illustration also depicts microbial detoxification, although in some instances the end product,

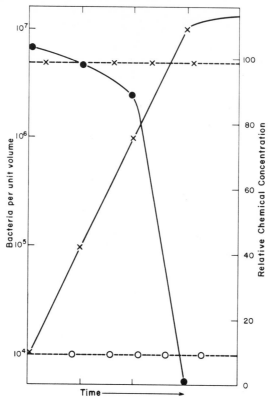

FIGURE 18.   Illustration of population changes and metabolism of a chemical modified by mineralizing and cometabolizing soil populations. The symbols in this figure are defined as: —●— mineralizable chemical, —×— mineralizing population, -- × -- cometabolized chemical, --○-- cometabolizing population. (Redrawn from Alexander, M., Biodegradation of chemicals and environmental concern, *Science*, 211, 133, 1981. With permission.)

such as nitrate, can be toxic. Until recently, microbiologists have left mineralization at this point. However, much recent evidence indicates that certain chemicals are subject to microbial conversion quite different from mineralization. Some of these chemicals are represented by well-known chlorinated hydrocarbons (e.g., DDT, 2, 4, 5-T, aldrin, and heptachlor) and some nonchlorinated molecules as well.[59] It is believed that some microorganisms can use certain chemicals as nutrients for energy but fail to use them to sustain growth. Thus, a new phenomenon is suggested that differs from mineralization called cometabolism or cooxidation. According to Alexander,[59] the populations presumably undertake transformations called cometabolism while growing on another cell supporting substrate. This phenomenon is characterized by a slow rate of substrate modification. The presumed inability of the populations to use the chemicals as growth substrates can contribute to the perpetuation of this condition of slow transformation rate that does not increase significantly with time as compared with typical mineralization. The two transformation rates are compared diagrammatically in Figure 18.

Because biodegradation is so important to a clean environment and to waste disposal in general, a few type reactions for transformation of chemicals are presented in Table   11 as originally compiled by Alexander.[59] In addition, he compiled type reactions for cleavage of chemicals of environmental importance (Table 12).   The type reactions listed have been reported as found only outside the cell in concentrations that were readily detectable from model ecosystems or, in some cases, by direct measurement of soils and water as labeled by the S or W.

Table 11
SOME CHEMICAL REACTIONS ILLUSTRATING TRANSFORMATIONS THAT HAVE ENVIRONMENTAL IMPLICATIONS

Category	Reaction[a]	Example[b]
Dehalogenation	$RCH_2Cl \rightarrow RCH_2OH$	Propachlor (S,M)
	$ArCl \rightarrow ArOH$	Nitrofen (S)
	$ArF \rightarrow ArOH$	Flamprop-methyl (S)
	$ArCl \rightarrow ArH$	Pentachlorophenol (S,M)
	$Ar_2CHCH_2Cl \rightarrow Ar_2C = CH_2$	DDT (M)
	$Ar_2CHCHCl_2 \rightarrow Ar_2C = CHCl$	DDT (G,W,M)
	$Ar_2CHCCl_3 \rightarrow Ar_2CHCHCl_2$	DDT (G,S,M)
	$Ar_2CHCCl_3 \rightarrow Ar_2C = CCl_2$	DDT (W,S,M)
	$RCCl_3 \rightarrow RCOOH$	N-serve (S), DDT (W,M)
	$HetCl \rightarrow HetOH$	Cyanazine (S)
Deamination	$ArNH_2 \rightarrow ArOH$	Fluchloralin (S)
Decarboxylation	$ArCOOH \rightarrow ArH$	Bifenox (S)
	$Ar_2CHCOOH \rightarrow Ar_2CH_2$	DDT (M)
	$RCH(CH_3)COOH \rightarrow RCH_2CH_3$	Dichlorfop-methyl (S)
	$ArN(R)COOH \rightarrow ArN(R)H$	DDOD (S)[c]
Methyl oxidation	$RCH_3 \rightarrow RCH_2OH$ and/or $\rightarrow$ $RCHO$ and/or $\rightarrow RCOOH$	Bromacil (S), diisopropylnaphthalene (G), pentachlorobenzyl alcohol (S,M)
Hydroxylation and ketone formation	$ArH \rightarrow ArOH$	Benthiocarb (S), dicamba (W)
	$RCH_2R' \rightarrow RCH(OH)R'$ and/or $\rightarrow$ $RC(O)R'$	Carbofuran (S), DDT (S,W,G)
	$R(R')CHR'' \rightarrow R(R')CHOH(R'')$	Bux insecticide (S)
	$R(R')(R'')CCH_3 \rightarrow$ $R(R')(R'')CCH_2OH$	Denmert (S)
β oxidation	$ArO(CH_2)_nCH_2CH_2COOH \rightarrow$ $ArO(CH_2)nCOOH$	ω-(2,4-Dichlorophenoxy)-alkanoic acids (S,M)
Epoxide formation	$RCH = CHR' \rightarrow RCH^{0-}CHR'$	Heptachlor (S,M)
Nitrogen oxidation	$R(R')NR'' \rightarrow R(R')N(\rightarrow O)R''$	Tridemorph (S)
Sulfur oxidation $=S$ to $=0$	$RSR' \rightarrow RS(O)R'$ and/or $\rightarrow$ $RS(O_2)R'$	Aldicarb (S,M)
	$(AlkO)_2P(S)R \rightarrow (AlkO)_2P(O)R$	Parathion (S,M)
	$RC(S)R' \rightarrow RC(O)R'$	Ethylenethiourea (S)
Sulfoxide reduction	$RS(O)R' \rightarrow RSR'$	Phorate (S)
Reduction of triple bond	$RC \equiv CH \rightarrow RCH = CH_2$	Buturon (S,M)
Reduction of double bond	$Ar_2C = CH_2 \rightarrow Ar_2CHCH_3$	DDT (W,M)
	$Ar_2C = CHCl \rightarrow Ar_2CHCH_2Cl$	DDT (W,M)
Hydration of double bond	$Ar_2C = CH_2 \rightarrow Ar_2CHCH_2OH$	DDT (W,M)
Nitro metabolism	$RNO_2 \rightarrow ROH$	Nitrofen (S)
	$RNO_2 \rightarrow RNH_2$	Pentachloronitrobenzene (S,M), Sumithion (S,W,M)
Oxime metabolism	$RCH = NOH \rightarrow RC \equiv N$	Aldicarb (S,M)
Nitrile/amide metabolism	$RC \equiv N \rightarrow RC(O)NH_2$ and/or $\rightarrow$ $RCOOH$	Bromoxynil (S,M), Dichlobenil (S,W)

[a] Abbreviations: R, organic moiety; Ar, aromatic; Alk, alkyl; Het, heterocycle.

[b] Reaction demonstrated in sewage (G), microbial culture (M), soil (S), or natural waters (W).

[c] 3-(3',5'-Dichlorophenyl)-5,5-dimethyloxazolidine-2,4-dione.

3. Recalcitrant Molecules

Some organic chemicals may persist for very long periods of time in natural ecosystems. They may not be available as a substrate for microorganisms, or if degraded at all, the process is so slow it is not readily measurable. Such chemicals are labeled recalcitrant

Table 12
EXAMPLES OF SOME ENVIRONMENTALLY IMPORTANT CHEMICAL
REACTIONS INVOLVING CLEAVAGE

Substrate	Reaction	Example
Ester	RC(O)OR' → RC(O)OH	Malathion (M), phthalates (W,M)
Ether	ArOR → ArOH	Chlomethoxynil (S), 2,4-D(S,W,M)
	ROCH$_2$R' → ROH	Dichlorfop-methyl (S)
C-N bond	R(R')NR" → R(R')NH and/or → RNH$_2$	Alachlor (S,M), trimethylamine (G,M)
	RN(Alk)$_2$ → RNHAlk and/or → RNH$_2$	Chlorotoluron (S), trifluralin (S,M)
	RNHCH(R')R" → RNH$_2$	Imugam (S)
	RNH$_2$CH$_2$R' → RNH$_2$	Glyphosate (S,W)
Peptide, carbamate	RNHC(O)R' → RNH$_2$ and/or HOOCR'	Benlate (S,M), dimethoate (S)
	R(R')NC(O)R" → R(R')NH + HOOCR"	Benzoylprop-ethyl (S)
NOC(O)R	RCH = NOC(O)R → RCH = NOH	Aldicarb (S,M)
C-S bond	RSR' → ROH and/or HSR'	Benthiocarb (S), Kitazin P (S,M)
C-Hg bond	RHgR' → RH and/or Hg	Ethylmercury (S,M), phenylmercuric acetate (S,M)
C-Sn bond	R$_3$SnOH → R$_2$SnO → RSnO$_2$H	Tricyclohexyltin hydroxide (S)
C-O-P	(AlkO)$_2$P(S^a)R → AlkO(HO)P(S^a)R and/or → (HO)$_2$P(S^a)R	Gardona (S), malathion (S,M)
	ArOP(S^a)(R)R' → ArOH and/or HOP(S^a)(R)R'	Diazinon (S), parathion (S,M)
P-S	RSP(O)(R')OAlk → HOP(O)(R')OAlk	Hinosan (M), Kitazin P (M)
Sulfate ester	RCH$_2$OS(O$_2$)OH → RCH$_2$OH and/or HOS(O$_2$)OH	Sesone (S,M)
S-N	ArS(O$_2$)NH$_2$ → ArS(O$_2$)OH	Oryzalin (S)
S-S	RSSR → RSH	Thiram (S,M)

Note: References for these reactions are provided by Alexander (1981) who originally compiled these chemical reactions.[59]

a   Sulfur or oxygen.

molecules. A few recalcitrant chemicals are toxic. Insolubility in water, acids, and alkalies is partly responsible for their persistence, but certain phenomena of the natural environment such as anaerobosis, high acidity, high salt, and other factors that reduce microbial activity may contribute further to their resistance to biodegradation. Both cometabolism and the recalcitrant molecules add a discouraging aspect to waste biodegradation as earlier believed. Further basis research requires urgent attention since biodegradation has become such a critical factor in waste disposal and pollutant control.

III. CHEMICAL CHARACTERISTICS

A. Chemical Properties

Despite the overlap between chemical and biological reactions and between chemical and physical reactions in soil, the objective here is to identify some predominantly chemical characteristics and to show how closely one soil characteristic can influence another. For example, microbial reactions most often begin and develop as a result of enzyme activity originating from the microorganism, yet, certain similar or identical reactions may occur as a result of chemical activity not involving microbes. Also, microbes may initiate chemical reactions independent of enzymes as a consequence of some substrate change.

Another example is the relationship between chemical and physical reactions of absorption, adsorption, and sorption involving soil particle surfaces. Chemical equilibria in soils can be shifted from one direction to another or increased or decreased, in the ongoing rate, as a result of change in physical properties of the soil.

B. Chemical Equilibria

The possibility or probability that a reaction will or will not take place can be predicted by knowing the associated chemical equilibria.[60] At any specific time a great number of chemical reactions occur in the soil. They proceed at various rates of speed ranging from those that attain equilibrium immediately to those that react so slowly they may never achieve equilibrium. Certain reactions are readily reversible with both forward and backward action possible. These always will be in a state of transition. The position will depend on the composition of the system with respect to reactants and reaction products present at any one time and the nature of environmental factors present at that time, such as temperature. Although equilibrium reactions have not been used extensively by the soil scientist in application to agriculture, equilibrium offers a real opportunity in all industries, including that of waste disposal, as a reference point for predicting which chemical reactions can take place, regardless of actual rates.[60] Equilibrium constrants may be expressed in terms of concentration or activity depending upon the units in which reactants and products are expressed. Each expression has certain advantages and disadvantages. An example of equilibrium reactions is provided for aluminum minerals and complexes in Table 13. Activity constants are peferred by the scientist concerned with pure systems where specific ionic and molecular species are known and measurable. However, in soils of mixed species and in mixed systems, activities cannot be measured with acceptable accuracy. The expression of concentration constants, therefore, are preferably used by the soil chemist although they change with ionic strength. Corrections must be applied when systems change or different systems are used. The chemical characterization of soils with respect to chemical equilibria is well described[1,60-63] and, therefore, will not be presented here except to draw attention to this important field as it influences pollutant reactions to soil and soil modification.

C. Ion Exchange

Many of the concepts of ion exchange evolve from chemical equilibrium relationships just discussed. Yet, the soil has charge properties that attract ions from the soil solution to its solid surfaces that are unique. Negative charges predominate over positive charges on most soil solids. These charges exercise a great impact on the internal soil environment. The unusual role of the soil in the control management of wastes must consider ion exchange as wastes react chemically with the soil. The nature and quantity of the negative and positive charges affect water behavior, attenuation of metal pollutants, movement and transformation of organic wastes, pesticides and solvents, and stability of municipal, industrial and biological wastes.

Soils contain a mixture of constant-charge clay minerals (crystalline) and pH-dependent-charge surfaces. The constant-charge clays dominate over the surface charges in temperate region soils, whereas the pH-dependent-charge surfaces of oxides and hydrous oxides dominate in intensively weathered soils of the tropical climates. Mixtures result from (a) edge effects on crystalline clay minerals, (b) isomorphous substitution on site vacancies in Si-Al co-gels, or (c) oxide coatings or inner layers on crystalline clay minerals in most soils.[64]

A combination of forces bind ions (cations and anions) to the soil solids ranging from electro-static to covalent, with related bonding energies. Specific sorption contrasts to ion exchange and occurs when covalent bonding dominates and represents a property of specificity for certain cations and anions. Ion exchange, as the name implies, is reversible but is less for specific sorption than those for electrostatic bonding.

A brief comment is desirable here to clarify what is meant by chemical bonds. Figure 19 diagrams some prominent chemical bonds encountered in the soil.[1] The common table salt, sodium chloride, provides an example of the strong ionic bond as Coulomb's law applies. Other chemical bonds are explained by Jenny.[1] The van der Waals bonds of cohesion may become involved with (a) neutral atoms and molecules that weakly attract each other

Table 13
EQUILIBRIUM REACTIONS OF ALUMINUM MINERALS AND COMPLEXES AT 25°C

Reaction	log K°
Oxides and Hydroxides	
$0.5\gamma\text{—}Al_2O_3(c) + 3H^+ = Al^{3+} + 1.5H_2O$	11.49
$0.5\alpha\text{—}Al_2O_3 \text{ (corundum)} + 3H^+ = Al^{3+} + 1.5H_2O$	9.73
$Al(OH)_3(amorp) + 3H^+ = Al^{3+} + 3H_2O$	9.66
$\alpha\text{-}Al(OH)_3(bayerite) + 3H^+ = Al^{3+} + 3H_2O$	8.51
$\gamma\text{-}AlOOH(boehmite) + 3H;^+ = Al^{3+} + 2H_2O$	8.13
$Al(OH)_3 \text{ (norstrandite)} + 3H^+ = Al^{3+} + 3H_2O$	8.13
$\gamma\text{-}Al(OH)_3(gibbsite) + 3H^+ = Al^{3+} + 3H_2O$	8.04
$\alpha\text{-}AlOOH(diaspore) + 3H^+ = Al^{3+} + 2H_2O$	7.92
Sulfates	
$Al_2(SO_4)_3(c) = 2Al^{3+} + 3SO_4^{2-}$	20.84
$Al_2(SO_4)_3 \cdot 6H_2O(c) = 2Al^{3+} + 3SO_4^{2-} + 6H_2O$	3.45
$KAl_3(SO_4)_2(OH)_6(alunite) + 6H^+ = K^+ + 3Al^{3+} + 2SO_4^{2-} + 6H_2O$	3.04
Hydrolysis	
$Al^{3+} + H_2O = AlOH^{2+} + H^+$	−5.02
$Al^{3+} + 2H_2O = Al(OH)_2^+ + 2H^+$	−9.30
$Al^{3+} + 3H_2O = Al(OH)_3^o + 3H^+$	−14.99
$Al^{3+} + 4H_2O = Al(OH_4^- + 4H^+$	−23.33
$Al^{3+} + 5H_2O = Al(OH)_5^{2-} + 5H^+$	−34.24
$2Al^{3+} + 2H_2O = Al_2(OH)_2^{4+} + 2H^+$	−7.69
Complexes	
$Al^{3+} + F^- = AlF^{2+}$	6.98
$Al^{3+} + 2F^- = AlF_2^+$	12.60
$Al^{3+} + 3F^- = AlF_3^o$	16.65
$Al^{3+} + 4F^- = AlF_4^-$	19.03
$Al^{3+} + 3NO_3^- = Al(NO_3)_3^o$	0.12
$Al^{3+} + SO_4^{2-} = AlSO_4^+$	3.20
$Al^{3+} + 2SO_4^{2-} = Al(SO_4)_2^-$	1.90
$2Al_3^+ + 3SO_4^{2-} = Al_2(SO_4)_3^o$	−1.88
Other Equilibria	
$CaSO_4 \cdot 2H_2O(gypsum) = Ca^{2+} + SO_4^{2-} + 2H_2O$	−4.64
$CaF_2(fluorite) = Ca^{2+} + 2F^-$	−10.41
$Al^{3+} + 3e^- = Al(c)$	−86.02

Lindsay, W. L., *Chemical Equilibrium in Soils*, John Wiley & Sons, New York, 1979, 24, 449. With permission.

if they are dipolar with unsymmetrical internal distributions of protons and electrons and with (b) inert organic chains as in certain hydrocarbons and soil humus. Ions attract dipoles and may surround themselves with a cluster of water molecules (Figures 20 and 21). The H^+ ion is highly important in soils since it holds ions together such as two oxygen or one oxygen in a hydrogen bond. The electron-pair bond or covalent bond is one of the strongest.

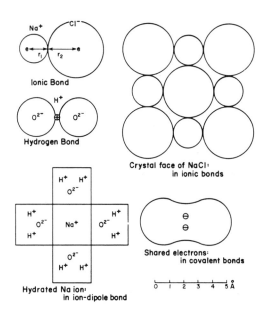

FIGURE 19. The prominent chemical bonds encountered in the soil solution. (Redrawn from Jenny, H., *The Soil Resource, Origin, and Behavior*, Springer-Verlag, New York, 1980, 377. With permission.)

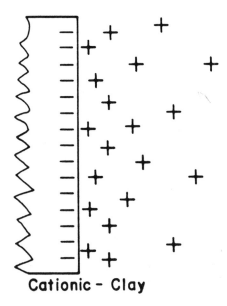

FIGURE 20. Electric double layer showing a negative wall of oxygen ions and positive swarms of exchangeable cations.

Two atoms share two electrons in this instance, and are represented most characteristically among carbon atoms in organic molecules. Several of these bonds may act simultaneously with metal ions such as Cu^{2+} and Zn^{+2}.[64]

Charge properties of soil solids range from those ubiquitous negative charges of cation exchange capacity (CEC_c), variable charge cation exchange capacity (CEC_v), total negative charge (CEC), positive charge or anion exchange capacity (AEC), and anion sorption capacity (ASC).[65] Cation exchange capacity (CEC_c) of a soil is a measurement mostly used in the

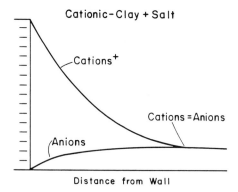

FIGURE 21. Concentration profile of cations and anions in relation to distance from negative wall.

Table 14
ION EXCHANGE CAPACITIES OF VARIOUS CLAY MINERALS[66]

Mineral	Cation exchange capacity (me/100 g)	Anion exchange capacity (me/100 g)	Cation exchange to Anion exchange (c/a)	Equilibrium pH
Nontronite	87	13	6.7	2.4
Bentonite	62	15	4.1	3.1
Illite	21	9	2.3	4.5
Kaolinite	27	43	0.63	4.7
Halloysite	4	7.7	0.52	4.3

agricultural industry and related to crop production. CEC_c is of transitory importance for waste disposal management. It usually represents exchange reactions of the major cations Ca, Mg, Na, and K. Since exchange cations on the soil solids are associated with negative charge sites through largely electrostatic bonding, they are subject to interchange with cations in the soil solution (Figure 20). CEC_c also is associated with the hydrous oxides of Fe, Al, and Mn as well as the aluminosilicates as represented by the soil clay minerals.

Cation exchange is found by displacing the cations from soil with various neutral salt (Na, K, and NH_4) solutions of known concentrations. The most commonly used solution is $1N$ NH_4OA_c. The cation exchange capacity (CEC) is an estimate of the amount of NH_4^+ held on the soil. The exchangeable bases are considered to be those contained in the extract. Soils from arid and semiarid lands require some modification of the standard CEC procedure in an attempt to prevent dissolution of $CaCO_3$ and $CaSO_4 \cdot 2H_2O$ and certain indigenous salts from contributing to the sum of exchangeable cations.

Anion exchange in soils has been given small attention compared with that of cations primarily because the level of exchange is relatively low at the pH ranges of most natural soils. Some soils, particularly acid soils, exhibit true anion exchange, however. The anions most often involved are NO_3^-, Cl^-, SO_4^{2-}, and $PO_4^{(n)}$. Whereas CEC ranges between 0.5 and 50 meq/100 g soil, anion exchange capacity ranges between 0 and about 2.6 meq/100 g.[63] Most soils exhibit anion exclusion because the net negative charge of soil colloid repels the anions in solution in the vicinity of the surfaces. Yet, the net negative charge of a colloidal clay system decreases with increasing acidity.[65] Kaolinite and allophane groups, for example, demonstrate positive charges at low pH values. Anion adsorption and exchange, therefore, take place at these positive charge sites. Table 14 from Schoen[66] provides data to make a comparison of cation and anion exchange capacities of clay minerals at very low soil pH values.

The conventional determinations of exchangeable cations and cation exchange capacity as a tool in the prediction of soil behavior for the attenuation of hazardous pollutants from wastes is not very helpful for evaluating soil attenuation of metals. First, in displacement (especially by 1 N NH_4OA_c), ions in addition to those in truly exchangeable positions may contribute to the extraction solution (such as Ca^{2+} from lime and gypsum). The solution of a great number of soil minerals, compounds, and complex inorganic and amorphous substances and salts may take place. Second, the displacement of cations with any of the monovalent cations now used does not appear to represent or describe the soil exchange capacity for heavy metals. The figures reported for exchangeable heavy metals using NH_4^+, for example, as the replacing ion may be misleading.[67-71] Zn exchange capacity cannot be expected to be defined by NH_4^+ exchange capacity. Third, cations that become stored temporarily at the liquid-solid interface will again be released upon change in composition of the liquid phase either by dilution or by ion concentration. Retention, thus, is not permanent as is desired for waste control. Fourth, the CEC is not the only factor influencing the amount of temporary ion storage. The concentration of constituents (salts, chelates, acids, and bases) in waste leachates, waste solutions, and solvents may actually totally offset the magnitude of the original soil exchange capacity in systems where the usual soil solution has been altered.

Despite these characteristics, CEC roughly correlates with attenuation of some heavy metals and certain other hazardous pollutants since many of the same soil characteristics that influence metal attenuation also influence the magnitude of CEC. There is a close relationship between particle size distribution (clay content) of a soil and CEC. The surface area of a soil and/or clay content appear to be more reliable properties describing attenuation prediction than CEC.[72-75]

D. Transfer of Protons and/or Electrons

A reaction in which an electron transfer takes place is called an oxidation-reduction process. This was discussed in terms of biological characteristics earlier; we now go into more fundamental details of the chemical reactions. An element or substance giving off electrons is being reduced (e.g., $Fe^{+++} + e^- \leftrightarrow Fe^{++}$; or $H^+ + 1/4\,O_2 + e^- = 1/2\,H_2O$; reduction of O_2). An element or substance accepting an electron is being oxidized. The transfer of electrons need not include oxygen. The intensity of the oxidizing or reducing action of a system is determined by a standard electrode placed in the solution system. The potential difference between the electrode and solution is called redox potential. The transfer of electrons from donor to acceptor concerns the solubility of reaction products formed when wastes and waste leachates, unnatural to the land, make contact with the natural soil body. Electron transfer is involved with every slight change in gaseous composition and movement through soil.

The redox reaction, e.g., transfer of electrons and protons from donor to acceptor, can be written chemically:

$$n_B A_{ox} + n_A B_{red} \leftrightarrow n_B A_{red} + n_A B_{ox} \tag{2}$$

equivalent to the proton reaction. The reaction can be split into the redox couples:

$$A_{ox} + n_A e^- = A_{red}, \quad K_{red}^\circ = \frac{(A_{red})}{(A_{ox})(e^-)^n} \tag{3a}$$

$$B_{ox} + n_B e^- = B_{red}, \quad K_{red}^\circ = \frac{(B_{red})}{(B_{ox})(e^-)^n} \tag{3b}$$

These half reactions are equivalent to the reactions involving the transfer of protons. Relative electron activity can also be described:

$$pe = -\log(e^-)$$

analogous to proton activity:

$$pH = -\log(H^+).$$

Strong oxidizing conditions associate with large pe values, and reducing conditions associate with small values of pe.

In the aqueous systems of soils, the dissociation of water into gaseous H_2 or O_2 imposes redox limits on soils. On the reducing side, Lindsay[5] provides the reactions:

$$H_2O + e^- \rightarrow \tfrac{1}{2} H_2(gas) + OH^-$$
$$\underline{H^+ + OH^- \rightarrow H_2O}$$

$$H^+ + e^- \rightarrow \tfrac{1}{2} H_2(gas) \tag{4}$$

which yields the equilibrium equations:

$$K^\circ = \frac{(H_2)(gas)^{1/2}}{(H^+)(e^-)} \tag{5}$$

or

$$\log K^\circ = \tfrac{1}{2} \log H_2(gas) - \log(H^+) - \log(e^-)$$

where the units of the gas phase are partial pressure.

The corresponding K° then equals unity (log K° = 0) for standard state conditions. This sets the electron activity (e^-) at unity for the standard hydrogen half-cell reaction, and since K° is 1, then

$$pe + pH = -\tfrac{1}{2} \log H_2(gas) \tag{6}$$

and when H_2 (gas) = 1 atm at P = 1 atm and T = 273 K, pe + pH = 0

On the oxidizing side, Lindsay, 1979,[5] describes the redox limit of the aqueous system by:

$$H^+ + e^- + \tfrac{1}{4} O_2 (gas) \rightarrow \tfrac{1}{2} H_2O \tag{7}$$

giving the equilibrium expression

$$K^\circ = \frac{(H_2O)^{1/2}}{(H^+)(e^-)(O_2(gas))^{1/4}} \tag{8}$$

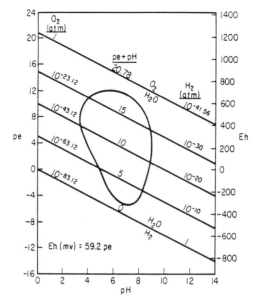

FIGURE 22. Equilibrium redox relationship of aqueous systems. The inscribed area was adapted from Baas Becking et al. (1960) and is representative of most soils. (From Lindsay, W. L., *Chemical Equilibrium in Soils*, John Wiley & Sons, New York, 1979, 25. With permission.)

Because in dilute aqueous systems the activity of water is near unity, the log form of the equilibrium expression yields:

$$-\log(H^+) - \log(e^-) - \tfrac{1}{4} \log O_2(gas) = 20.8 \tag{9}$$

or

$$pe + pH = 20.78 + \tfrac{1}{4} \log O_2(gas)$$
$$\text{(when } O_2(gas) \text{ is 1 atm, pe + pH = 20.8)} \tag{10}$$

The equilibrium redox relationships of aqueous systems including the pe, pH relationships of aqueous systems including the pe, pH and the more familiar Eh have been diagrammed by Lindsay[5] and reproduced in Figure 22. This figure also includes an inscribed area, as adapted from Baas Becking et al.,[76] which is representative of most soils. The broad representation of the dynamic chemical equilibria occurring in soils with the solution as the focal point is summarized in Figure 23.[46]

E. Soil pH (Acidity and Alkalinity)

The pH as measured by the glass electrode in soil saturated with water roughly indicates the acidity (H^+ ion activity) and alkalinity (OH^- ion activity) of soil. The pH is defined as the negative logarithm of the H^+ ion concentration. Thus pH = log $1/[H^+]$ with H^+ ion

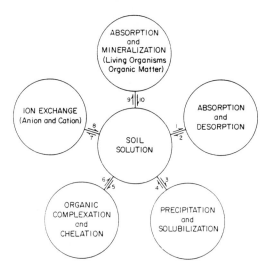

FIGURE 23.   Illustration of the dynamic chemical equilibria occurring in the soils with solution at the focal point. (From Cope, C. B., Fuller, W.H., and Willetts, S. L., *The Scientific Management of Hazardous Wastes*, Cambridge University Press, 1983, 310. With permission.)

concentration in moles per liter. The pH of pure water is 7 or 0.0000001 mol/ℓ. More simply:

d			i	
e	a	pH 7 = 0.0000001 moles per liter	n	a
c	c	pH 6 = 0.000001 moles per liter	c	c
r	i	pH 5 = 0.00001 moles per liter	r	i
e	d	pH 4 = 0.0001 moles per liter	e	d
a	i	pH 3 = 0.001 moles per liter	a	i
s	t	pH 2 = 0.01 moles per liter	s	t
i	y	pH 1 = 0.1 moles per liter	i	y
n			n	
g			g	

Soils usually occupy the middle ranges of the pH scale and higher plants (crops) have adapted well to soil pH levels which usually range between 4 to 8.4 as shown in Figure 24.   Many plants and economic crops are highly sensitive to soil pH and selectively are adapted to narrower ranges. This also is true for microorganisms and other biological life indigenous to soil habitats. Thus pH measurements provide a valuable tool for biologists and soil scientists.

Since soil is a highly buffered medium, true acidity can be estimated if the CEC and the degree of saturation of the exchange complex with Al (reserve acidity) become known in addition to pH (active acidity). Also, the pH at the surface of soil particles (pH_s) is more acid than the soil solution beyond the particle surface or the bulk (pH_b). Organic matter, also, including the volatile acids often occurring in dilute aqueous solid waste leachates provides buffer capacity to soils and organic containing disposals that undergo biodegradation.

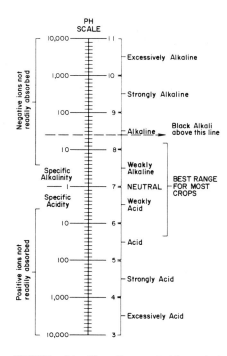

FIGURE 24. The pH range (acidity and alkalinity) for soils.

IV. PHYSICAL CHARACTERISTICS

A. Physical Properties

The physical characteristics of the soil are equally important as biological and chemical characteristics for pollutant attenuation.[6,16,46,55,75] Some prominent physical soil characteristics and processes that influence movement and retention of pollutants in soils are: (a) texture — sand, silt, and clay, (b) surface area, (c) soil structure, (d) kind of clay mineral, (e) stratifications of materials (f) compaction, (g) restrictive layers, (h) hydraulic conductivity, (i) shrinking and swelling, and (j) drying and wetting — as well as freezing and thawing.[77-79] All of the factors that influence soil permeability play an important role in pollutant attenuation by soil. Relationships between physical, chemical, and biological processes are diversely interrelated and difficult to identify.[1,5,6,8] This emphasizes the complexity of the soil as a waste treatment system and the problems in the identification of specific mechanisms of pollutant attenuation and of placing them into a strictly physical, biological, or chemical classification.

B. Particle Size and Surface Area

Texture refers to particle size distribution as sand silt, and clay.[12] Texture and the related surface area of the soil particles clearly dominate the migration rate of trace and heavy metals through soils.[75,78] Sand is found to be negatively correlated to heavy metal ion attenuation and clay (< 2 μm) positively correlated. Surface area of the particles increases greatly as the particle size becomes smaller (Table 15). Clay, the inorganic colloid of < 2 μm sizes, forms the principal site of soil reactions and activity. Both primary and secondary minerals constitute the clay-sized particles. Secondary minerals usually dominate. They form by the alteration of pre-existing minerals as a result of weathering. Clay minerals vary considerably in composition and properties.[80] From a practical standpoint they can be separated on the basis of their capacity to expand when wetted or shrink upon drying as

Table 15
THE RELATIONSHIP OF SURFACE AREA TO PARTICLE SIZE

Diameter of sphere	Textural name	Volume per particle (1/6 πD^3)	Number of particles in $\pi/6$ cc	Total surface area (cm²)
1 cm	Gravel	$1/6 \pi (1)^3$	1	3.14
0.1 cm (1 mm)	Coarse sand	$1/6 \pi (1/10)^3$	1×10^3	31.42
0.05 cm (0.5 mm or 500 μ)	Medium sand	$1/6 \pi (5/100)^3$	8×10^3	62.83
0.01 cm (0.1 mm or 100 μ)	Very fine sand	$1/6 \pi (1/100)^3$	1×10^6	314.16
0.005 cm (0.05 mm or 50 μ)	Coarse silt	$1/6 \pi (5/1,000)^3$	8×10^6	628.32
0.002 cm (0.02 mm or 20 μ)	Silt	$1/6 \pi (2/1,000)^3$	125×10^6	1,570.8
0.0005 cm (0.005 mm or 5 μ)	Fine silt	$1/6 \pi (5/10,000)^3$	8×10^9	6,283.2
0.0002 cm (0.002 mm or 2 μ)	Clay	$1/6 \pi (2/10,000)^3$	125×10^9	15,708
0.0001 cm (0.001 mm or 1 μ)	Clay	$1/6 \pi (1/10,000)^3$	1×10^{12}	31,416
0.00005 cm (0.0005 mm or 500 mμ)	Clay	$1/6 \pi (5/100,000)^3$	8×10^{12}	62,832
0.00002 cm (0.0002 mm or 200 mμ)	Colloidal clay	$1/6 \pi (2/100,000)^3$	125×10^{12}	157,080
0.00001 cm (0.0001 mm or 100 mμ)	Colloidal clay	$1/6 \pi (1/100,000)^3$	1×10^{15}	314,160
0.000005 cm (0.00005 mm or 50 mμ)	Colloidal clay	$1/6 \pi (5/1,000,000)^3$	8×10^{15}	628,320

Baver, L. D., Gardner, W. H., and Gardner, W. R., *Soil Physics*, John Wiley & Sons, Inc., New York 1972, 498. With permission.

well as size. Montmorillonitic types expand and shrink extensively, whereas kaolinite types do so only minimally.

Soil properties usually are characterized by the material that passes a 2 mm screen although larger sizes are present in many soils. Cobbles and stones are not uncommon (Table 16), and receive attention in the classification system. The soil scientist refers to the completely dispersed particles as textural separates of sand, silt, and clay. In combinations these form textural classes such as silty clay, sandy loam, etc. (Figure 25). Alternate schemes exist in the literature contributing to some confusion of terminology (Figure 26). The Unified Soil Classification System (USCS) is used primarily by engineers. Soil type criteria in the USCS system is based on particle (grain) size and response to physical manipulation at various water contents. The AASHO system is the old pre-1938 U.S. Department of Agriculture system. The new USDA system serves agricultural requirements and land management. It is based both on physical and chemical properties of soils that are quantitatively measurable as they exist in the field.

The present USDA system of soil classification extends beyond particle size distributions. It is continually being refined and is generally accepted by soil scientists. The nomenclature is used in most of the current U.S. literature. However, the distribution of soil grain or particle size and associated modifiers (stony, gravelly, mucky, diatomaceous, and micaceous) of both systems may be compared directly.

The size ranges for the USDA and the USCS particles designates are listed in Table 16. The USDA uses such textural terms as sandy loam, silty loam, silty clay, etc., whereas,

Table 16
U.S. DEPARTMENT OF AGRICULTURE (USDA) AND
UNIFIED SOIL CLASSIFICATION SYSTEM (USCS)
PARTICLE SIZES

USDA		USCS	
Particle	Size Range (mm)	Particle	Size Range (mm)
Cobbles	76.2—254	*Cobbles*	>76.2
Gravel	2.0—76.2	*Gravel*	4.76—76.2
coarse gravel	12.7—76.2	coarse gravel	19.1—76.2
fine gravel	2.0—12.7	fine gravel	4.76—19.1
Sand	0.05—2.0	*Sand*	0.074—4.76
very coarse sand	1.0—2.0		
coarse sand	0.5—1.0	coarse sand	2.0—4.76
medium sand	0.25—0.5	medium sand	0.42—2.0
fine sand	0.1—0.25	fine sand	0.074—0.42
very find sand	0.05—0.1		
Silt	0.002—0.05	*Fines*	<0.074
		(Silt and Clay)	
Clay	<0.002		

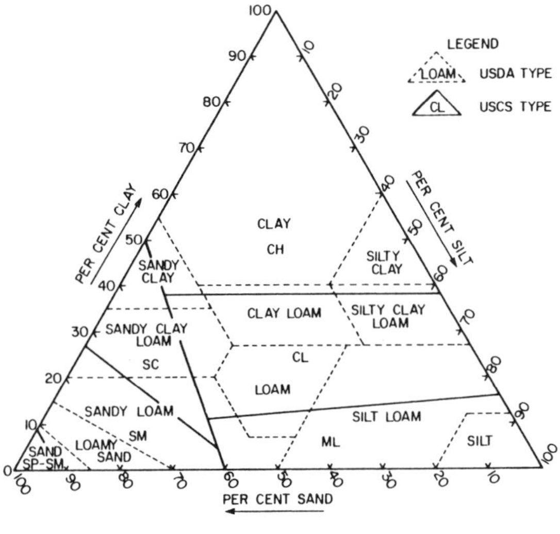

FIGURE 25. Soil texture classification of soils comparing the USDA and USCS systems based on the percentage of different sizes of particles they contain. (From Meyer, M. P. and Knight, S. J., Trafficability of Soils, Soil Classification, U.S. Army Engineers Waterways Experiment Station, Tech. Memo., 3 Suppl. 16., Vicksburg, Va, 1961.)

the USCS soil types refer to GC clay gravel and SC clayey sand descriptions but both are based on relative amounts of different sized particles in a soil. Figures 25 and 26 provide diagrammatic comparison between the USDA and USCS systems with respect to texture. Table 17 correlates the two systems of particle size classification with verbal detail. This correlation is not precise because texture is a high-level criterion in the USDA system.[79] A soil of a certain texture can be classified into only a limited number of the 15 USCS soil types, whereas, in the USDA system, soils of the same texture can be found throughout the

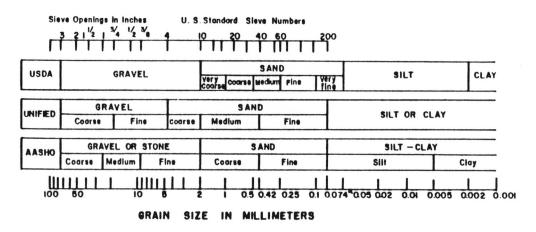

FIGURE 26.   Soil texture chart showing soil particle sizes and related classes as given by three separate systems involved nationally and internationally.

Table   17
CORRESPONDING USCS AND USDA SOIL CLASSIFICATION DESCRIPTION[a]

Unified Soil Classification System (USCS) Soil Types	Corresponding United States Department of Agriculture (USDA) Soil Textures
1. GW	Same as GP — gradation of gravel sizes not a criterion.
2. GP	Gravel, very gravelly[b] sand less than 5% by weight silt and clay.
3. GM	Very gravelly[b] sandy loam, very gravelly[b] loamy sand very gravelly[b] silt loam, and very gravelly[b] loam.[c]
4. GC	Very gravelly clay loam, very gravelly sandy clay loam, very gravelly silty clay loam, very gravelly silty clay, very gravelly clay.[c]
5. SW	Same — gradation of sand size not a criteria.
6. SP	Coarse to find sand; gravelly sand[d] (less than 20% very fine sand).
7. SM	Loamy sands and sandy loams (with coarse to find sand), very fine sand; gravelly loamy sand and gravelly sandy loam.[d]
8. SC	Sandy clay loams and sandy clays (with coarse to fine sands); gravelly sandy clay loams and gravelly sandy clays[d]
9. ML	Silt, silt loam, loam, very fine sandy loam.[e]
10. CL	Silty clay loam, clay loam, sandy clays with <50% sand.[e]
11. OL	Mucky silt loam, mucky loam, mucky silty clay loam, mucky clay loam.
12. MH	Highly micaceous or diatomaceous silts, silt loams — highly elastic.
13. CH	Silty clay and clay.[e]
14. OH	Mucky silty clay.
15. PT	Muck and peats.

[a]   From U.S. Soil Conservation Service, 1979.
[b]   Also includes cobbly, channery, and shaly.
[c]   Also includes all textures with gravelly modifiers where $>1/_2$ of total held on No. 200 sieve is gravel size.
[d]   Gravelly textures included if less than $1/_2$ of total held on No. 200 sieve is of gravel size.
[e]   Also includes all of these textures with gravelly modifiers where $>1/_2$ of the total soil passes the No. 200 sieve.

10 soil orders and 43 suborders due to differences in chemical properties of the climatic areas in which they formed.

C. Soil Structure

Soil structure describes the arrangement of soil particles into larger units   (Figure 27). Thus, soil structure can control the rate of fluid flow through soil, water-holding capacity, solute retention, permeability, and may exert both favorable and unfavorable characteristics to soil surrounding disposal sites. For example, a highly cemented granular structure of a

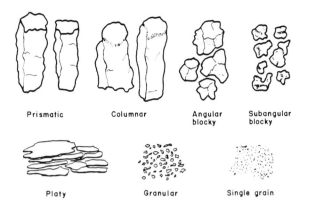

Prismatic   Columnar   Angular blocky   Subangular blocky

Platy   Granular   Single grain

FIGURE 27. Diagrams of some naturally occurring and observable soil structure.

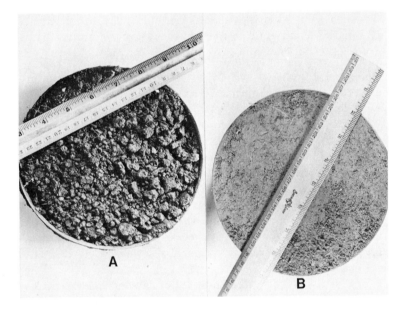

FIGURE 28. The good crumb structure of soil A was destroyed to demonstrate that poor soil structure of sample B can be associated with poor water penetration. (From Fuller, W. H., Soil modification to minimize movement of pollutants from solid waste operations, *Crit. Rev. Environ. Cont.*, 9, 222, 1980.)

dried Oxisolic (tropical) clay may behave like coarse sand particles even though the ultimate particle size is clay. If polluting constituents channel through the soil rapidly with such little diffusion into the aggregates that retention, and attenuation is poor. Soils of little structural stability and those whose particles are highly dispersed allow only slow penetration of fluids, if at all (Figure 28). [79,81] An example of the relationship of unstable structure to water is seen in example B of Figure 28. The structure in sample A became destroyed by excessive raindrop action resulting in sample B. Soil B is puddled at the surface with a tight seal extending over the entire surface area. The dispersed soil pores were closed to water penetration and excessive runoff took place. Soil structure may be dispersed by chemical means in addition to the physical means just described. When sodium (and certain other monovalent ions) exceeds about 15% of the cation-exchange capacity, some soils become dispersed upon wetting and water movement slows or virtually stops. Divalent ions such as calcium can correct the dispersed condition by replacing sodium and the restored soil becomes flocculated. The salts in waste streams, therefore, must be known and managed according to the objective desired for the most efficient and effective land disposal.

Soils form structural patterns of all sizes from primary and secondary dispersed soil particles (Figure 28). The pore spaces within these structures represent two major configuration classes: (a) micropores located within the granules and (b) macropores located between granules. The structures vary in stability and, therefore, failure patterns. Structure is defined by the USDA Soil Conservation Service[79] as "the aggregation of primary soil particles which are separated from adjoining aggregates by surfaces of weakness." The single aggregate is called a ped. Soil structures are described and classified on a basis of (a) shape and arrangements of peds, (b) size of peds, and (c) durability of peds as related to type, class, and grade (Table 18).

Soils composed mostly of clay size particles may provide as many problems in waste disposal as porous sands. They crack on wetting and drying, and freezing and thawing. They may be very impervious and puddle water, establish water tables, and encourage runoff and erosion. Most often soils having a balance of sand, silt, and clay may provide a better base for disposal purposes. This is certainly true for land treatment of wastes. Water stability, like texture, is of prime consideration in all site selection criteria.

D. Kind of Clay Mineral

Clay minerals form a more stable internal structure than original minerals and are represented by several different and distinct configurations as described earlier. Internal substitution of a variety of elements give each of the types further differences in reactive characteristics (Table 15). These clay minerals can be identified and broadly quantified by X-ray analysis.[80] Because the surface areas exposed to ion reactions differ, they also differ in physico-chemical behavior as exemplified by cation and anion exchange properties. Because they expand and hydrate differently in water, they exert different physical properties to soils such as shrinking and swelling. Kaolinite (1/1 nonexpanding lattice), for example, swells when wetted and shrinks when dried, but less than smectite (montmorillonite) types that possess a 2/1 expanding lattice structure. Therefore, the kind of clay mineral dominating the soil influences Atterberg limits, plasticity, compressive strength, liquid limits, compaction, and volume change.

Soils seldom possess a single pure (unsubstituted or substituted) type of clay mineral. Mixtures of several forms are encountered more frequently, therefore, correlations between kinds of clay mineral and attenuation of pollutants such as heavy metals are infrequently recognized.[75-77] Moreover, attenuation correlations may be shifted due to variations in the presence of variable concentrations of the hydrous oxides of Fe, Al, and Mn in soils that play a significant role in chemical and physical reactions. The importance of clay mineral types should not, however, be underestimated, particularly at extreme climatic temperature ranges, which show particularly good correlations between kind of clay mineral types and extent of attenuation. Differences between clay mineral types are highly important since shrinking and cracking create failures as a barrier to movement and migration of pollutants that are designed to be retained. The influence of different kinds of clay minerals may be minimized also by the dominance of variables in particle size distribution, pH, and soluble salt and organic matter in soils, to name a few factors.[81-85] There is no doubt, that should all other factors in a soil be equal, differences in kind of clay mineral can be an important factor in pollutant attenuation.

E. Pore Size Distribution and Porosity
1. Total Porosity

The total porosity (f) of a soil is related to bulk density (ρ_b) and particle density (ρ_s) by:

$$f = 1 - \rho_b/\rho_s \qquad (11)$$

2. Bulk Density

Bulk density may be determined by several methods, several which make use of undisturbed soil cores or clods. The core method of known soil volume includes interclod cavities whereas individual clods do not. Immersion most often is in water using a thin coating of paraffin wax or other water proofing material before dunking. A third method is by excavation.

3. Particle Density

Particle density is the true density defined as the ratio of the mass of the soil dried at 105 C to the volume of the solid particles. Normally, ρ_s is determined by a water or air picnometer. Fortunately, the value is in a fairly narrow range (2.4 to 2.7 g cm^{-3}) for a wide range of minerals.

4. Pore-size Distribution

Pore size distribution is most often measured using a hanging water column or pressure plate apparatus. This is discussed in the soil water section and is equivalent to finding the soil water characteristic (moisture-tension) relationship.

Evaluations can also be made by means of the pressure-intrusion method of Diamond,[84] in which mercury is forced into pores of the dried soil. This method is most suitable for course grained soils. A desorption, capillary-condensation method by Vomocil[85] is used for fine-grained soils usually with water as the permeating fluid.

Pores may be classified into macropores and micropores when aggregates are distinct. The macropore class represents predominantly interaggregate cavities and micropores, intraaggregate cavities. Air and water migrate primarily between aggregates or peds in the macropores and within aggregates in micropores. This separation is, of course, only descriptive but can be useful in attempts to explain pollutant-attenuation discrepancies between soils of similar textural classes. Permeability, as related to clay mineral type and amount of clay, varies greatly depending on the dominance of one class over the other.

Fluid movement is determined more by pore size distribution than by total porosity. This is most evident where the range of sizes is broad as in well-aggregated soils, but where the range of pore sizes is narrow in sand (large pores) or dispersed clay (small pores) the differences are not often of practical significance. In waste disposal, attention must be given to alteration of the natural flow processes, caused by the soil penetration of the waste fluids. Plugging of pores with suspended solids, microbial growth and biological debris alter the pore sizes and lessen the total pore volume. The permeability also may be altered by organic solvents.[82,83] The permeability of two clays, for example, can be enhanced by contact with certain neutral-polar, neutral-nonpolar, and basic solvents.

Pore size distribution and porosity is represented in such physical characteristics as compaction, cementation, and stratification. They relate so closely that, in the final analysis, they all affect pollutant permeability in the same way.

5. Compaction

Soils may be compacted naturally, by tilling or by heavy equipment and vehicular traffic. Compaction relates to other soil properties of bulk density, porosity, and pore size. Bulk density is the mass per unit bulk volume of soil. Porosity describes the total volume of pore space in a unit bulk volume of soil. Normally, total porosity is higher for finer textured soil. In disturbed soils porosity, density, and pore size vary according to the way the soils are compacted. Figure 29 shows the effect of tractor wheels on soil compaction and resulting lack of plant growth.

Table 18
THE CLASSIFICATION OF SOIL STRUCTURE ACCORDING TO USDA SOIL SURVEY STAFF[79]

A-Type: Shape and arrangement of peds

Platelike. Horizontal axes longer than vertical. Arranged around a horizontal plane.	Prismlike. Horizontal axes shorter than vertical. Arranged around vertical line. Verticles angular.		Blocklike-Polyhedral-Spheroidal. Three approximately equal dimensions arranged around a point			
			Blocklike-Polyhedral. Plane or curved surfaces accommodated to faces of surrounding peds.		Spheroidal-Polyhedral. Plane or curved surfaces not accommodated to faces of surrounding peds.	
	Without rounded caps	With rounded caps	Faces flattened; vertices sharply angular	Mixed rounded, flattened faces; many rounded vertices	Relatively nonporous peds	Porous Peds

B-Class: Size of peds

Size of peds	Platy	Prismatic	Columnar	Blocky	Subangular blocky	Granular	Crumb
1. Very fine or very thin	1 mm	10 mm	10 mm	5 mm	5 mm	1 mm	1 mm
2. Fine or thin	1—2 mm	10—20 mm	10—20 mm	5—10 mm	5—10 mm	1—2 mm	1—2 mm
3. Medium	2—5 mm	20—50 mm	10—20 mm	10—20 mm	10—20 mm	2—5 mm	2—5 mm
4. Coarse or thick	5—10 mm	50—100 mm	50—100 mm	20—50 mm	20—50 mm	5—10 mm	
5. Very coarse or very thick	10 mm	100 mm	100 mm	50 mm	50 mm	10 mm	

C-Grade: Durability of peds

C-Grade: Durability of peds

0. Structureless	No aggregation or orderly arrangement.
1. Weak	Poorly formed, nondurable, indistinct peds that break into a mixture of a few entire and many broken peds and much unaggregated material.
2. Moderate	Well-formed, moderately durable peds, indistinct in undisturbed soil, that break into many entire and some broken peds but little unaggregated material.
3. Strong	Well-formed, durable, distinct peds, weakly attached to each other, that break almost completely into entire peds.

FIGURE 29.   Compaction of soil caused by tractor wheels and resulting
lack of plant growth. (From Fuller, W. H., *Management of Southwest Desert
Soils*, The University of Arizona Press, Tucson, 1975, 17.)

6. Cementation

Soils may be cemented by a number of materials to the extent fluid behavior becomes
highly erratic, causing anaerobic conditions and undesirable water tables to develop. Natural
soils under humid climates may possess layers of soil cemented with hydrous oxides of iron,
aluminum and to a lesser extent, manganese. Silicates may also act as cementing agents.
In arid and semiarid climates, cemented layers of native limestone (caliche) and gypsum
abundantly occur in various degrees of purity. Soil particles cemented with either or both
(Figure   30) alter water movement. Highly alkaline soils of arid climates may become very
dense from infiltration of soluble salts and/or silicates called "gatch".[81] All of these indurated
(restrictive) layers have a profound affect on land disposals and should be avoided for many
reasons.

7. Stratification

Stratification defines the layering of soil and geologic materials into horizontal lenses,
usually of different soil textures of sand silt, and clay, of variable thicknesses as illustrated
in Figure 31.   The heterogeneous nature of stratified soils is a deterrent to securing pollutant
retention and provides serious problems in pollutant migration. Predictions of the rate and
extent of movement through the soil cannot be made with confidence. Recent observations
of hazardous solvent migration in deep soils, delegate against disposal where soils are highly

FIGURE 30.   Gypsum and lime veinings cement soil particles together. (From Fuller, W. H., *Management of Southwestern Desert Soils,* The University of Arizona Press, Tucson, 1975.)

FIGURE 31.   Stratifications of sand, silt, and clay in Vinton loamy fine sand. (From Fuller, W. H., *Management of Southwestern Desert Soils,* The University of Arizona Press, Tucson, 1975, 71.)

stratified and underground lenses or benches create unpredictable heterogeneous conditions. Variable pore size distributions as defined by these soil structures, therefore, affect many of the important physical processes in the soil.

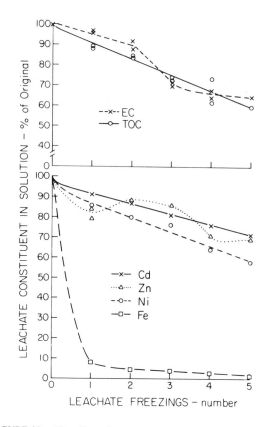

FIGURE 32. The effect of successive freeze-thaw action on the solubility of Cd, Ni, Zn, Fe, and total organic carbon and salt on MSW
landfill-type leachate. (From Fuller, W. H. and Alesii, B. A., Behavior
of municipal solid waste leachate. II. In Soil, *J. Environ. Sci. Health,*
A14(7), 559, 1979. Reprinted by courtesy of Marcel Dekker, Inc.)

F. Hydration and Dehydration

There are certain dominant physical processes related to dehydration such as (a) shrinking
and swelling, (b) drying and wetting, and (c) freezing and thawing that deserve special
attention, independent of water flow characteristics, because of their specific physical mechanism for pollutant retention in soil.

Dehydration of the soil by drying at temperatures above freezing or below freezing favors
the retention in place of pollutants in liquid fluids. Upon rehydration some of the constituents
return into solution, but not all, because of the hysteretic phenomenon. The dry-wet and
freeze-thaw effects on municipal solid waste landfill were found[56,87] to markedly affect the
concentration of all soluble components. Upon rewetting and thawing, the dehydrated insoluble components only partially redissolved (see Figure 32). Repeated dehydration cycles
continued to remove constituents from solution. Thus, a mechanism was revealed that is
desirable from a standpoint of stabilizing pollutant migration, since the more of a component
that precipitates and remains insoluble, the less it is available to move through the soil to
underground water sources. In cold regions, freezing can be important for some level of
pollutant control.

The tendency of the soil to shrink upon drying, and swell upon wetting is often overlooked
in waste management operation (Figure 33). Soils dominated by an expanding lattice-type
clay (e.g., montmorillonite) are more prone to shrink and swell than those that do not have
an expanding-lattice type clay (e.g., kaolinite). Shrinking and swelling influence movement

FIGURE 33.   Highly fractured soil from excessive swelling and shrinking and dispersion when wet. (From Fuller, W. H., *Soils of the Desert Southwest,* The University of Arizona Press, Tucson, 1975, 32.)

of pollutants through soils both favorably and unfavorably. Dehydration favors retention of soluble constituents from liquids, whereas concurrent cracking and fissure formations open the soil for rapid movement of solids, slurries, and liquids to greater depths as the fresh waste stream again deposits. Soil cracks, wherever they occur in soils used for waste disposal, whether in landfills, lagoons or land treatment. This rapid movement used for waste disposal represents soil failure.

V. SOIL WATER

Soil water is described both according to the amount present and the associated energy level. The amount present is the water content; the energy level is the soil water potential. Measurements and the appropriate parameters reflect which mode of description is of interest. The potential components are defined for a system in static equilibrium, but the results are useful for describing a dynamic system. In fact, a primary use of soil water potential is to evaluate magnitude and direction of flow. Several textbooks provide comprehensive discussions of soil water.[88-93]

A. Water Content
Water content describes how much water is present. Alternative definitions include

θ_m = (Mass of water in sample)/(Dry mass of soil in sample)

θ_v = (Volume of water in sample)/(Apparent volume of sample)

S = (Volume of water in sample)/(Total pore space)

θ_{LR} = (Volume of water)/(True volume of solids)

$d_w = \theta_v D$ (D = depth increment of profile)

The first four alternatives are interrelated and are easily shown to satisfy

$$\theta_m = (\rho_w/\rho_b)\theta_v$$

$$S = \theta_v/\theta_s$$

$$\theta_{LR} = (\rho_s/\rho_b)\theta_v$$

where ρ_w, ρ_b, ρ_s and θ_s are the density of water, bulk (apparent) density, particle (true) density and saturated water content ($=$f), respectively. In addition to being expressed as fractions, θ_m, θ_v, S, and θ_{LR} can be expressed on a percentage basis also. When used without qualification, most commonly "water content" refers to θ_m. This also is called water content by weight or simply gravimetric water content. However, the volumetric water content θ_v is also quite commonly used, especially in soil physics.

The most common method for evaluating water content is by sampling and drying. To find θ_m, a sample is weighed at the field wetness, then dried in an oven (e.g., 105 °C for 24 hr), reweighed and the value calculated. The principal advantage of the method is its simplicity, ease of calibration and the fact that it is the fundamental standard against which other methods are evaluated. Problems can arise — although usually relatively minor — regarding what is really "dry" and with destruction of organic matter or easily destroyed minerals. The gravimetric method gives sufficient information to calculate θ_m but an independent measurement of the bulk density is needed to find θ_v or S. Another consideration for some applications is that the method is destructive and repeated measurements at exactly the same point in the soil profile cannot be accommodated.

A second method for evaluating soil moisture is by neutron thermalization (or neutron scattering). A neutron moisture meter consisting of a probe and counter is required. The probe is lowered into an access tube (see Figure 34), and emits high (1 to 15 MeV) neutrons from a radioactive source, normally Ra-Be or Am. The probe also has a detector which is connected back to the counter to be read by the operator. The amount of slow neutrons present is in accordance to loss in kinetic energy by elastic collisions with the various nuclei present in the soil. The nuclei of hydrogen are the most effective in slowing the fast neutrons, so if water is the preponderant source of hydrogen, then the count of slow neutrons is highest with high water contents. The volume sampled is approximately a sphere of soil whose radius is 40 cm for dry soil to 10 cm for wet conditions. An advantage of the method is that measurements can be made over and over at the same site and readings can be made deep in the profile without excavating or boring each time. Precise calibrations are site specific and laborious. Fortunately, differences with time at the same locations are not overly dependent on an exact calibration. The most common access tubes are aluminum or steel. Repeatability is on the order of 0.01 for θ_v. More details are in the aforementioned texts and elsewhere.[93-95]

Many other methods exist for evaluating water content. These include gamma attenuation and methods based on dielectric constants. In addition, measurements of soil water potential

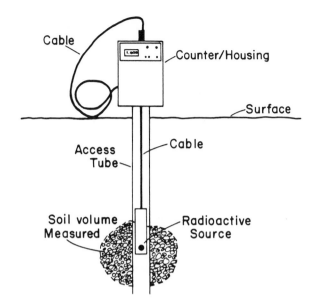

FIGURE 34. Neutron probe apparatus for determining water content.

— such as with moisture resistance blocks — can be used to deduce water content and will be discussed later.

B. Soil Water Potential

The soil water potential is defined formally as the work necessary to transfer a unit quantity of water from a standard reference state where it is zero to the situation of interest. The potential is influenced by:

- elevation
- matric effects
- liquid pressure
- air pressure
- solutes present
- temperature
- overburden

as well as possibly other effects. The numerical value is arbitrary to the extent that the reference condition must be defined. Another source of ambiguity is that the "unit quantity" can be either a volume, mass, or weight and often the units. are interchanged.

Water transfer occurs as a result of potential differences — with movement from regions of higher or lower potential. The "potential energy" is for static conditions as contrasted with "kinetic energy" due to inertia (motion) as would be important, for example, for water in a swiftly running river. Only for very fast intakes such as ponded water recharging a coarse sand would inertial forces be of consequence.

For the purposes of this discussion, the soil water potential ϕ_T is simplified to

$$\phi_T = z + h + \pi \tag{12}$$

where z is elevation expressing gravitational effects, h the pressure head reflecting positive liquid pressure for which it is positive or matric effects for which the soil water is unsaturated

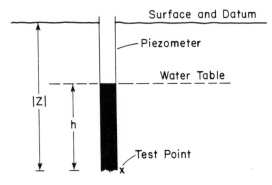

FIGURE   35.   Piezometer tube for measuring positive pressure
head. (Sketch depicts static conditions.)

and h is negative. The third component "π" is the osmotic or solute component. The air
pressure, temperature, and overburden effects are neglected in Equation 12. The above
definition of potential is the energy per weight.

Equation (12) may also be written as

$$\phi_T = H + \pi \tag{13}$$

where "H" the hydraulic head combines the effects of gravity and pressure head. When
the osmotic effects are negligible, only the hydraulic head remains. For the vast majority
of flow problems, only the hydraulic head is assumed significant and water moves according
to the hydraulic gradient. The osmotic effects are most often evaluated using a soil extract
although its combined effect with matric potential can be evaluated directly with soil
psychrometers.

1. Measuring Gravitational Potential

Of the potential components, the gravitational part is by far the simplest. For all conditions,
z is defined with respect to a reference elevation. If the test location is above the reference
elevation, z is positive. If the test location is below the reference elevation, z is negative.
The definition of the reference is arbitrary. The soil surface may be a convenient reference,
a surveying marker may be convenient or the lowest point to be measured may be suitable.
The choice of the zero point, of course, does not affect differences in potential of two
locations, only the individual values.

2. Measuring Pressure Head — Saturated Conditions and Positive Pressures

For water saturated (water-logged) conditions the water pressure is normally greater than
the adjacent atmosphere. The pressure potential may be evaluated by a piezometer tube as
in Figure 35. A piezometer is a pipe or tube with impermeable walls open on each end.
The height of rise in the piezometer tube is the pressure head h, a positive quantity. The
corresponding "work" from the basic definition is the effort necessary to transfer a unit
weight of water at atmospheric pressure and at the test elevation into the soil water. Of
course, to evaluate the hydraulic head H = h + z, it is necessary to define a reference
level (datum) for the gravitational component. If the soil surface is chosen as the zero point,
the gravitational component is negative and equal in magnitude to the depth. The hydraulic
head H is also negative for the example in the figure.

3. Measuring Pressure Head — Unsaturated Conditions, Low Suctions

For unsaturated conditions, the pressure head h will be negative and, in general, much

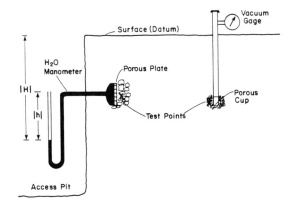

FIGURE 36. Tensiometers for measurement of negative pressure heads under moderately wet conditions.

more difficult to evaluate than for saturated conditions. For these conditions, the pressure head is also called the matric potential.

If h is in the low suction range of 0 to 800 cm, then a tensiometer can be used for the measurement. A simple tensiometer is sketched on the left in Figure 36 and consists of a porous plate and manometer which extends laterally from the test location out to an adjacent access pit. The porous plate allows water to move freely to or from the manometer in order to equilibrate with the soil water, but does not allow air transfer. The affinity of the soil for the water is measured by the negative pressure head h. Also shown is the hydraulic head H, the sum of h and z. The hydraulic head is also negative for this example. On the right of Figure 36 is a more common type of tensimeter which is designed to operate from above the soil surface. The principles are the same, but the soil water pressure is measured by a vacuum gage which obviates the necessity of the access trench alongside. Normally the gage is expressed as centibars "suction". If the height of the gage is "d" above the test depth, then the pressure head expressed as "cm" is

$$h = -(\text{Reading in centibars})/(1022) + d \qquad (14)$$

For laboratory work, a hanging water apparatus as shown in Figure 37A is routinely used. The pressure head h is simply the negative of net length of water below the porous plate. The column can be only water or it can be water and another immiscible liquid, commonly mercury. If a volumetric burette is added as shown, the pore-size distribution can be evaluated by noting volumes of water drained for given changes in h. The effective radius r_e of pores in equilibrium with a given pressure head h is from the well-known "capillary-rise" equation

$$h = (2\sigma \cos \alpha)/(\rho_w g r_e) \qquad (15)$$

with σ the surface tension of water, α the contact angle, ρ_w the density of water, and g the gravitational acceleration constant. Just as for the tensiometers, the hanging water system as in Figure 37 is limited to a range of approximately h = −800 cm or greater (wetter), the reasons for which follow in the next section.

4. Measuring Pressure Head — Drier Conditions

For drier conditions, tensiometers and hanging water columns cannot operate. The systems shown in Figures 36 and 37A have at some point, water under a tension greater than or

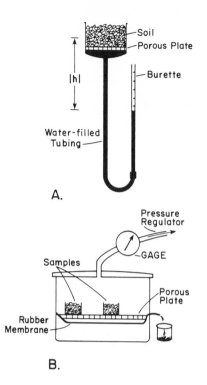

FIGURE 37.   Apparatus for measuring negative pressure head in the laboratory. (A)
Water-filled tubing, and (B) pressure plate.

equal to |h|. As |h| becomes large the tension eventually approaches one atmosphere. Thus, the pressure is reduced and approaches a complete vacuum. As a practical matter, problems develop at about h = −800 cm. The definition of soil water remains valid for the drier conditions, however, the conceptual model for describing soil water must include not only capillary forces and tension effects, but more general attractive forces of water to the soil. Even though "h" no longer equals a physical pressure, it is still referred to as a pressure head (as well as simply the matric potential).

The most common method for determining matric potential under drier conditions is by inference from a measured water content value. The method is dependent upon knowing the soil water charactertistic relationship for the specific location of the determination. The soil water characteristics are most often determined by a pressure plate apparatus as sketched in Figure 37B. A soil sample (disturbed or undisturbed) is placed on a porous plate, saturated with water, and then subjected to a positive gas pressure. The pressure is maintained at a constant known level and is used to force the water from the sample until equilibrium is attained. Then the water content is determined, normally by removing the sample and determining gravimetrically (some small systems are designed so the whole apparatus can be weighed before and after water is removed). The procedure is repeated for several pressures and results in corresponding points defining θ_v vs. h as shown in Figure 38.   The relationship of h to the air pressure is

$$h = -(Gage\ Pressure)/(\rho_w g) \qquad (16)$$

where h, gage pressure, water density ρ_w and gravitational acceleration constant g are in consistent units to give h as a length.

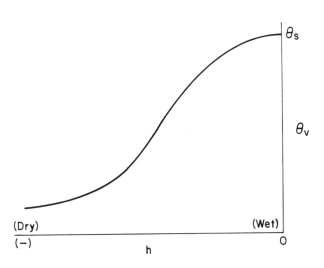

FIGURE 38. Soil water characteristic, i.e., θ_v vs. h.

In defining the soil water characteristic the wet range can be by a "hanging water column" and the drier range by a pressure plate. The pressure plate apparatus is used with gage pressures up to 15 or 30 atmospheres routinely and can be used up to 100 atmospheres if built strongly enough. For the wetter range, soil structure is generally critical and undisturbed cores are preferred. For the drier conditions, say h less than -5000 cm, disturbed samples are generally acceptable. In addition to soil structure, the effects of hysteresis can also be a factor. Hysteresis occurs in that a soil is generally wetter when drained than when wetted to the same value of h.

Methods exist for evaluating matric potential in situ more directly under the drier conditions. Soil psychrometers are commercially available and measure combined pressure head and osmotic potential over the approximate range of 2000 to 10^6 cm of water. The psychrometer measures relative humidity (for very high humidities) which is easily related to the h $+$ π. Also, devices are available based on heat dissipation from a porous absorber. For further discussion and references, see Hillel.[7]

5. Notations and Alternative Expressions of Soil Water Potential

Terminology and communications regarding soil water potential is often ambiguous and confusing. Sources of confusion are that there are several terms used, such as potential, energy, tension, heads, and suction; there are several effects involved — gravity, matric, salt, etc., there are several systems — English, c.g.s., and SI; there are both gage and absolute pressures involved; and the energy may be per unit mass, weight, or volume. First with regard to terms, the following are noted:

- tension — a positive number equal to $-h$ for unsaturated conditions or can be in other units than head.
- suction — a positive number equal to $-h$ for unsaturated conditions which is more properly called matric suction — sometimes is $-\pi$ and called osmotic suction. It is usually expressed as a length.
- head — soil water potential per unit weight or can also be simply a pressure converted to a length.

Commonly used units are

- energy/unit weight — any length measurement

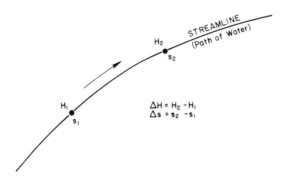

FIGURE 39. Streamline depicting imaginary path followed by water through soil.

- energy/unit volume — any pressure units, such as pascal, dynes/cm², p.s.i., atmospheres, bars, can be a gage or absolute pressure
- energy/unit mass — ergs/g or joules/kg

Some equivalences although not dimensionally the same are

1 std. atmosphere is equivalent to 10.2 m of H_2O
is equivalent to 1.013 bar
is equivalent to $(1.013)(10)^5 Pa$

Also whether a pressure is "gage" or "absolute" sometimes requires clarification. Absolute pressure can never be physically less than zero and is measured from a perfect vacuum. Gage pressure is always evaluated relative to another pressure, the most obvious case being relative to the local atmosphere. A common design for a pressure gage is a diaphram which is deflected when gas pressure on the inside is different from the local atmospheric pressure on the outside.

C. Soil Water Dynamics

Water moves from regions of higher to lower soil water potential. Any of the individual potential components can contribute, but emphasis in this section will be on gravitational and pressure head components. For unsaturated conditions the pressure head will be negative and identical to the matric potential.

For flow along a streamline, the value of the hydraulic head at coordinates s_1 and s_2 can be taken as H_1 and H_2 as in Figure 39. The energy of the water at s_1 and s_2 is

$$H_1 = h_1 + z_1 + v_1^2/2g + C$$

$$H_2 = h_2 + z_2 + v_2^2/2g + C + \Delta H \tag{17}$$

For most flow problems the inertial terms $v^2/2g$ may be safely neglected. For an idealized fluid ΔH would be zero and Equation (17) would follow Bernoulli's law. However, for a real fluid there is an overall consumption of energy dissipated as work against friction — consequently ΔH needs to be included.

The hydraulic gradient i is defined by

$$i = \lim_{\Delta s \to 0} (\Delta H/\Delta s) = dH/ds \tag{18}$$

Table 19
PERMEABILITY CLASSES AS USED BY
U.S. SOIL CONSERVATION SERVICE

Class	K (cmh^{-1})	k (cm^2)	
Very slow	<0.125	<3	$(10)^{-10}$
Slow	0.125—0.5	3—15	$(10)^{-10}$
Moderately slow	0.5—2.0	15—60	$(10)^{-10}$
Moderate	2.0—6.25	60—170	$(10)^{-10}$
Moderately rapid	6.25—12.5	170—350	$(10)^{-10}$
Rapid	12.5—25.0	350—700	$(10)^{-10}$
Very rapid	>25.0	> 700	$(10)^{-10}$

and is dimensionless. In general "i" is a vector quantity and has an associated direction as well as a magnitude. For simplicity, the discussion here is with "i" as a scalar only, whose value is always assumed opposite the direction of flow. The dynamics of water flow for saturated flow will be discussed next, followed by that for unsaturated flow.

1. Saturated Flow

For saturated flow, the pressure head component h will be positive corresponding to a positive hydrostatic pressure. The flow rate along any streamline (cf. Figure 39)   is given by Darcy's law:

$$q_D = -Ki = -K(\Delta H/\Delta s) \qquad (19)$$

with K being the hydraulic conductivity (units L/T). The Darcian flow rate q_D is the flow volume per unit cross-sectional area per unit time and has units of velocity. The q_D is also referred to as Darcian velocity or flux density. Sometimes it is also referred to as "velocity" or "flow rate", but these terms need to be clearly defined in context to avoid ambiguity and confusion with the closely associated convective (pore water) velocity.

The origin of Darcy's law was based on empirical observations for sand filters. Numerous attempts at a derivation on more fundamental terms have met with only partial success.[96] The best known conditions for which Darcy's law is not valid are for turbulent, non-laminar flow associated with very high velocities and for very slow velocities in certain clay materials — especially those which shrink and swell.

Darcy's law may alternatively be expressed in terms of the intrinsic permeability k:

$$q_D = -(\rho gk/\eta)i \qquad (20)$$

with ρ and η the fluid density and dynamic viscosity dimensions (ML^{-1}T^{-1}), respectively. The dimensions of k are L^2. If the fluid and solid were totally non-interactive, then k would be a property only of the porous media and independent of the fluid — whether the fluid be water or alcohol or air. Of course, the soil interacts with all fluids to some extent, but nevertheless, the intrinsic permeability k determined for a given soil for various fluids would tend to be closer to each other than the hydraulic conductivities. With regard to terminology, "permeability" is used qualitatively for both hydraulic conductivity and intrinsic permeability, probably more often for hydraulic conductivity. The differences in hydraulic conductivity are expected to be large for the same soil when fluids have vastly differing viscosities, such as water and glycol.

Table 19   gives permeability classes as used by the Soil Conservation Service. Values are given both for hydraulic conductivity of water and for intrinsic permeability. Either K or k can be used to indicate the ease with which a fluid passes through the soil.[100]

2. Measuring Hydraulic Conductivity — Saturated Flow

Methods for determining hydraulic conductivity are by Boersma[97] and Amoozegar-Fard and Warrick.[98] Briefly, some of the more common methods are

Laboratory methods
  constant head
  falling head
In situ methods
  auger hole
  piezometer
  pit bailing method
  reverse auger hole
  infiltration ring
  double tube method

A typical laboratory setup for a constant head permeater is shown in Figure 40A. A constant head of water (h_2) is maintained on the soil surface at Position 2 on a vertical soil column. The water pressure at the lower surface (Position 1) is atmospheric as is allowed to drip freely. Defining the reference level for the gravitational potential at the base of the column gives

$$H_1 = 0$$
$$H_2 = h_2 + L \tag{21}$$

with L the length of the column. The hydraulic gradient is

$$i = (h_2 + L)/L \tag{22}$$

and the Darician velocity is

$$q_D = Q/(At) \tag{23}$$

where Q is the volume of flow collected in time t and A the cross-sectional area. By Equation (23), the hydraulic conductivity follows:

$$K = |q_D/i| = QL/[At(h_2 + L)] \tag{24}$$

The procedure is to establish a constant head, measure Q for the time t and calculate K. Similar systems can be used with horizontal or vertically upward flow.

A falling-head apparatus is depicted in Figure 40B. The lower conditions are the same as for the constant head system, but the upper head varies from H_2 at time zero to H_2^* at time t. The hydraulic gradient thus varies from H_2/L to H_2^*/L from which K can be shown to be

$$K = (aL/At) \ln(H_2/H_2^*) \tag{25}$$

with "a" is the cross-sectional area of the supply tube. The flow amount Q need not be measured directly. A special case is when a = A which is often convenient. Rather than measure H_2 and H_2^* directly, only H_2 and the time for the water to all enter the soil are

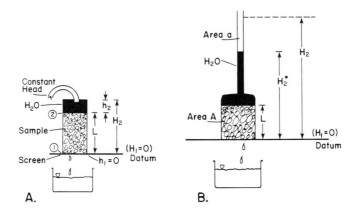

FIGURE 40. Permeameters for measuring hydraulic conductivity. (A) Constant head, and (B) falling head apparati.

sufficient. If the drop from H_2 to H_2^* is not excessive, an average gradient and the steady-state formula Equation (24) give reasonable results.

Of the in situ determinations, the auger-hole, piezometer, and pit-bailing methods require shallow water tables. The basic idea is to excavate below the water table, remove the water and measure how fast the hole refills. For the auger hole the excavation is a cylindrical hole from the surface to below the water table, usually on the order of 10 to 20 cm in diameter. For the piezometer method, a solid-walled pipe is extended below the water table and a small cavity excavated below its end. For the pit-bailing method, the cavity can be a larger pit, such as with a tractor mounted scoop or "back hoe".

The remaining in situ methods do not require a shallow water table, but water must be added to the system. In each case, an initial stage is required in order to reach a steady flow. Infiltration rings approximate the hydraulic conductivity, provided the horizontal flow is negligible, sufficient time has elapsed and the soil is homogeneous. For the reverse auger hole method, water is added to a cylindrical cavity and the intake rate measured. This is a version of the commonly used septic-tank suitability test. The double-tube method is a more elaborate technique, designed to attain one-dimensional flow for the measured volume.

In general, hydraulic conductivity values are highly variable. For several different studies in agricultural fields, Warrick and Nielsen[99] found coefficients of variation the order of 100%. Even for "replicate" samples a considerable variation can occur. Values are obviously influenced by where and how the sample is taken. For some cases an "undisturbed" sample or an in situ method is preferred, although if the site is totally reworked before disposal or very deep samples are needed, disturbed samples are justified. With regard to vertical flow, stratification — whether natural or man-made — requires special consideration. In all of the methods, sufficient time must be allowed to reach steady state, although in some cases volumes continue changing over long periods of time.

3. Unsaturated Flow

For unsaturated conditions, water is under tension. Thus, movement occurs through capillaries and along surfaces or in a vapor form. As a consequence, flow is much slower and more tortuous than for saturated flow. Darcy's law may be assumed of the form

$$q_D = K(\theta_v)i \qquad (26)$$

where $K(\theta_v)$ denotes the hydraulic conductivity varies with water content θ_v. The generalization of Darcy's law to unsaturated flow is primarily for convenience, although it has

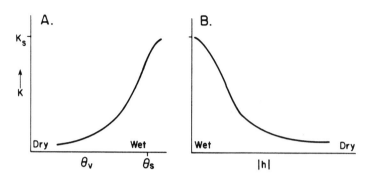

FIGURE 41.   Unsaturated hydraulic conductivity as a function of (A) water content, and (B) suction.

proved reasonably satisfactory for many problems. Equation (26) along with a continuity requirement results in the well-known Richards' Equation:

$$\partial\theta/\partial t = \nabla \cdot (K\nabla H) \tag{27}$$

where "∇" is the vector gradient operator. When combined with suitable initial and boundary conditions, Richards' equation may be solved in order to determine water status through space and time. As the relationship is non-linear, the solution is difficult and often only by numerical methods.

A typical unsaturated-hydraulic-conductivity function K as a function of θ_v is shown in Figure 41. The values can change by several orders of magnitude as the water-filled portion of the soil becomes less. Thus, it is often more convenient to plot K on a logarithmic scale. Of course, water status may also be expressed in terms of pressure level h (or the suction, $-h$) and Figure 40B shows this type of a relationship.

In general, methods for determining unsaturated hydraulic conductivity function are much more tedious than the corresponding saturated values. Methods are discussed by Klute (1965)[100] and Hillel (1982)[7] including laboratory and field methods. Shortcuts have been proposed to evaluate the conductivity based on pore-size distribution and capillary-bundle theory. These are reviewed somewhat by van Genuchten.[101] Several alternatives have been used including that by Mualem of the form[102]

$$K/K_s = \theta^{0 \cdot 5}\left[\int_0^\theta dx/h(x)\right]^2\left[\int_0^1 dx/h(x)\right]^{-2} \tag{28}$$

where $h(\theta)$ is the soil water characteristic relationship and θ is a reduced water content. Mualem compared K values calculated for 45 soils with measured values with generally favorable results.

4. Convective Velocity

In addition to the Darcian velocity, the convective (pore-water) velocity is also often of interest. This is the velocity of translation on a macroscopic basis. For example, if a non-interacting pollutant is moving with the water, how fast does the front move through the soil profile? This is shown schematically as Figure 42. Suppose water is moving vertically downward from 0 at a steady Darcian velocity of q_D. In time t the total volume of fluid passing z = 0 is

$$Q = q_D At \tag{29}$$

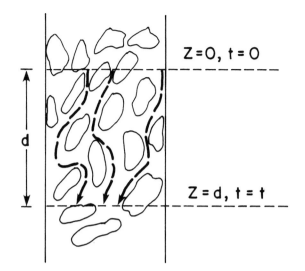

FIGURE 42. Microscopic pathlines of soil water through porous medium. Travel is from z = o to z = d plane in time t.

where A is the cross-sectional area considered. If the advance is along a planar front, this same volume will occupy z = 0 to d where

$$Q = \theta_v Ad \qquad (30)$$

As the displacement is over a distance d in time t, the corresponding velocity is

$$q_c = d/t$$

or

$$q_c = q_D/\theta_v \qquad (31)$$

This relationship is valid for both saturated or unsaturated conditions. For saturated conditions θ_v is the porosity of the system. The same relationship also holds for 2- or 3-dimensional flow.

The convective velocity will always be greater than the Darcian velocity as θ_v is necessarily less than unity for porous material. In fact, generally q_c will be considerably more than twice q_D especially for unsaturated conditions. Similarly if the cumulative discharge is known, the associated depth of displacement for a nonreactive material will be several times the cumulative amount added, even when perfect "piston displacement" occurs without channeling, macro-pore effects or other factors which speed up translation.

Additionally there is a "microscopic" velocity associated with porous media flow. This is the velocity within the pore and along the tortorous path between the soil particles and is greater in magnitude than even q_c.

REFERENCES

1. **Jenny, H.,** *The Soil Resource, Origin and Behavior,* Springer-Verlag, New York, 1980, 377.
2. **Fuller, W. H., Korte, N., Niebla, E. E., and Alesii, B. A.,** Contribution of the soil to the migration of certain common and trace elements, *Soil Sci.,* 122, 223, 1976.
3. **Clark, F. W.,** The data of geochemistry, *U. S. Geol. Survey, Bull.,* 770, 15, 1924.
4. **Dixon, J. B. and Weed, S. B.,** *Minerals in Soil Environment,* 1971, Soil Science Society America, Madison, Wis., 1977, 948.
5. **Lindsay, W. L.,** *Chemical Equilibrium in Soils,* John Wiley and Sons, New York, 1979, 24, 449.
6. **Fuller, W. H.,** Investigation of Landfill Leachate Pollutant Attenuation by Soils, 1978, EPA-600/2-78-157, U. S. Environmental Protection Agency, Cincinnati, Ohio, 1978, 239.
7. **Hillel, D.,** *Introduction to Soil Physics,* Academic Press, Inc., New York. 1982, 385.
8. **Letey, J., Stolzy, L. H., and Kempero, W. D.,** Soil aeration, in *Irrigation of Agricultural Lands,* Hagan, R. M., Haise, H. R., and Edminster, T. W., Eds., Amer. Soc. Agron., Madison, Wis., 1967, 934.
9. **Currie, J. A.,** Soil Respiration, in Soil Physical Conditions and Crop Production Tech. Bull. 29, Min. of Agr., Fisheries and Food, Her Majesty's Stationery Office, London, 1975, 35.
10. **Connell, R. Q.,** Soil aeration and compaction in relation to root growth and soil management, *Adv. Appl. Biol.,* 2, 1, 1977.
11. **Soil Science Society America,** *Glossary of Soil Science Terms,* Soil Sci. Soc. Am., Madison, Wis., 1978 34.
12. **Soil Survey Staff,** *Soil Taxonomy, U. S. Department of Agriculture, Soil Conservation Service Agricultural Handbook No. 436,* Washington, D. C., 1975, 2.
13. **Leighton, M. M. and MacClintock, P.,** The weathering mantle of glacial tills beneath original surfaces in north-central United States, *J. Geol.,* 70, 267, 1962.
14. **Rhue, R. V.,** *Quaternary Landscapes in Iowa,* Iowa State University Press, Ames, 1969, 255.
15. **Fuller, W. H.,** *Soils of the Desert Southwest,* The University of Arizona Press, Tucson, 1975, 102.
16. **Fuller, W. H.,** Movement of Selected Metals, Asbestos, and Cyanide in Soil: Application to Waste Disposal, EPA-600/2-77-020, U. S. Environmental Protection Agency, Cincinnati, Ohio, 1977, 243.
17. **Fuller, W. H.,** *Management of Southwestern Desert Soils,* The University of Arizona Press, Tucson, 1975, 195.
18. **Korte, N. E., Fuller, W. H., Niebla, E. E., Skopp, J., and Alesii, B. A.,** Trace element migration in soils: desorption of attenuated ions and effects of solution flux, in *Residual Management by Land Disposals,* Fuller, W. H., Ed., EPA 600/9-76-015, U. S. Environmental Protection Agency, Cincinnati, Ohio, 1976, 280.
19. **Korte, N. E., Niebla, E. E., and Fuller, W. H.,** The use of carbon dioxide to sample and preserve natural leachates, *J. Water Pollut. Control Fed.,* 48, 959, 1976.
20. **Martin, J. P. and Focht, D. D.,** Biological properties of soils, in *Soil for Management of Organic Wastes and Waste Waters,* Elliott, L. F. and Stevenson, F. J., Eds., American Society of Agronomy, Madison, Wis., 1977, 115.
21. **Clark, F. E.,** Bacteria, in *Soil Biology,* Burges, A. and Raw, F., Eds., Academic Press, Inc., New York, NY, 1967, 49.
22. **Richards, L. A.,** *Diagnosis and Improvement of Saline and Alkaline Soils,* USDA Agricultural Handbook No. 60, U. S. Government Printing Office, Washington, D. C., 1954, 80.
23. **Rao, H. G.,** Effects of green-manuring on red loamy soils freshly brought under swamp paddy, *J. Indian Soil Sci. Soc.,* 4, 225, 1956.
24. **Ganem, I., Hassan, M. N., Rhadr, M., and Tadros, Y.,** Manganese in soils: II. Effect of adding organic materials on the UAR, *J. Soil Soc.,* 11, 125, 1971.
25. **Olomu, M. O., Racz, G. J., and Cho, C. M.,** Effects of flooding on the Eh, pH, and concentration of iron and manganese in several Manitola soils, *Soil Sci. Soc. Amer. Proc.,* 37, 220, 1973.
26. **Hemstock, G. A. and Low, P. F.,** Mechanisms responsible for retention of manganese in the colloidal fraction of soil, *Soil Sci.,* 76, 331, 1953.
27. **Ponnamperuna, F. N.,** The Chemistry of Submerged Soils in Relation to the Growth and Yield of Rice, Ph.D. thesis, Cornell University, Ithaca, N. Y., 1955.
28. **Ponnamperuna, F. N., Tianco, E. M., and Log, T. A.,** The iron hydroxide systems, *Soil Sci.,* 103, 374, 1967.
29. **Ponnamperuna, F. N., Log, T. A., and Tianco, E. M.,** Redox equilibria in flooded soils: II. The Manganese oxide systems, *Soil Sci.,* 108, 48, 1969.
30. **Cheng, B. T.,** Dynamics of soil manganese, *Agrochemical,* 17, 84, 1973.
31. **Collins, J. F. and Buol, S. W.,** Effect on fluctuations in the Eh-pH environment on iron and/or manganese equilibria, *Soil Sci.,* 110, 111, 1970.
32. **Collins, J. F. and Buol, S. W.,** Patterns of iron and manganese precipitation under specific Eh-pH conditions, *Soil Sci.,* 110, 157, 1970.

33. **Hannapel, R. J., Fuller, W. H., Bosman, S., and Bullock, J. S.,** Phosphorus movement in a calcareous soil: I. Predominance of organic forms of phosphorus in phosphorus movement, *Soil Sci.,* 97, 350, 1964.

34. **Hannapel, R. J., Fuller, W. H., and Fox, R. H.,** Phosphorus movement in a calareous soil: II. Soil microbial activity and organic phosphorus movement, *Soil Sci.,* 97, 421, 1964.

35. **Fuller, W. H. and Hardcastle, J. E.,** Relative absorption of strontium and calcium by certain algae, *Soil Sci. Soc. Amer. J.,* 31, 772, 1967.

36. **Hardcastle, J. E. and Fuller, W. H.,** Relative absorption of calcium and strontium by some desert soil fungi, *Chemosphere,* Oxford, England, 2, 59, 1974.

37. **Fuller, W. H., Hardcastle, J. E., Hannapel, R. J., and Bosma, S.,** Calcium-45 and Strontium-89 movement in soils, and uptake by barley plants as affected by Ca(Ac)$_2$ and Sr(Ac)$_2$ treatment of the soil, *Soil Sci.,* 101, 472, 1966.

38. **Fuller, W. H. and L'Annunziata, M.,** Movement of algae- and fungal- bound radiostrontium as chelate complexes in a calcareous soil, *Soil Sci.,* 170, 223, 1968.

39. **Hodgson, J. F.,** Chemistry of the micronutrient elements in soil, *Advance Agron.,* 15, 119, 1963.

40. **L'Annunziata, M. F. and Fuller, W. H.,** The chelation and movement of $Sr^{89} - Sr^{90}(Y^{90})$ in a calcareous soil, *Soil Sci.,* 105, 311, 1968.

41. **Mortensen, J. L.,** Complexing of metals by soil organic matter, *Soil Sci. Soc. Am. Proc.,* 27, 179, 1963.

42. **Schnitzer, M. and Skinner, S. I. M.,** Organo-metal interactions in soils: 7 stability constants of Pb^{++}, Ni^{++}, Co^{++}, and Mg^{++} fulvic acid complexes, *Soil Sci.,* 103, 249, 1967.

43. **Artiola-Fortuny, J. and Fuller, W. H.,** Phenols in municipal solid waste leachates and their attenuation by clay soils, *Soil Sci.,* 133, 218, 1982.

44. **Sposito, G.,** Soil Chemistry, in *McGraw-Hill Encyclopedia of Science and Technology,* Vol. 12, 4th ed. Lapedes, D. N., Ed., McGraw-Hill Book Co., NY, 1977, 491.

45. **Sommers, L. E.,** Chemical composition of sewage sludges and analyses of their potential use as fertilizer, *J. Environ. Qual.,* 6, 225, 1977.

46. **Cope, C. B., Fuller, W. H., and Willetts, S. L.,** *The Scientific Management of Hazardous Wastes,* Cambridge University Press, Cambridge, England, 1983, 263.

47. **Schnitzer, M. and Kodoma, H.,** Reactions of minerals with soil humic substances, in *Minerals in Soil Environment,* Dixon, J. B. and Weed, S. B., Eds., Soil Sci. Soc. Amer., Madison, Wis., 1977, 741.

48. **Schnitzer, M.,** Reactions between fulvic acid, a soil humic compound and inorganic soil constituents, *Soil Sci. Soc. Am. Proc.,* 33, 75, 1969.

49. **Chian, E. S. K. and DeWalle, F. B.,** Evaluation of Leachate Treatment: Vol. I., Characterization of Leachate, EPA-600/2-77-186a, U. S. Environmental Protection Agency, Cincinnati, Ohio, 1977a, 210.

50. **Chian, E. S. K. and DeWalle, F. B.,** Evaluation of Leachate Treatment, Vol. 2, Biological and Physical-Chemical Processes, EPA-600/2-77-186b, U. S. Environmental Protection Agency, Cincinnati, Ohio, 1977b, 245.

51. **Artiola-Fortuny, J. and Fuller, W. H.,** Humic substances in landfill leachates: I. Humic acid extraction and identification, *J. Environ. Qual.,* 11, 663, 1982.

52. **Sommers, L. E., Nelson, D. W., and Silveira, D. J.,** Transformations of carbon, nitrogen, and metals, in soils treated with waste materials, *J. Environ. Qual.,* 8, 287, 1979.

53. **Stevenson, F. J. and Ardakani, M. S.,** Organic matter reactions involving micronutrients, in *Micronutrients in Agriculture,* Mortvedt, J. J., Giordano, P. M., Lindsay, W. L., Eds., Soil Sci. Soc. Am., Madison, Wis., 1972, 79.

54. **Fuller, W. H., Alesii, B. A., and Carter, G. E.,** Behavior of municipal solid waste leachate: I. Composition variations, *J. Environ. Sci., Health,* A14, 461, 1979.

55. **Fuller, W. H., Amoozegar-Fard, A., Niebla, E. E., and Boyle, M.,** Behavior of Cd, Ni, and Zn in single and mixed combinations in landfill leachates, in Land Disposal: Proc. 7th Ann. Res. Symp., Shultz, D., Ed., U. S. EPA-600/9-81-0026, U. S. Environmental Protection Agency, Cincinnati, Ohio, 1981, 18.

56. **Fuller, W. H. and Alesii, B. A.,** Behavior of municipal waste leachate: II. In Soil, *J. Environ. Sci. Health,* A14, 559, 1979.

57. **Flaig, W.,** The chemistry of humic substances, in *The Use of Isotopes in Soil Organic Matter Studies,* Pergamon, New York, 1966, 103.

58. **Allison, F. E.,** *Soil Organic Matter and Its Role in Crop Production,* Elsevier Scientific Publishing Co., Inc., New York, NY, 1973, 1.

59. **Alexander, M.,** Biodegradation of chemicals of environmental concern, *Science,* 211, 132, 1981.

60. **Brady, N. C.,** *Nature and Properties of Soils,* 8th Ed., John Wiley and Sons, Eds. New York, 1974, 1, 16.

61. **Dowdy, R. H., Ryan, J. A., Volk, V. V., and Baker, D. E.,** Chemistry in the Soil Environment, ASA Special Pub. No. 40, Am. Soc. Agron., Madison, Wis., 1981, 259.

62. **Kittrick, J. A.,** Mineral equilibria and the soil system, in *Minerals in Soil Environment,* Dixon, J. B. and Weed, S. B., Eds., Am. Soc. of Agron., Madison, Wis., 1977, 1.

63. **Kenney, D. R., and Wildung, R. E.,** Chemical properties of soil, in *Soils for Management of Organic Wastes and Waste Waters,* Elliott, L. F. and Stevenson, F. J., Eds., Am. Soc. of Agron., Madison, Wis., 1977, 75.

64. **Gast, R. G.,** Surface and colloid chemistry, in *Minerals in Soil Environments,* Dixon, J. B. and Weed, S. B., Eds., Soil Sci. Soc. Am., Madison, Wis., 1977, 27.

65. **Mehlich, A.,** Charge properties in relation to sorption and desorption of selected cations and anions, in *Chemistry in the Soil Environment,* Dowdy, R. H., Ed., Am. Soc. of Agron., Madison, Wis., 1981, 47.

66. **Schoen, U.,** Kennzeichung von Tonen Durch Phosphatbindung und Kationenumtausch Z. *Pflanzenernahr Dungung u. Bodenk,* 63, 17, 1953.

67. **Leeper, G. W.,** *Managing the Heavy Metals on Land,* Marcel Dekkar, New York, 1978, 121.

68. **Steengjerg, F. T.,** Undersgelser over Manganindholdet i dansk Jord. 1. Det ombyttelige Mangan [The Mn content of Danish Soils: 1. The exchangeable Mn] *Tidsskr Planteavl,* 39, 401, 1933.

69. **Crooke, W. M.,** Effect of soil reaction on uptake of nickel from a serpentine soil, *Soil Sci.,* 9, 113, 1956.

70. **Jones, H. R.,** Waste disposal control in the fruit and vegetable industry, in *Pollut. Techno. Rev. No. 1,* Noyes Data Corporation, Park Ridge, N.J., 1973, 261.

71. **James, R. O. and Healy, T. M.,** Adsorption of hydrolysable metal ions at the oxide water interface: I. Co(II) adsorption of SiO_2 and TiO_2 as model systems, *J. Colloid and Interface Sci.,* 40, 42, 1972.

72. **Fuller, W. H.,** The importance of soil attenuation for leachate control, in Waste Management Technology and Resource and Energy Recovery, Proc. of the 5th Natl. Congr., Dallas, Dec. 7 to 9, 1976; co-sponsored by National Solid Waste Management Assoc. and the U. S. Environmental Protection Agency, EPA SW-22P, Washington, D.C., 1977, 239.

73. **Fuller, W. H.,** Soil-Waste Interactions, in *Disposal of Industrial and Oil Sludges by Land Cultivation,* Shilesky, D. M., Ed., Resource Systems, and Management Association, Ocean City, N.J., 79, 1978.

74. **Korte, N. E., Skopp, J. M., Niebla, E. E., and Fuller, W. H.,** A baseline study of trace metal elution from diverse soil types, *Water Air Soil Pollut.,* 5, 449, 1975.

75. **Korte, N. E., Skopp, J. M., Fuller, W. H., Niebla, E., and Alesii, B. A.,** Trace element movement in soils: Influence of soil physical and chemical properties, *Soil Sci.,* 121, 350, 1976.

76. **Baas Becking, L. G., Kaplan, I. R., and Moore, D.,** Limits of the natural environment in terms of pH and oxidation-reduction potentials, *J. Geol.,* 68, 243, 1960.

77. **Fuller, W. H.,** Soil modification to minimize movement of pollutants from solid waste operations, *Critical Rev. Environ. Contr.,* 9, 213, 1980.

78. **Fuller, W. H.,** Liners of natural porous materials to minimize pollutant migration, EPA-600/S2-81-122, U. S. Environmental Protection Agency, Cincinnati, Ohio, 1981, 5.

79. Soil Survey Staff, *Soil Survey Manual,* USDA Soil Conservation Service, Handbook No. 18, Washington, D.C., 1951, 8.

80. **Jackson, J. L.,** Soil clay mineralogy analysis, in *Soil Clay Mineralogy,* Rich, C. I. and Kunze, G. W., Eds., University North Carolina Press, Chapel Hill, 1963, 245.

81. **Fuller, W. H.,** Desert soils, in *Desert Biology,* Brown, G. W., Ed., Academic Press, Inc., New York, NY, 1974, 32.

82. **Brown, K. W. and Anderson, D.,** Effect of organic chemicals on clay liner permeability, in Proc. of the 6th Ann. Res. Symp., Shultz, D., Ed., EPA-600/-80-010, U.S. Environmental Protection Agency, Cincinnati, Ohio, 1980, 123.

83. **Anderson, D. and Brown, K. W.,** Organic leachate effects on the permeability of clay liners, in *Land Disposal: Hazardous Waste,* 7th. Ann. Res. Symp. Proc., EPA-600/9-81-002b, U.S. Environmental Protection Agency, Cincinnati, Ohio, 1981, 119.

84. **Diamond, S.,** Pore size distribution in clays, in *Clays, Clay Miner.,* 18, 7, 1970.

85. **Vomocil, J. A.,** Porosity, in *Methods of Soil Analysis,* Part 1., Black, C. A., Ed., Am. Soc. Agron., Inc., Madison, Wis., 1965, 299.

86. **Blake, G. R.,** Bulk Density, in *Methods of Soil Analysis,* Part 1., Black, C. A., Ed., Am. Soc. Agron., Inc., Madison, Wis., 1965, 374.

87. **Bitterli, R.,** Freezing and Drying: Effects on the Solubility of Municipal Solid Waste Leachate Constituents, M.S. thesis, The University of Arizona Library, Tucson, 1981, 36.

88. **Hillel, D.,** *Soil and Water,* Academic Press, Inc., New York, 1971, 288.

89. **Hillel, D.,** *Fundamentals of Soil Physics,* Academic Press, Inc., New York, 1980, 413.

90. **Baver, L. D., Gardner, W. H., and Gardner, W. R.,** *Soil Physics,* John Wiley and Sons, Inc., New York, 1972, 498.

91. **Hanks, R. J. and Ashcroft, G. L.,** Applied Soil Physics, Springer-Verlag, New York, 1980, 159.

92. **Kirkham, D. and Powers, W. L.,** *Advanced Soil Physics,* John Wiley and Sons, Inc., New York, 1972, 534.

93. **Gardner, W. H.,** Water content, in *Methods of Soil Analysis,* Part 1., Black, C. A., Ed., Monograph 9, Amer. Soc. Agron., Madison, Wis., 1965, 82.

94. **Nakayama, F. S. and Reginato, R. J.,** Simplifying Neutron Moisture Meter Calibration, *Soil Sci.,* 1982, 133, 48.
95. **McGown, M. and Williams, J. B.,** The water balance of an agricultural catchment, I. Estimation of evaporation from soil water records, *J. Soil Sci.,* 31, 217, 1980.
96. **Sposito, G.,** General criteria for the validity of the Buckingham-Darcy flow law, *Soil Sci. Soc. Amer. J.,* 44, 1159, 1980.
97. **Boersma, L.,** Field measurement of hydraulic conductivity below a watertable, in *Methods of Soil Analysis,* Part I., Black, C. A., Ed., Monograph 9, Amer. Soc. Agron., Madison, Wis., 1965, 222.
98. **Amoozegar-Fard, A. and Warrick, A. W.,** 1984, Field measurement of saturated hydraulic conductivity, in *Methods of Soil Analysis,* Part I, Klute, H., Ed., Amer. Soc. Agron., Madison, Wis., 1984. (Accepted for publication)
99. **Warrick, A. W. and Nielsen, D. R.,** Spatial variability of soil physical properties in the field, in *Applications of Soil Physics,* Hillel, D., Ed., Academic Press, Inc., New York, 1980, 319.
100. **Klute, H.,** Laboratory Measurement of Hydraulic Conductivity of Saturated Soil, in *Methods of Soil Analysis,* Part I., Black, C. A., Ed., Monograph 9, Amer. Soc. Agron., Madison, Wis., 1965, 234.
101. **van Genuchten, M. Th.,** A closed form equation for predicting the hydraulic conductivity of unsaturated soils, *Soil Sci. Soc. Amer. J.,* 44, 892, 1980.
102. **Mualem, Y.,** A new model for predicting the hydraulic conductivity of unsaturated porous media, *Water Resource. Res.,* 12, 513, 1976.

Soil as a Waste Utilization System

Chapter 2

WASTE UTILIZATION ON CULTIVATED LAND

I. RATIONALE AND SCOPE

The land is a gigantic biodigestion system developed over millions of years. This natural system digests the animal and plant wastes to become part of the soil. The soil continues to be the primary means of mass disposal. The role of the land (soil) for recycling wastes, particularly organic wastes and waste waters, receives increasing attention because of the quantity and variety of wastes and variations in methods of disposal. Land disposal sites, however, present a serious potential threat to the quality of surface and underground waters as well as to the soil itself. The best methods of disposing hazardous and solid wastes on land have yet to emerge into safe and sound management practices.

As a consequence of the accelerating threat of pollution to the environment and our incomplete knowledge of ways to stabilize pollutant movement throughout the environment, land treatment technology is being actively sought. Land treatment is particularly noted in the Resource Conservation and Recovery Act (94:580), as a viable waste management option.[1] The proposed regulations issued by the U.S. EPA's Office of Solid Waste is given as one of the land disposal technologies for management of hazardous wastes. As defined by the U.S. EPA, "land treatment" implies that the land or soil is used as a medium to treat hazardous waste.[1] A land treatment facility is defined as "that portion of a facility at which hazardous waste is applied onto or incorporated into the soil surface". The U.S. EPA considers land treatment as "a viable waste management option" for selected solid and hazardous wastes not available through any of the other proposed options.

In this discussion, "land treatment" is considered in a broad interpretation. In addition to hazardous and solid waste management, it will include land and soil management of organic wastes and waste waters, some of which are not usually considered hazardous or in a solid form. Thus, land treatment for this discussion has been divided into two parts; (a) utilization (resource) and (b) treatment (waste). The general title could easily be "Land Treatment/Utilization". The waste management technology is broadened, therefore, to include such waste materials as are generated by animals, food processing, leather industries, forest products, and mine tailings, and organic waste waters that are not necessarily "hazardous" or "solids". Such wastes have received attention as soil builders, fertilizers, and/or soil conditioners by agriculturalists.[2-5] For this discussion, land utilization will be limited to management associated with direct waste resource application to the land without unusual interruption of agronomic crop production. Animal manures, municipal sewage sludges, and sewage waters are good examples of resource materials that have been considered as wastes resources. Land utilization, therefore, serves dual objectives, one of waste disposal and a second of valuable resource.

Land treatment as used here involves application of hazardous industrial wastes to the land when no crop is being grown because of an incompatibility with successful plant establishment and/or economic crop production. This does not preclude the use of the land or soil after a period of biodegradation that would render the remnants of the waste compatible for plant establishment and economic production. For convenience, land treatment/utilization will be separated into (a) waste disposal on cultivated agricultural land (crop land) and (b) waste disposal on uncultivated land (pasture, range, and forest land). The wastes designated for land treatment, also, should be considered as potential resources when managed in a way to provide improvement in the productive quality of soil and/or direct nutritional benefit to crops grown on the soil. The soil has almost unlimited capacity to accept large quantities of waste and transform them through microbial activity (biodegradation) into essential nu-

Table 1
PARTITIONING OF MATERIALS IN THE MUNICIPAL
WASTE-TYPE LANDFILL

	U.S.[9]	Massachusetts[9]	California[9]	Arizona[84]
Paper	36.8	35.0	43.0	47.5
Newspaper	(7.2)	(7.8)		
Metals	8.8	9.2	7.0	6.1
Ferrous	(7.8)	(8.3)	(6.0)	(5.0)
Nonferrous	(1.0)	(0.9)	(1.0)	(1.1)
Glass	9.4	18.6	9.0	5.8
Plastic	3.5	4.1	2.0	2.0
Cloth, rubber, leather	3.9	5.2	4.0	8.2
Wood	3.4	1.1	4.0	0.5
Food waste	15.6	5.9	6.0	14.7
Yard and garden	17.3	0.5	19.0	12.3
Miscellaneous	1.3	19.6	6.0	1.1
Ash and soil	NR	NR	NR	4.1
Total	100.0	100.0	100.0	100.0

Note: NR means not reported

trients for plant utilization. However, wastes that would permanently destroy the quality of the land for growth of some kind of vegetation should not be applied.

Land treatment must be understood, planned, and managed with the same degree of attention provided any other process operation. The complexities of waste- soil and natural process interactions, must be understood if land treatment is to be an acceptable practice and develop on a sound technical basis. An appreciation and knowledge of suitable land treatment site locations is the first step in adequate pollution control and protection of the surrounding environment.

Planning for land treatment/resource utilization, therefore, begins with (a) the development of a data base for location (site), soil, and waste characteristics, (b) the identification of the limiting characteristics of the land, soil hydrology, and waste to be involved, and (c) the level of constraints necessary to protect the health of mankind and environment from undesirable deterioration. The mechanics of operation follow for each location, recognizing that no two locations will be exactly the same. Because such variation is also shared by wastes, it is most convenient to discuss waste disposal here under our definition of utilization.

The land utilization programs of waste disposal may be grouped according to similar kinds of materials to be applied and placed into obvious categories with some examples identified as follows:

- Agricultural   Animal generated waste (manures, meat processing, and tanning)
Crop production (crop residues, orchard material)
Forest residues (mill waste as bark, sawdust, and forest fuels)
Food processing (cannery waste, waste waters, and whey)
- Municipal   Solid waste of residential and commercial collections not associated with a product (Table 1)
Sewage sludges and waters
- Industrial   Wastes from industrial manufacturing and production
- Institutional   Schools, colleges, hospitals, and health centers (this will overlap municipal in small communities)

Table 2
MAXIMUM ALLOWABLE
SLUDGE-METAL ADDITIONS TO
PRIVATELY OWNED FARMLAND[85]

Metal addition (kg/ha) to soil with a
cation exchange capacity (meg/100 g) of:

Metal	Less than 5	5 to 15	Greater than 15
Pb	500	1000	2000
Zn	250	500	1000
Cu	125	250	500
Ni	50	100	200
Cd	5	10	20

U.S. Environmental Protection Agency, 1975, 10.

II. WASTE RESOURCES FOR CULTIVATED LAND

Certain constraints are required for waste materials and waste waters to be used on land where food and fiber plants are established and grow productively. Some of the essential constraints are as listed:

- Plant and crop growth quality
  Waste materials should be:

  free of toxic properties at efficient loading rates. For example, Table 2.2 identifies maximum sludge metal additions to privately owned farmland.
  free of excessive concentrations of heavy metals.
  free of excessive concentrations of common salts of Na, K, Ca, and Mg that would inhibit or limit growth.

- Loading rates
  knowledge of optimum loading rates to insure maximum crop yields without leaving an excess of substances that would migrate to underground waters.

- Land application
  application and/or incorporation must not require unusual or additional farm practices or excessive energy needs over and beyond expected return.

- Amenability for biodegradation
  there must be assurance the waste will biodegrade within a reasonable period of time and leave no toxic or adverse residue.

- C/N ratio
  The waste material should have a favorable C/N ratio of 30 or less or, if low in N, add fertilizer.

- Benefits to crop
  benefits to plant growth may include plant nutrients, soil conditioning, mulching, and water utilization and behavior.

- Cost
  cost of hauling, applications, and incorporation should not exceed the economy of return.

- Off-site pollution
  susceptibility to wind, water and/or odor problems should be minimal.

- Soil properties
  should not adversely affect soil physical properties of infiltration, structure, percolation, and aeration for economic crop production.

- Health factors
  Provide no health hazard to the food chain, humans, or animals.

The key to successful waste utilization on cultivated land where farm practices are already established is (a) to avoid requirements for appreciable extra work or energy use and (b) to assure crop improvement (better production than if the waste was not used). Wastes adapted for use on cultivated land, therefore, must classify as resources (not wastes) and fit into the ongoing agricultural practices as well as be sufficiently beneficial to crop production that their use will prove economically feasible. A good example of what is expected here is the use of animal manures. Despite the historical benefits attributed to animal manures, they share in all the constraints listed above. Should a manure source not meet the requirements of the constraint, it must be reconsidered only as a waste and not a resource and receive different disposal treatment. The same is true of any material to be used on cultivated land. Thus, a waste may be considered as a resource out of place, low in quality, quantity, or simply lacking in qualities necessary for adaptation to present management technology. This concept finally brings us to the main purpose: soil modification to minimize movement of pollutants from industrial wastes, municipal solid wastes, and wastewaters. Some examples of solid wastes (Table 3), wastewaters, and sludges (Table 4) that are suitable for cultivated lands (resource utilization) are provided.

III. SOIL MANAGEMENT OF SOLID WASTES

The utilization of agricultural wastes is well covered in scientific publications stemming from state agricultural experiment station research as supported by many private, state, and federal (e.g., USDA, EPA) sources and published in journals.[2] There is no purpose served by repeating these numerous, in-depth research reviews except briefly to provide some ready references. One book entitled "Soils for Managemement of Organic Wastes and Waste Waters" reports the proceedings of a national symposium on this and some related topics. It is published by Soil Science Society of America, American Society of Agronomy, and Crop Science Society of America, Madison, Wis.[2] Although this publication has in excess of 600 pages, only the usual agricultural and municipal wastes, such as manures, sewage effluents, and sludges that have been studied for many years, are included and discussed at any depth. The same is true for "Guidelines for Manure Use and Disposal in the Western States".[6] These subjects are brought up to date and the literature is well reviewed up to 1975. A publication by the U.S. EPA entitled "Beef Cattle Feedlot Site Selection" is a good example of the land treatment/utilization approach.[7]

The approach to be taken in this chapter, which will be more related to the general objective, is to review briefly the the methods of land application and soil incorporation of industrial wastes on cultivated land. It is this area of land treatment that requires innovations and the development of new technologies to accommodate wastes other than the conventional agricultural or municipal sewage sludges and effluents.

**EXAMPLES OF SOLID WASTES THAT HAVE BEEN APPLIED
TO CULTIVATED LAND GROWING CROPS[a]**

Waste (resource)	Crops
Manures	
Animal (all sources)	Edible and non-edible
Plant (green manures)	Edible and non-edible
Municipal waste	
Sewage water effluent	Pasture, cotton, grains
Sewage slurries (3 to 10% solids)	Pasture, cotton, grains
Sewage sludge (thick slurry or dry)	Pasture, cotton, grains
Solid waste (untreated)	Pasture (alfalfa)
Solid waste (pretreated—compost)	Home gardens
Paper mill (pulp and paper mfg.)	
Waste waters	Alfalfa, grass pasture
Slurries	Alfalfa, grass pasture
Hardboard	
Paper board	
Composted and/or biodegraded	Wheat, corn beans
Cannery (tomatoes, corn, etc.)	Small grain
Tannery dust and slurry	Cultivated pasture
Dairy products residues	Cultivated pasture
Animal slaughter waste (e.g., blood meal)	Home gardens, yards
Mine tailings (processed)	All land

[a] Compiled by Fuller, W. H., from various sources

Table 4
**ESTIMATED QUANTITIES OF INDUSTRIAL WASTEWATERS AND
SLUDGES SUITABLE FOR LAND CULTIVATION**

Industry	Waste type	Waste Quantities[a]		
		1975	1980	1985
Food and kindred products				
Meatpacking	Wastewater[b]	150	170—190	190—230
Poultry dressing plants	Wastewater[b]	87	98—110	110—140
Dairy products	Wastewater[b]	45	51—57	58—70
Fruits and vegetables	Wastewater[b]	280	320—350	360—440
Cane sugar	Wastewater[b]	68	79—87	89—110
Malt beverages	Wastewater[b]	87	98—110	110—140
Textile mill products				
Textile finishing	Secondary wastewater treatment sludge[c]	735	840—920	940—1,160
Lumber and wood products		$7.2—7.5 \times 10^6$	$8.8—9.1 \times 10^6$	$10.8—1.1 \times 10^6$
Wood preserving	Wastewater[b]	1.9	2.2—2.3	2.5—2.9
Paper and allied products	Primary wastewater treatment sludge[c]	1.4×10^6 $1.7 \times 10^{6-}$	1.6×10^6 $1.9 \times 10^{6-}$	1.8×10^6 $2.1 \times 10^{6-}$
Chemicals and allied products				
Organic fibers, noncellulosic	Secondary wastewater treatment sludge[c]	4,800	7,300	9,800
Pharmaceuticals	Waste mycelium[c]	6.3×10^4	8.1×10^4	1.0×10^5
Soap and other detergents	Wastewater[b]	12—14	17—18	20—22

Table 4 (continued)
ESTIMATED QUANTITIES OF INDUSTRIAL WASTEWATERS AND SLUDGES SUITABLE FOR LAND CULTIVATION

Industry	Waste type	Waste Quantities[a]		
		1975	1980	1985
Organic chemicals	Wastewater treatment sludges[c]	5.5×10^6	6.8×10^6	8.5×10^6
Petroleum refining and related industries				
Petroleum refining	Non-leaded product tank bottoms[c]	4.1×10^4	5.1×10^4	6.2×10^4
	Waste bio sludge[c]	4.0×10^4	5.0×10^4	6.2×10^4
Petroleum refining (continued)	API separator sludge[c]	3.4×10^4	4.2×10^4	5.2×10^4
	Dissolved air flotation float[c]	2.9×10^4	3.7×10^4	4.5×10^4
	Slop oil emulsion solids[c]	1.7×10^4	2.1×10^4	2.6×10^4
	Crude tank sludge[c]	390	490	610
Leather and leather products				
Leather tanning and finishing (vegetable)	Wastewater[b]	2.5	2.3	2.2
	Secondary wastewater treatment sludge[c]	820	740	680
Total wastewater		735	840—920	940—1,160
Total sludge		$7.2—7.5 \times 10^6$	$8.8—9.1 \times 10^6$	$10.8—1.1 \times 10^6$

[a]   Waste quantities for 1975 based on ref. 1 to 15 and contacts with industry representatives and current researchers. Values for 1980 and 1985 assume the 1975 relationship between production and waste generation, and are based on industry-specific production projections (if available).
[b]   Wastewater quantities are given in units of millions of cubic meters per year.
[c]   Sludge quantities are given in units of metric tons per year on a dry weight basis.

Phung, T., Barker, L., Ross, D., and Bauer, D., U.S. Environmental Protection Agency, 1978, 205.

Since agriculture is the historic leader in effective economic and environmental quality acceptance for waste disposal on lands, it is understandable that management for land utilization may well begin on the farm. Management practices of the land, water, soil, fertilizers, and crop production in response to the kind of waste significantly determine the extent of biological degradation, attenuation, and environmental impact of the waste disposal system. Emphasis in this chapter will center on those waste utilization management practices already successfully in place in which chemical constituents in agricultural, municipal, and industrial wastes are biodegraded on cultivated land.

A. Procedure for Land Utilization of Wastes

A great number of decisions are required in land treatment of wastes. A wide range of knowledge is necessary to finalize the disposal successfully with a minimal of adverse impact on the environment. Loehr et al.[4] suggest a design approach, as illustrated in Figure 1, be considered before proceeding directly to land application in order to help overcome this problem. This general step-wise diagram in an engineering procedure for waste application is further divided into two levels to make the material easier to assimilate. Level I is suggested as an aid in making initial feasibility decisions as to whether land application of waste

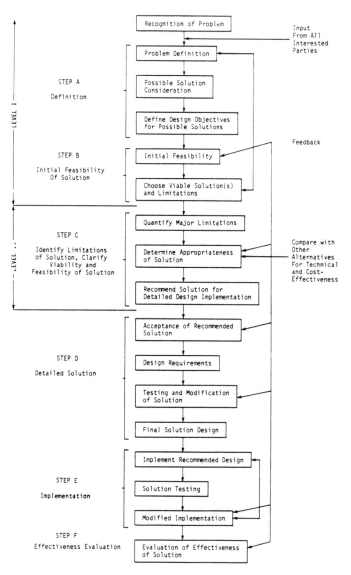

FIGURE 1.   Illustration of the general steps involved in an engineering design procedure. (From Loehr, R. C., Jewell, W. J., Novak, J. D. Clarkson, W. W., and Friedman, G. S., *Land Application of Wastes,* Van Nostrand Reinhold Co., New York, 1979, chap. 1. With permission.)

represents a viable management alternative (Figure 2).   Level II is suggested to permit potential limiting factors to be examined in detail (Figure 3).   From these considerations, then, final recommendations may be made. Although these designs were developed for sludges and wastewaters, they also contain features useful for land applications of solid and hazardous wastes.

B. Placement of Solid Waste

Land treatment/utilization requires that waste applications be on top of the soil surface or incorporated to a shallow depth. It does not include deep burial.

1. Surface Placement

Surface placement of solid organic materials includes (a) mulching, (b) trash-farming, (c) stubble mulching, (d) land farming (as in oil and oil wastes), (e) strip mulching, and (f) vertical mulching. The loading rate from point-sources is shallow and allows for agricultural

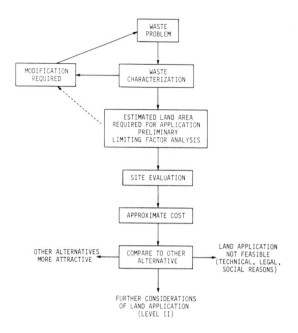

FIGURE 2.   Level I design procedure for land application of waste.
(From Loehr, R. C., Jewell, W. J., Novak, J. D., Clarkson, W.
W., and Friedman, G. S., *Land Application of Wastes,* Van Nos-
trand Reinhold Co., New York, 1979, chap. 1. With permission.)

cropping. Deeper layering or stacking comes under the heading of composts. Temporary
piling or stacking of organic residues for compost production is a means of insuring controlled
partial decomposition with the resulting material being well adapted to surface application.
Surface placement of solid waste has certain advantages and disadvantages for crop production, namely:

Advantages of surface placement
- Moisture conservation in the soil surfaces compared with bare surface
- Temperature modification of the soil surface
- Less nitrogen demand from the soil
- Lower cost of application
- Energy conservation because of nonincorporation power requirements
- Can aid in preventing soil, water, and wind erosion to the land

Disadvantages of surface placement
- More susceptible to soil, water, and wind erosion
- Greater social nuisance in the form of odors, volatilization of chemicals, dust and/or
  trash blowing and microbial aerosol depending on the nature and particle size of the
  waste
- May be unsightly and attract vermin, flies, birds, and rodents
- Loss of certain plant nutrients to the atmosphere and erosion
- More available for surface runoff and pollution of surface waters
- Loading rate more critical to control
- Greater problems with disease spreading
- Potentially toxic or hazardous constituents more of a problem when surface applied
- Loss of water source by evaporation compared with subsurface

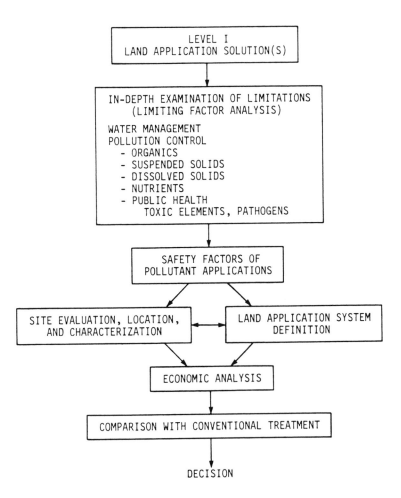

FIGURE 3.   Generalized activities completed during level II analysis. (From Loehr, R. C., Jewell, W. J., Novak, J. D. Clarkson, W. W., and Friedman, G. S., *Land Application of Wastes*, Van Nostrand Reinhold Co., New York, 1979, chap. 1. With permission.)

Some of the disadvantages can be reduced acceptably or eliminated by pretreatment of some wastes.

Surface treatment is less expensive than subsurface incorporation because of the greater power demand for pulling equipment through the soil. A great number of surface spreaders can be used such as tank trucks, honey guns, rear-end truck spreaders, and blade and bulldozers, as have been used for farm manure spreading.[8,9] During the 1940s the USDA Soil Conservation Service and Midwest farmers developed a waste utilization program to conserve water and reduce soil and water erosion by not plowing-under the old crop residues before reseeding. Thus, a waste utilization program was developed out of need for minimum tillage and mulch to protect the soil. Crops were seeded through the trash or grain stubble. Soils highly subject to wind and water erosion because of loose textures and structure, and sloping topography benefit most from this type of surface waste utilization. This technique can be adapted to external wastes brought in provided they are compatible with the crop being grown and will remain in place against the eroding forces of wind and water. Similarly, wastes can be placed on pastures where the sod will hold them from moving. Also, vertical

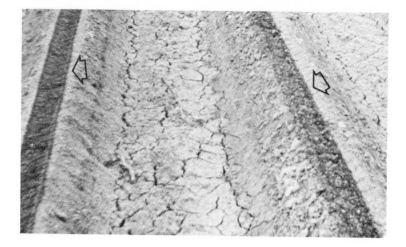

FIGURE 4.   Asphalt mulch strip over the seed row. (From Fuller, W. H., Management of Soils of the Southwest Desert, University of Arizona Press, Tucson, 1975, 195. With permission.)

mulching, where organic wastes and residues are placed in a trench, combines the advantages of both surface and subsurface application.[10]

Surface applications of solid wastes may play an important role in soil renovation and reconstitution by supplying organic matter to aid in water penetration and water-holding capacity. Good examples of this are the recovery of mine tailings and mine waste lands treatment. As further research is undertaken, soil surface application of industrial and hazardous waste can be shown to become more effective and provide a resource for land use rather than a waste for disposal only. The potential for land treatment has not been fully researched. For example, asphalt and petroleum mulches have been shown to be effective for retaining moisture and increasing the soil temperature, resulting in earlier seedling emergence, better stands and earlier yields.[11-13] The research, which was only in a preliminary stage (Figure 4) was discontinued in 1973 because of the energy crisis. We now anticipate a resumption of research based on petroleum wastes rather than the higher quality asphalts.

2. Soil Incorporation

Most solid wastes, as defined by land treatment, are well adapted to incorporation into the soil. There is less chance of trash scatter by wind and water as compared with surface application. The greater contact with the soil also aids the rate of decomposition, allows less interference with planting operations, affords better opportunity for water infiltration, and is aesthetically more acceptable.

The selection of a method and equipment for soil incorporation depends on the specific nature of the waste and crop to be grown. For example, municipal refuse should be subdivided and the large pieces of metal and glass removed. A large number of shredders and particle size reducing machines are available to accomplish almost any size objective. The organic materials may then be placed on the land surface, incorporated, or composted. Composting is designed primarily for bulk reduction and pretreatment to reduce the hazard of toxic substances and disease organisms the waste may contain. Such wastes as originate from the pesticide and pharmaceutical industry, and strong acids and caustics may require pretreatment. Lime is inexpensive and readily available to use to neutralize acids and aid in attenuation of heavy metals. Organics may require considerable chemical pretreatment before they can safely be added to the land. In some instances, aqueous dilution may be used to mitigate the adverse effects of toxic pollutants on the soil microorganisms. The waste can then be land treated safely. Pretreatment for land-applied solid and hazardous waste materials is so highly specialized for each waste that generalities here can be of little value to the reader. Furthermore, the cost-effectiveness of each waste varies considerably and must be programmed on a case-by-case basis.

The nature of the waste has a great influence on the kind of equipment used for application and kind of soil incorporation required. Agriculturalists use a large number of different kinds of machinery in manipulation of the soil for specific objectives. Most farm tillage equipment is suitable for land treatment with a variety of wastes. Incorporation requires more sophisticated equipment (e.g., the farm cultivation tools which include the plow, disc, harrow, and Rototiller®) than surface spreading. The units appear in all sizes and power ranges. Vertical mulching, where organic trash, sawdust, compost, etc., are placed into a narrow slit trench and compacted, requires special equipment.[10] The great versatility of the Rototiller® makes it most popular for refuse incorporation. Wheel tractors, truck tractors, and refuse blades also are widely accepted for refuse-soil management.

A newer innovation is the subsurface injectors for sludges and waste waters. The attractiveness of the soil injector is due to the capability of precise placement of slurries and sludges of up to 8% solids and the convenience of wastewater placement. Other advantages claimed relate to (a) conservation of water and other natural resource materials, (b) favorable placement of nutrients directly into the most active root zone, (c) improved soil physical condition due to subsurface organic matter incorporation, and (d) minimization of human and animal disease and vermin problems. Land treatment of wastes can be designed to reclaim wasteland, e.g., eroded farm land, over-grazed pastures, saline and alkali soils, and other disturbed land. The placement of organic wastes in the surface soil allows for aerobic decomposition, thus eliminating gas formation and production of concentrated potentially hazardous leachates so commonly found with landfill-type disposals. The added depth of soil available for attenuation of metals and biodegradation by microorganisms clearly distinguishes land treatment from other methods of excavated (landfill) burials. Sludges that contain noxious, volatile, or odorous components may have the air pollution problems corrected by shallow injection into soil surfaces. Depths of injection can be adjusted from just below the surface to as much as 0.5 m. A minimum depth is usually selected since the power (energy) demand is less and heterogeneities in the soil profile that may cause problems are avoided. Injection, however, does not possess the advantages of a rototiller for soil mixing and homogeneous distribution of the waste with the soil. Loading rates also differ. The Rototiller® has the greater advantage here. Rototillers® have the capacity of homogeneously mixing tons of solid waste into the soil to depths up to 0.5 m and in widths of 2.4 m where soils are not compacted.

The disk or disk-plow combines surface and subsurface incorporation by mixing waste with the soil after it has been spread. Mixing may be necessary for some wastes to promote biodegradation, reduce excessive volatilization, and minimize odors. Although disk-incorporated waste is not preferred over rototilled-waste, where rototillers are not available, it may not be economical to buy special equipment. The standard farm along with a bulldozer can be effective and efficient in managing the waste in a reasonable homogeneous mix. Also, the choices of disks allow for a wide range of soil manipulations which are capable of accommodating a wide variety of solid and hazardous wastes. Disks classify into three major types: disk-plows, disk-harrows, and disk-tillers.

C. Solid Waste Loading Rates

Each of the major kinds of equipment used in land treatment has a different capacity to suitably incorporate solid waste at a given frequency of pass-over the land. However, the capacity to mix wastes with the soil varies more with the bulkiness than any other single factor. Some of these differences can be modified by the degradability of the solid waste as related to frequency of application. Application frequency, in turn, depends on the land use, cropping pattern, climatic factors, and economics. The most uncertain of the above variables is the economics. Cost of land application systems for waste waters and municipal slurries has been reviewed by Loehr et al.[4] It is not repeated, partly because it does not apply particularly well to solid and hazardous wastes considered here and partly because there is so little actual data on solid and hazardous waste loading rate responses in the different soil types. It has been pointed out by Phung et al.,[9] however, that the economics of optimum loading rates of solid wastes to cultivated land are based on (a) the lowest loading rate that will economically enhance crop yield and (b) the highest loading rate that the land can accept without evidence of environmental damage. Other factors in the economics are dependent on (c) cost of application and incorporation, (d) period of time the land is not producing, and (e) benefits to crop growth and yield. The economics must still make provision for nonproliferation of pollution in the food chain and groundwater above all other objectives in any disposal management plan.

Loading rates also will be dependent on certain practical factors as summarized:

- The microbiological reactivity of the waste
- The chemical properties of the waste
- The physical properties of the waste
- Site topography features
- Site climatological characteristics
- Soil chemical, physical, and microbiological properties
- Land use
- Cropping plans
- Crop nutrient requirements and sensitivities
- Public reaction and sentiment

1. Animal Manures and Municipal Sludges

Nearly all of the loading rate data for solid wastes, up to 1982, are centered on animal manures and municipal sludges. Manure loading in Western U.S.A. for cultivated land varies with the temperature and rainfall of the region (Table 5).[6]   Lesser amounts are applied over a longer period of time in cold (less than 6000 annual degree days) than in warm (more than 6000 annual degree days) climatic regions. Also, in dry land regions (less than 76 cm annual precipitation) less manure is recommended than in humid regions (more than 76 cm annual precipitation). Manures may be applied at approximate rates as listed in Table   5.

Municipal sewage sludges have been applied to cultivated land at about half the loading rates of cattle manure. When they are not from a highly industrialized community and are relatively free of heavy metals, suitable loading rates appear to be similar to those of livestock manures on a dry-weight basis.[14-15]

Municipal sewage sludges vary as much in composition and presence of potential polluting constituents within selected climatic regions, as among different regions, although some real regional trends have been established. Sludges are drier and higher in dirt, and sand in arid and semiarid than in humid regions. In warmer climates, such as the Southwest, decomposition may proceed rapidly and extend annually for a longer period of time than in the temperature humid climates. The type of pollutants and quality reflects the presence of local industrial disposal into the sewage facilities more than any other single factor since strictly municipal sludges throughout the U.S.A. differ only moderately. The nonindustrialized Tucson, Arizona, sludge, for example, contains considerably fewer heavy metals (As,

Table 5
ANIMAL MANURE USE ON CULTIVATED LAND[a]

Region	Manure (dry weight basis)	
	Optimum	Maximum loading
Warm irrigated regions		
Livestock manure	34—56 t/ha	80 t/ha
Poultry manure	11—22 t/ha	(to salt tolerant crops)
Cold irrigated regions		
Livestock manure	34—45 t/ha	50 t/ha
Poultry manure	9—20 t/ha	(to salt tolerant crops)
Dryland regions		
Livestock manure	6.7—11 t/ha	10 t/ha
Poultry manure	half of irrigated region	(to salt tolerant crops)
Humid regions		
Livestock manure	34—45 t/ha	34—45 t/ha
Poultry manure	11—22 t/ha	11—22 t/ha
Slurry (5% cattle solids)	4 cm depth in 30 days	4 cm depth in 30 days

[a]   Modified from Meeks et al., 1975

Be, Cd, Cr, Cu, Hg, Ni, Pb, and Zn) than the industrially influenced Chicago sludge.[14-15] Of lesser importance to sludge quality is the method of handling, efficiency of the treatment system, and modernization of the system.[16] The quality of municipal sludge is emphasized, because like all other residues and wastes, the loading rates and capability of maintaining soil productivity depend largely on this characteristic. Municipal sludges usually contain a large proportion of soil particles. Table 6 was developed to roughly compare the composition of the lithosphere for a 10 mile depth, to soils, and further to city sludges. Again, the variability of sludges is as great as that of soil. The significance of the data, therefore, lies more in a broad perspective in composition than in precise mathematical comparisons. Certain trends in composition can be real and appear quite obvious in such constituents as organic carbon (organic matter), nitrogen, phosphorus and sulfur, and in heavy metals, Cd, Cr, Cu, Hg, Ni, Pb, and Zn.

2. Municipal Solid Waste

Municipal solid waste may be fortified with municipal sewage sludge in land treatment. For example, solid municipal waste was applied by Cottrell to Sagehill sand at rates of about 0, 150, 298, and 598 t/ha of dry matter with sewage sludge (2% solids) at about 190 liters per metric ton of solid waste.[17] Yields of alfalfa and grass were depressed at loading rates of solid waste above about 298 t/ha. Optimum rates were nearer 150 t/ha dry matter where supplemental nitrogen fertilizer was added.[18] All materials were incorporated by rototilling to a depth of 15 cm.

Municipal solid waste added to soils directly with a limited sorting and shredding at Davis, Calif. provided considerable problems with glass, cans, metal, and plastic when cropped.[19] Loading rates ranging from 112 to 896 t/ha (dry weight basis) were difficult to maintain on an annual basis because of accumulated bioresistant residues. Volk emphasizes the need for shredding to facilitate soil incorporation.[20]

Since municipal solid waste is 40 to 60% paper, an experiment on waste paper (shredded paper and cardboard combined) is of interest.[8] Three sizes of shreddings (0.6 to 3.8, 10 to 15, and 31 cm) were incorporated at rates of 45 and 448 t/ha to a sandy and clay soil. Depths of incorporation by a soil stabilizer ranged from 0 to 48 cm. Incorporation was good, but cost was high. With further study, this research could be more meaningful, since cost

Table 6

**A COMPARISON IN THE CHEMICAL PROPERTIES OF MUNICIPAL
SEWAGE SLUDGES FROM VARIOUS CITIES WITH THE
LITHOSPHERE AND SOILS**

Constituent		Lithosphere[a] 10 miles depth	Soils[b] (total)	Seven[c] States	Phoenix[d] 1977	From all USA[e,f]
			selected aver.	mean	ann. mean	median
pH	%	ng	ng	nt	6.5	ng
Organic C	%	0.95	2.0	31.0	20.1	ng
Total N	%	ng	0.14	3.2	1.8	3.3
NH$_4$-N	µg/g	ng	ng	0.74	ng	ng
NO$_3$-N	%	ng	ng	646	102	ng
Total P	%	0.12	0.06	1.9	1.5	2.3
Total S	%	0.06	0.07	1.1	1.7	ng
Ca	%	3.6	1.37	5.0	3.4	3.9
Fe	%	5.1	3.80	1.4	2.4	1.1
Al	%	8.1	7.10	0.5	0.5	0.4
Na	%	2.8	0.63	0.35	0.16	0.2
K	%	2.6	0.83	0.30	0.40	0.3
As	µg/g	5.0	5.0	43.1	10.0	4.0
Mg	µg/g	21,000	5,000	5,414	5,200	4,000
Ba	µg/g	430	430	576	400	ng
B	µg/g	10	10	109	2.5	20
Zn	µg/g	80	50	2,997	4,120	2,200
Cu	µg/g	70	30	1,308	712	700
Ni	µg/g	100	40	440	72	50
Cr	µg/g	200	100	3,280	450	400
Mn	µg/g	900	600	390	295	ng
Cd	µg/g	0.2	0.06	101	50	10
Pb	µg/g	16	10.0	1,656	254	500
Hg	µg/g	0.1	0.03	1,007	9	3
Co	µg/g	40	8	5.3	8	ng
Mo	µg/g	2.3	2.0	27.7	9	5
Cd/Zn	%	0.25	0.12	5.7	0.12	ng

Note: ng = not given.

[a] From Lindsay, 1979.[95]
[b] From Lindsay, 1979.[95]
[c] From McCalla et al., 1977.[88]
[d] From Fuller and Tucker, 1977.[16]
[e] From Sommers, L. E., 1977.[94]
[f] From Page and Chang, 1975.[96]

effectiveness may be altered by proper nitrogen application and timing of cropping.

In a similar study, Polson applied refractory metal processing waste as a slurry or dried material to Dayton silty clay loam at rates of 0, 11, 22, 56, and 112 t/ha.[21] Rye grass yields were not significantly changed by waste product additions, up to 112 t/ha, and were similar to those of commercial farms not using the refractory waste. The dried refractory waste was crushed and mixed into the top 15 cm with a Rototiller®.

3. Compost

Composting is dependent on microbiological processes for converting organic solid wastes into a more stable, humus-like material, the chief uses of which are to supply plant nutrients and for soil conditioning. The process is aerobic and involves both mesophiles and thermophiles. The principal constraints, therefore, are microbiological activity and genetic traits.[22]

Table 7
SELECTED ELEMENTS IN 42 DAY OLD MUNICIPAL
SOLID WASTE COMPOST FROM THE TVA
JOHNSON CITY SITE

	Percent dry weight (average)		
Element	Containing sludge (3%—5%)	Without sludge	Range (all samples)
Carbon	33.07	32.89	26.23—37.53
Nitrogen	0.94	0.91	0.85— 1.07
Potassium	0.28	0.33	0.25— 0.40
Sodium	0.42	0.41	0.35— 0.51
Calcium	1.41	1.91	0.75— 3.11
Phosphorus	0.28	0.22	0.20— 0.34
Magnesium	1.56	1.92	0.83— 2.52
Iron	1.07	1.10	0.55— 1.68
Aluminum	1.19	1.15	0.32— 2.67
Copper	<0.05	<0.03	
Manganese	>0.05	<0.05	
Nickel	<0.01	<0.01	
Zinc	<0.005	<0.005	
Boron	<0.0005	<0.0005	
Mercury	not detected	not detected	
Lead	not detected	not detected	

U.S. Environmental Protection Agency, 1971.

Composting results in volume reduction and certain nutrient concentration as related to well known processes of controlled biodegradation of organic matter. Agriculturalists and gardeners are familiar with composts.[11] They have used them as a mulch, soil conditioner, and plant nutrient source. The composition varies with the kind of residue used in the production (Tables 7 and   8).   In fact, almost any waste organic material can be made into compost. Some of the commonly known materials are

- Home garden trash, lawn clippings, leaves, shrub and tree prunings, vegetable trimmings, and kitchen wastes
- Straw and waste hay
- Animal manures with and without straw and bedding
- Wood — sawdust, bark, small chips
- Paper and cardboard
- Municipal solid waste (MSW)
- Food processing waste — cannery and dairy

Compost originating from plant wastes represents a resource that benefits plant growth (crop production) attributable to:

- Promotion of a favorable soil structure
- Promotion of favorable moisture relation, water movement, and water conservation
- Promotion of a good tilth, and thereby a working and good root penetration
- Contribution to ion exchange colloids, thus aiding in plant nutrition by holding nutrient elements against loss by leaching

Table 8
ANALYSES OF COMPOSTED REFUSE AND FRESHLY
SHREDDED REFUSE

Constituent	Japan[a]	Florida[b]	Tennessee[c]	Washington[d]	Ontario[e]
			% dry-weight basis		
Total N	1.0	1.2	1.3	0.78	0.57
P	0.24	0.26	0.26	0.16	0.08
K	0.98	0.38	0.97	0.27	0.314
Ca	5.1	1.30	4.6	1.97	0.850
Mg		0.07	0.5	0.11	0.209
Na			0.6	0.26	0.187
Fe				0.51	
S			0.5		
Total C	28		27	48	
Organic C					37
			μg/g dry-weight basis		
Mn		130		242	250
B		25		58—62	
Cu		125	100	96	28
Zn		250	1,500	715	400
Pb					200
Cd					7
Cr					31

[a]   Composted municipal refuse.
[b]   Composted municipal refuse.
[c]   Compost with up to 20% sewage sludge.
[d]   Freshly shredded refuse >2 μm.
[e]   Freshly shredded unsorted municipal refuse.

McCalla, T. M., Peterson, J. R., and Lue-Lunin, C., *Soils for Management of Organic Wastes and Wastewater*, American Society of Agronomy, Madison Wis., 1977, 11. By permission of the American Society of Agronomy.

• Addition of humus which acts as a slowly available food source for microorganisms to produce soil structure forming substances

Compost of municipal solid waste in the field of waste disposal and land treatment/ utilization is not only technically feasible, but also essential to the health and welfare of the society. Composts prepared in a biodigester and deep windows at elevated temperatures by aerobic microbial decomposition processes, undergo partial sterilization whereby pathogenic bacteria, fungi, and virus are eliminated and weed seeds, insects, and insect eggs are destroyed.[22,24]

Loading rates of most composts of plant and municipal solid waste origins applied to crop land are similar to those of sewage sludges and cattle manure. When nitrogen fertilizer has not been added to the composted material during processing, nitrogen needs must be calculated and applied along with the compost when put onto land. The same kind of equipment used to incorporate manures and sewage sludges works well with composts. Loading rates of well rotted municipal solid waste compost range from 30 to 75 t/ha. As much as 100 t/ha of MSW compost was applied to an Arizona irrigated soil with 100 kg/ha of N fertilizer with excellent results on winter lettuce. The compost, however, was irrigated once and the land allowed to stand fallow 180 days before bedding-up and seeding.

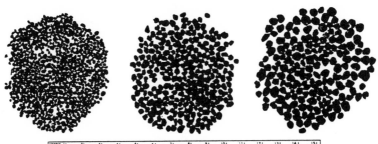

FIGURE 5. Pellets of biodigested and composted municipal solid waste used as fertilizer and soil conditioner.

Composted municipal solid waste at loading rates of 9 and 18 t/ha increased the yield of corn in Georgia soils the first year with some beneficial residual effects carrying over the second year in low fertility soils.[25] Larger loading rates of municipal compost ranging from 0, 23, 46, 82, 246, and 327 t/ha containing 20% sewage sludge were effective in improving forages of sorghum, common bermuda grass, and corn.[26] Municipal compost is used beneficially in Europe for control of erosion and runoff.[27] Also, nutritional benefit to small grain crops is attributed to the N content. The yield benefit roughly parallels the level of N added in the waste, a factor in loading rate.

Pelleted MSW compost has additional advantages over loose compost.[28] The pellets handle easily, are clean and odorless. Wind will not blow them away and they may be made into any size desirable (Figure 5) for home use, particularly if they are fortified with extra nitrogen and phosphorus fertilizer. Considerable processing must be put into MSW compost to remove large pieces of glass, metals, and plastic which may be offensive to the agricultural grower or home gardener. Because composting may not always return a profit and is an added cost to the agriculturalist, compost must be viewed as a waste disposal as well as a resource.

4. Oily Wastes

Optimum and maximum loading rates for oil waste disposal by soil cultivation on agricultural land has yet to be determined since the few quantitative stuides reported provided little or no crop evaluations.[29,30] Loading rates for land not intended for immediate crop production vary according to:

- Oil composition (e.g., % paraffins, aeromatics, asphaltine oils). The odor of most rapid decomposition is: paraffins > aeromatics > asphaltines.
- Soil temperature — warm soils biodegrade oils more rapidly than cold soils.
- Soil moisture — about 30% moisture is minimum required.
- Soil pH and capacity for neutralizing acids produced.
- Frequency of application.
- Number of tillings or soil mixing during period of biodegradation.
- Fertilizer requirements — primarily N and P.

Oily wastes vary greatly in composition and associated waste constituents.[31] The three major hydrocarbons may be grouped for convenience into (a) saturates or paraffins, (b) aeromatics, and (c) resins or asphaltines. Associated with oily wastes are a number of non-oily pollutants of salty water, acid, heavy metals (such as arsenic, lead, and zinc) and sulfur compounds including sulfides and oxides of sulfur, volatiles, and odors. Each oily waste source, therefore, has a large number of quality variables that must be evaluated in the selection of a suitable loading rate. For example, Kincannon[29] used a general loading rate of 10% by weight of oily material for the three oils — crude oil, Bunker C oil, and waxy oil. The loss of saturates, aeromatics, and resins for each was 36, 18, and 90; 56, 56, and 10; 8, 26, and O HC-type in weight-percentage, respectively. Bunker C oil decomposed more rapidly than the other two. The resin-asphaltines were biodegraded the least at all locations. Loading of initial 10% (by weight for 15 cm depth of soil) was difficult to apply and mix into the soil. Two loading rates at 5% were much more manageable. One must consider that loss by volatilization also varies for each oil fraction and must not be attributed wholly to biodegradation. Mixing may be accomplished successfully with a Rototiller®, disc, or barber-shank harrow depending on the viscosity of the oily waste. Huddleston and Cresswell[30] also found that initial oil additions of 10% in soil (12 cm depth) was difficult to manage with equipment but losses were greater over a given period of time than the 5% applications. No doubt the climatic factor is very important in loading rate selection. Higher loading rates can be made in warm than in cold climates. Snyder et al.[31] applied waste lagoon oil at a rate of 10,800 gal/a (10 ℓ/m^2) for a 15 cm soil tilling depth. Their suggested application rates for the climate of Salt Lake City, Utah, are reproduced in Table 9. They have not made recommendations for optimum or minimum levels for crop production land, as yet. Biodegradation was more rapid where lime and nitrogen were added. They accounted for about 80% of the oil loss during the warmer months. Calculated on a pure oil basis, the rate of loading was approximately equivalent to 7.5% per acre.

5. Paper Pulp Mill Waste

Paper mill sludge varies in composition, like most wastes, depending on the process, wood source, and amount of recycled paper incorporated in the final products (Table 2). According to Dolar et al.[32] paper mill sludge is relatively low in nitrogen, giving the sludge from primary treated process a C/N ratio ranging near 150 and from secondary treatment a ratio between 12:50. Sludges having a C/N ratio above about 20, require additional nitrogen to biodegrade at a maximum rate. Also, from a standpoint of crop production, the initial concentration of all the major nutrients (N, P, and K) in the paper sludges is low (Table 10). Because of the wide C/N ratio of primary paper mill sludge, N is immobilized. The waste may actually rob the soil of available N during the first year of decomposition in cultivated land. Aspitarte et al.[33] found that crops planted during this initial biodegradation period were adversely affected, presumably by lack of sufficient N to achieve maximum growth, although he had added fertilizer N to provide a final C/N ratio of 1:10. After the year, all crops receiving loading rates of 448, 896, and 1344 t/ha (at 150 C/N ratio) benefited from the sludge, with or without enough N additions to give a 1:10 C/N ratio.

Loading rates of 2.5% by weight, of secondary paper mill sludge, were found by Dolar et al.[32] to improve oat yields in five Wisconsin soils the first year. At a rate of 10%, by weight, however, oat yields decreased as compared with the controls. Unfortunately, the literature on loading rates of solid waste from the paper and pulp mills is very limited. Yet in order to make knowledgeable loading rates to cultivated land, certain information must be available as given in the beginning of this section such as biodegradation rate for the specific climatic conditions of land treatment.

6. Wood and Forest Waste

Wood and forest wastes may be readily adopted for home gardeners and home and industrial landscaping as well as for surface and subsurface incorporation into wastelands for land reclamation, rejuvenation, and repair. All tree parts do not have the same chemical composition or physical structure, nor do they decompose equally well. Unprocessed tree parts do not contain significant levels of heavy metals to be a threat to pollution of the food

Table 9

TEST APPLICATION RATES OF PETROLEUM OIL EMULSION OR SLUDGE WASTE TO SOIL TILLED TO A 12-CM DEPTH AS A LAND TREATMENT OPERATION[89]

Area Acres	Material	Amount of Material per Plot	Pure Ca(OH)$_2$[a] (kg)	C/N Ratio	Theoretical N (kg)	Urea 45% N (kg)	Theoretical[b] P (kg)	Application Rate
6.9	Emulsion	75,700 gal[c]	1200	20:1	5300	12,600	570	10,800 gal/6″/acre[e]
10.97	Emulsion	119,000 gal	1900	50:1	3310	7,370	330	10,800 gal/6″/acre
8.04	Emulsion	86,600 gal	1400	50:1	2420	5,360	240	10,800 gal/6″/acre
.73	Emulsion	7,884 gal	140	50:1	220	540	24	10,800 gal/6″/acre
0.96	Emulsion	10,368 gal	140	10:1	1455	2,700	120	10,800 gal/6″/acre
0.87	Emulsion	18,792 gal	250	20:1	1320	3,600	160	21,600 gal/6″/acre
0.91	Emulsion	19,656 gal	250	50:1	550	1,340	60	21,600 gal/6″/acre
1.08	Sludge	160 cu. yd.[d]	2200	20:1	3890	8,000	360	60 metric tons/6″/acre[f]
.96	Sludge	160 cu. yd.	2200	50:1	1400	3,200	144	60 metric tons/6″/acre
1.2	Sludge	190 cu. yd.	2200	50:1	1730	3,200	144	60 metric tons/6″/acre
1.02	Sludge	160 cu. yd.	2200	10:1	7340	16,000	720	60 metric tons/6″/acre
0.94	Emulsion	20,302 gal	350	20:1	1470	3,240	145	10,800 gal/6″/acre
0.84	Emulsion	18,144 gal	300	50:1	508	1,200	54	10,800 gal/6″/acre
0.84	Emulsion	18,144 gal	300	100:1	254	600	27	10,800 gal/6″/acre
app. 17	Sludge	1000 metric tons	17,100	20:1	30,000	67,000	3,000	60 metric tons/6″/acre

[a]   Includes approximately 125 kg Ca(OH)$_2$ per acre 6″ to stabilize soil pH.

[b]   Should be applied as commercially available potassium phosphate. Theoretical value should be multiplied by 10 to obtain approximate commercial weight of weight of fertilizer (0-10-0).

[c]   Multiply by 0.003785 to convert to m^3.

[d]   Multiply by 0.765 to convert to m^3

[e]   Multiply by 0.00935 to convert to m^3/15 cm/ha.

[f]   Multiply by 0.4047 to convert to metric tons/15 cm/ha.

Table 10

SELECTED ANALYSES OF PAPER MILL SLUDGE

Constituent	N	P	K	Ca	Mg
			%		
Paper mill sludge	2.33	0.50	0.74	1.53	0.15
Paper mill sludge	0.15	0.29	0.85	0.10	0.09
Paper mill sludge	0.27	0.16	0.44	1.59	0.10
Paper mill sludge	0.99	0.42	0.62	0.37	0.12
Paper mill sludge	0.62	0.29	0.52	0.85	0.09
Paper waste	0.37	0.19	0.58	0.98	0.09

Paper mill sludge data from Reference 32; paper waste data from Fuller, W. H., unpublished report to Southwest Forest Industry, Inc., Phoenix, 1981.

chain or underground waters. The greatest problem requiring adjustment with woody wastes is the C/N ratio (Table 11) as related to their use as a container medium for specimen plants, plant nutrient source, physical improvement of the soil or soil conditioner source of energy for microorganisms, or waste disposal. Except for leaves, the nitrogen is low providing a C/N ratio that exceeds 100.[33] Because mature wood resists rapid decomposition, the biodegradation of only a few tree woods (hardwoods) respond to N additions. Bark is more resistant to microbial attack than the woody materials but hardwood bark is more resistant than softwood.[34-36]

Loading rates of woody wastes depend somewhat on the particle size. Generally, the smaller the particle size, the greater permissible loading rate. Rates of application, again, depend on crops to be grown and timing. Loading rates do not depend on polluting or toxic constituents except in rare instances. California incense cedar wood, and white pine bark adversely affected pea seed germination and seedlings development at 1% or above, and the bark of California incense cedar and wood retarded pea (*Pisum sativum* L.) growth at 2% rate of addition. Red cedar wood and bark also retarded pea seedling growth at the 2% application rate.[33] Woody materials and bark should be applied to cultivated soils well in advance of planting (4 to 6 months, at least) with irrigation if rainfall is limited. Most woody materials excell as mulches. Loading rates may exceed those of animal manure because of the low soluble salt content.

Sawdust makes excellent plant container mix provided it is free of shavings and has been composted (moistened and kept damp) with soluble nitrogen and phosphate fertilizer for at least 2 warm months prior to mixing with soil and planting. A ratio of three parts composted sawdust to one part sandy clay loam soil or silty clay loam soil has provided excellent growing media.[37]

7. Cannery Wastes

Only a small part of cannery wastes occur as solids. Yet the disposal of this small quantity poses problems. The tendency to attract flies, insects, and vermin require such wastes to be incorporated into the soil.[38] Large applications of cannery sludge also pose a problem of incorporation, unless the loading rates are unusually light.[38] Where only one crop a year is grown, off-season applications can be made successfully, provided they are not placed on frozen soil or snow. Use of certain kinds of cannery wastes on pasture or forage land has been successful. For example, peach waste to Lakeland sand and Cecil sandy loam improved the productivity of coastal bermudagrass yields. Reed et al.[39] found tomato and fruit cannery wastes beneficial to wheat yields. Loading rates based on dry weight were not given. Suitable guidelines have yet to be developed.

8. Industrial Processing Wastes

Various industrial and miscellaneous processing wastes and their N, P, K, Ca, and Mg contents are listed in Table 12 for comparisons.[40] These natural material wastes offer few problems of disposal from a heavy metal or toxic pollutant standpoint. The main difficulty is associated with methods of incorporation and, in some instances, a very wide C/N ratio for the most desirable rate of biodegradation. Loading rates are similar to other natural plant residues as also the method of incorporation. Most of them will provide soil conditioning benefits.

9. Soil Amendments

Soil conditioning, through the addition of soil amendments, originated many years ago in the home garden and more recently spread to commercial agriculture.[41] Commercially, it is more common to arid and semiarid areas of the West as an aid in soil reclamation and to eliminate or counteract poor water movement in soils. Soil-conditioning also has roots

Table 11
THE CARBON AND NITROGEN COMPOSITION OF WOODS AND BARKS OF SOME TREE SPECIES[90]

Species	Source	Wood (%)			Bark (%)		
		C	N	C/N ratio	C	N	C/N ratio
Softwoods							
California incense cedar (*Libocedrus decurrens*)	California	51.1	0.097	526:1	51.8	0.038	1363:1
Redcedar (*Juniperus virginiana*)	Virginia	50.8	0.139	366:1	46.0	0.206	223:1
Cypress (*Taxodium distichum*)	North Carolina	50.3	0.057	883:1	47.3	0.324	146:1
Redwood (*Sequoia sempervirens*)	California	49.9	0.060	832:1	48.3	0.060	805:1
Western larch (*Larix occidentalis*)	Montana	48.6	0.180	270:1	49.9	0.161	310:1
Eastern hemlock (*Tsuga canadensis*)	Pennsylvania	48.5	0.106	458:1	51.1	0.060	852:1
Red fir (*Abies magnifica*)	California	48.2	0.227	212:1	49.1	0.259	190:1
White fir (*Abies concolor*)	California	44.8	0.045	996:1	51.8	0.135	384:1
Douglas fir (*Pseudotsuga menziesii*)	Oregon	48.1	0.051	943:1	52.7	0.041	1285:1
Engleman spruce (*Picea engelmanni*)	Montana	48.5	0.118	411:1	51.2	0.390	131:1
White pine (*Pinus strobus*)	Maine	48.3	0.087	555:1	51.5	0.101	510:1
Shortleaf pine (*Pinus echinata*)	Maryland	45.0	0.130	346:1	51.3	0.128	401:1
Loblolly pine (*Pinus taeda*)	South Carolina	48.7	0.068	716:1	50.9	0.082	621:1
Slash pine (*Pinus elliotti*)	Florida	49.2	0.050	984:1	52.1	0.056	930:1
Longleaf pine (*Pinus palustris*)	Louisiana	49.9	0.038	1313:1	50.2	0.092	546:1
Ponderosa Pine (*Pinus ponderosa*)	Oregon	45.1	0.052	867:1	51.8	0.048	1079:1
Western white pine (*Pinus monticola*)	Idaho	48.9	0.113	433:1	49.5	0.171	290:1
Lodgepole pine (*Pinus contorta*)	Montana	46.9	0.071	661:1	49.3	0.179	275:1
Sugar pine (*Pinus lambertiana*)	California	50.1	0.124	404:1	51.7	0.166	311:1
Hardwoods							
Black oak (*Quercus velutina*)	Illinois	47.3	0.070	676:1	44.5	0.102	436:1
White oak (*Quercus alba*)	Illinois	46.9	0.104	451:1	41.6	0.129	323:1
Red oak (*Quercus falcata*)	Louisiana	47.4	0.099	479:1	46.2	0.284	163:1
Post oak (*Quercus stellata*)	Louisiana	47.2	0.096	492:1	41.8	0.270	155:1

Table 11 (continued)
THE CARBON AND NITROGEN COMPOSITION OF WOODS AND BARKS OF
SOME TREE SPECIES[90]

Species	Source	Wood (%)			Bark (%)		
		C	N	C/N ratio	C	N	C/N ratio
Hardwoods							
Hickory (*Carya sp.*)	Arkansas	46.8	0.100	468:1	48.1	0.413	117:1
Red gum (*Liquidambar styraciflua*)	Arkansas	46.7	0.057	819:1	45.1	0.177	255:1
Yellow poplar (*Liriodendron tulipifera*)	North Carolina	47.1	0.088	535:1	47.6	0.351	136:1
Chestnut (*Castanea dentata*)	North Carolina	47.1	0.072	654:1	47.3	0.273	173:1
Black walnut (*Juglans nigra*)	South Carolina[a]	47.0	0.100	470:1	45.1	0.177	255:1
Averages		48.0	0.093	615:1	48.7	0.174	452:1

[a]   Bark from Maryland.

in multitudinous attempts to utilize wastes by disposal on the land to achieve some magical crop improvement. The desire to convert wastes to resources has a great appeal.

Conditioning is important to soils which have excesses of certain hazardous salts, poor internal drainage, high clay content, compacted layers, hardpans, or crusting problems. Soils adversely affected by low pH or highly acidic or high pH or highly alkaline conditions also respond to soil conditioning. The main purpose is to make the soil more favorable for water movement and aeration, to modify the soil pH for more favorable plant growth, and improve the soil structure with consequent better root growth and penetration. The purpose of conditioning is not to add nutrients, although it may favorably affect the nutrient status of the soil incidentally or indirectly.

The term soil amendment is used in this text, as distinguished from fertilizer sold in the market, to refer to a heterogeneous host of materials, organic as well as inorganic, which may influence biological, chemical, and physical conditions of the soil, although usually it refers to materials which primarily alter or change the soil's physical properties.[41] Soil amendments are academic orphans. They have received little scholarly attention except in association with reclamation, rate of water penetration, and excessive alkalinity and acidity.[42]

The function of a soil conditioning material is evaluated in terms of its favorable influence on:

- Soil moisture movement, retention, and availability
- Control of undesirable salt accumulation
- Soil temperature
- Soil pore space
- Soil microbial activity
- Nutrient availability
- Soil structure

Many hazardous solid wastes classify as soil amendments and, therefore, soil amendments qualify for attention in land treatment/utilization. For convenience, soil amendments and conditioners are grouped into general categories of like materials.[41]

Table 12
THE MOISTURE AND MAIN NUTRIENTS IN INDUSTRIAL AND MISCELLANEOUS WASTES (%)

Material	Details	Moisture	N	P	K	Mg	Ca	# of samples
Brewer's waste		6—94	0.9—3.2	0.04—0.18	0.83 (1)	—	0.02 (1)	2
Carbon (activated) waste		48	0.9	0.18	0.02	—	—	1
Castor meal		11	5.5	1.06	1.16	—	—	1
Chicory waste		—	1.5	—	—	—	—	1
Cockleshell		2	0.3	0.04	0.08	—	—	1
Cocoa waste		6—71 48	0.8—2.9 1.8	0.26—0.66 0.44	—	—	10.2 (2)	8
Coconut fiber/matting waste	Dust M	18	0.2—0.9 0.4	0.04	0.33—0.58 0.50	—	—	3
	Fiber M	(1)	0.1	—	—	—	—	1
Coffee waste	M	42—70 62	0.7—3.1 1.5	0.01—5.50 0.09	0.01—0.75	—	0.1	6
Coffee-chicory residue		66—67	0.8—1.7	0.26 (1)	0.01 (2)	0.04 (1)	—	2
Cotton cake	M	11	4.1—6.9 4.8 (3)	0.75 (2)	1.33—2.32 (2)	—	—	1
Felt waste		—	13.6					
Flax	Cavings & cleanings	6	—	1.58	7.64	—	—	1
	Residues from retting tanks (dried)	7	0.8	0.09	0.58	—	—	1
	Flax & jute dust	9						1
Ginger root		—	1.2 1.8	—	—	—	—	1
Glue factory waste		28—64 41	1.4—4.7 2.1	0.004—0.48 0.09	0.03—0.50 0.21	—	—	6

Table 12 (continued)

THE MOISTURE AND MAIN NUTRIENTS IN INDUSTRIAL AND MISCELLANEOUS WASTES (%)

Material	Details	Moisture	N	P	K	Mg	Ca	# of samples
Hemp waste		66—68 M (2)	0.7—1.5 / 0.9	—	—	—	—	4
Hops	Spent	9—87 M 73	0.6—5.7 / 1.1	0.09—1.50 / 0.13	0.005—2.16 / 0.08	0.12—0.16	—	9
Leather	Chamois dust	—	8.3	—	—	(2)	—	1
Leather	Hide meal, ground leather	19—68 M	0.4—7.5	0.04—0.17	0.02—0.25	—	—	4
Malt fiber		52 M	6.1 (7)	0.04	0.12	—	—	2
Paper waste		7 (1)	4.4—5.2	0.70—0.92	0.91—1.83	—	—	1
Sawdust		85	0.1	—	—	—	—	4
		4—66 M 48	0.1—0.9 / 0.2	0.004—0.22 / 0.03	0.03—1.16 / 0.06	—	—	
Tannery waste		6—85 M 55	0.1—14.1 / 5.3	—	—	—	—	10
Tobacco waste								
Woodshavings		71	0.4	0.04	0.75	—	—	1

ᵃ   Percent in material as received.

ᵇ   The range in composition and the median value M are given where possible; where the number of determinations differ from the total number of samples, the number is given in parentheses below the analysis.

Loehr, R. C., *Agricultural Waste Management: Problems, Processes and Approaches*, Academic Press, New York, N.Y., 1974, 51. With permission.

Table 13
THE AMOUNT OF VARIOUS AMENDMENTS
REQUIRED TO SUPPLY 100 POUNDS (45.4 kg) OF
SOLUBLE CALCIUM WHEN APPLIED UNDER
PROPER SOIL CONDITIONS

Amendment	Purity %	Pounds required to supply 1000 lbs. of soluble Ca
Gypsum ($CaSO_4 \cdot 2H_2O$)	100	4300
Sulfur (S)	100	800
Sulfuric Acid (H_2SO_4)	95	2600
Iron Sulfate ($FeSO_4 \cdot 7H_2O$)	100	6950
Lime-sulfur Solution (Ca-polysulfide)	24% or S[a]	3350

[a] Because Ca-polysulfide has indefinite chemical composition, purity is expressed in terms of sulfur (S) content.

Group I. Inorganic chemicals — sulfuric acid, sulfur, polysulfides, iron sulfates, and gypsum

Group II. Inert materials — sands, gravels, ground rock, exploded silicates, coal ash, fly ash, vermiculite, and glass wool

Group III. Materials of plant and animal origin — sawdust, forest debris, bark, paper pulp wastes, straw, woody stems, vines, branches, peat moss, and manures, composts, sewage sludges, dried blood, fish residues, guano, grass clippings, tannery and cannery wastes, and malt waste

Group IV. Synthetic organic compounds — surfactants (soaps), polyelectrolytes, and hydrolyzed polyacrylonitriles

Group V. Oils and oily wastes

The usual method of application is by spreading and working the amendment waste into the soil with the conventional agricultural tillage equipment to depths up to 15 cm depending on the material. Loading rates depend on the final result to be achieved in waste disposal, resource recovery, or improved crop production.

Group I includes inorganic chemicals that chemically react in the soil to provide soluble calcium for replacing excessive monovalent ions such as Na or K (Table 13).   They also flocculate the colloids and aid in the improvement of the soil structure, aeration, and pore size distribution.[43] Gypsum, sulfur (S), polysulfides, and sulfuric acid do not benefit all soils. Where water movement and pH are satisfactory for plant growth, and the soil does not need either Ca or Fe, these chemicals may not be beneficial to crop production. Crushed limestone is used on soils that have little calcium and already are acid rather than sulfur (S), sulfuric acid (H_2SO_4), and iron sulfate ($FeSO_4 \cdot 7H_2O$) that contain no Ca. These three acid-producing wastes are applied to calcereous soils or soils well supplied with calcium.

Certain mine tailing wastes may be reacted with sulfuric acid and used to correct copper, iron, and zinc deficiencies in grain sorghum (*Sorghum bicolar* L.) thus turning two wastes into useful resources when applied to loading rates of 2.24 t/ha to 15 cm.[44]

As a result of more stringent air pollution regulations, the sulfur oxides and sulfuric acid will appear more prominent as a surplus or waste. Land treatment appears as a valid alternative disposal method of this hazardous waste. Significant quantities may be used beneficially on western agricultural soils.[45] Loading rates of 2 to 6 t/ha (1 to 3 t/a) are applied to highly sodic soils for reclamation. Lesser amounts, ranging from 500 to 2000 kg/ha have been successful in improving vegetable crop production by making the P, Mn, Fe, or Zn available.[46] Massive loadings can be made to calcareous agricultural soils as a disposal sink according to Wallace.[47]

Industrial acid waste streams of nitric acid (HNO_3) and phosphoric (H_3PO_4) from automobile finishing may be neutralized with ammonia and the resulting supernatant soluble NH_4NO_3 used as a nitrogen source at the usual fertilizer loading rates. The heavy and hazardous metals precipitate out and are disposed as sludges in greatly reduced volume and solubility. Acids used on agricultural crops are most effective when applied in bands next to the crop or spotted where trace metals are most obviously lacking. Although this limits the disposal to low levels compared with general broadcasting, it is more cost-effective and permits successful continued use over a longer time.

Ferrous sulfate ($FeSO_4 \cdot 7H_2O$) reacts in soils to provide soluble iron for plants as well as acts as an acid soil conditioner. Loading rates for improving crop production must be modest if they are to be cost-effective. Soils will accept fairly large loading rates with no detriment to productivity.

Mine residues, certain mine tailings contain sulfides of Cu, Fe, Mn, and Zn that can be beneficial to crops growing on soils where these elements are poorly available.[44] Ferrous sulfide (pyrite) is the most prominent, but without pretreatment the level of available Fe is very low. Other iron-sulfur compounds are the double sulfides (pyrrhotite) and multi-sulfides (polysulfide). Some of these mine wastes can be placed directly onto the land at loading rates that vary from a few to many tons per hectare depending on the texture, pH, and crop selection. Coarse soils will tolerate greater loading rates than fine textured soils.

Pyrrhotite, unlike pyrite, oxidizes relatively rapidly by the action of the native soil microorganisms and will, therefore, provide immediate soil conditioning. Copiapite, a naturally occurring oxidation product of iron pyrite, oxidizes further and quite rapidly, to provide available Fe to plants in soils deficient in soluble iron.

Iron sulfates are byproducts of copper mining. They (ferrous and ferric iron) are quite soluble in water, but are readily oxidized to the hydrous ferric oxides when added to soils. The need for the iron sulfur compounds for better crop production in western states is quite limited in relationship to the supply of waste; but land treatment possibilities, as a disposal option, appear to be favorable in the West where calcareous soils dominate.

Group II, the more inert sand, gravel, stone, and crushed rock characteristic of this group, is used in rapidly growing quantities for landscaping and soil erosion control purposes (Figure 6).   Land treatment is highly specialized as are loading rates. Should it be desired to alter the texture (particle size distribution) of soil, for example, by as little as 10% for the upper 30 cm, it would take an equivalent of approximately 224 metric tons of material per hectare. Viewing it in another way, large disposals can be made with minimum effect on the natural soil characteristics favoring this means of waste treatment. Rock mulches are being used effectively as "surface armor" for soil and water erosion control of sloping landscapes (Figure 6).   All sizes of stones and gravel have been used successfully in landscaping of urban areas, roads, and highway medians.

Exploded silicate (perlite) waste can be applied to agricultural land with no significant detrimental effects at relatively high rates, however the wind hazard in redistributing the light material can be a problem even when mixed into the soil. Potted greenhouse plants grow quite favorably in perlite and soil mixes.

Coal ash, fly ash, and carbon black have been applied to soils with some success for special purposes as mulch. Their main use has been related to raising the soil temperature for better seed germination. Incorporation is more suitable for these as waste residues to prevent their redistribution by wind and water. Loading rates vary with the special uses and offer small problems compared with most other wastes compatible with land treatment.

Fly ash usually is alkaline in nature. Because of this, fly ash is attractive in neutralizing acid soils.[48] Other desirable uses are to improve the aeration, waterholding capacity, and

FIGURE 6.   Rock mulch covering sloping land as an erosion control practice.

provide certain trace metal nutrients to problem soils. Again, to be effective, their use must be highly selective. On the other hand, relatively large quantities may be applied to fine-textured soils with few or no detrimental effects. Jastrow et al.[48] found an alkaline fly ash to be helpful in soil reclamation of coal-refuse-affected land. The neutralizing capacity appears to be quite variable but generally less than lime.[49] Chang et al.[50] found that additions greater than 10% by volume to acid soils decreased the hydraulic conductivity and Jones and Amos[51] found both an increase and no change in plant available water depending on the original texture of the soil. Maximum loading rates to agricultural land has not been studied, since farmers are reluctant to apply such inert materials to their soils. Disposal on nonagricultural land, however, is well established.

Vermiculite has been established as a favorable medium for correcting the adverse effects of soil crusting and surface drying for better seed germination and seedling stand establishment. Mixing of vermiculite with soil to improve water holding capacity and aeration is well known. Loading rates for fine textured soils as well as for sands are only limited by its small supply as a waste material.

Group III — considerable attention has already been given to the most important materials classified in this group, i.e., manures, composts, municipal sludges, and forest products. The loading rates for cultivated land for some common crop residues depend on the C/N ratio of the material (Table 14). The C/N ratios can be adjusted by adding fertilizer nitrogen (as ammonium nitrate, urea, etc.) to the residue or soil to prevent nitrogen deficiencies in the succeeding crop. This is more important to double than single cropping. In general, loading rates of plant and animal wastes are well developed for the different regions of the U.S.A. by state agricultural experiment station research.

Group IV — certain wastes from the synthetic fiber industries have appeared as soil granulators or structure stabilizers.[52-54] Organic chemical, hydrolyzed polyacrylonitriles, and vinyl acetate malic acid compounds have been shown to be effective soil conditioners

Table 14

**GROUPING OF COMMON CROP RESIDUES ON
THE BASIS OF THEIR C/N RATIOS**

Residues	C	N	C/N ratio
	—%—		
High in both C and N	40	2.5	16/1
Legume residues			
Legume green manures			
Young cereal green manures			
Low C and high N	30	3.0	10/1
Composted strawy manure			
Animal feces			
High C and low N	40	0.5	80/1
Straw			
Stubble			
Strawy manure			
Mature cereal green manures			
Leaves, sawdust, or wood shavings			
Low C and low N	32	0.8	40/1
Composted straw			

although they do not represent a significant waste now as at one time in their history. Moreover, the quantities usable in agricultural production are very limited, and disposal as a waste to agricultural land is very limited. Decomposition of these residues, like plastic, is slow. Wetting agents also fall into the same classification. Surfactant appear to be short-lived and can be applied only infrequently because of cost. Usually, the disposals are confined to problem soils of slow water penetration because of the claims for improved water movement.

Application rates relating the organic solid waste (manures, composts, dried sludges, forest wastes, sawdust, etc.) as percent of amended soil in terms of percentage with depth of amended soil in inches are provided in Table 15 for convenience. Loading rates thus may be calculated by two different methods.

IV. SOIL MANAGEMENT OF LIQUID WASTES

Waste utilization on cultivated land should be manageable and the user of the liquid waste must realize some benefit. Soil management of organic wastewaters must include, as an important feature in the objectives, not only the protection of the environment by its practice, but realize actual improvement of the local environmental conditions, even though the latter cannot always be achieved immediately. The soil management of wastewaters applied to cultivated land, in addition to the obvious factors important to application, should consider the (a) degree of renovation, (b) best water reuse, (c) best land use, (d) pollution of underground water sources, and (e) long time effect on land productivity. To know what impact the management program is having on the environment, a monitoring program should be estabished.

General factors that influence management of liquid wastes should be emphasized briefly early in this section with a warning that operation practices in land treatment of wastewaters, even on cultivated land, are so dependent on local conditions that a detailed design of operations probably would inhibit rather than advance progress in this important disposal option. As a point of departure for the discussion, some general factors influencing good management practices should be identified. They are listed briefly as:

Table 15
SOIL AMENDMENT VOLUME TO ADD FOR VARIOUS
DEPTHS OF TREATMENT

Amendments as percent of amended soil	Depth of amended soil (inches)[a]						
	3	4	5	6	7	8	9
	(Cubic yards per 1000 square feet)[b]						
5	0.46	0.61	0.77	0.93	1.08	1.23	1.39
10	0.93	1.23	1.54	1.85	1.16	2.47	2.78
15	1.39	1.85	2.32	2.78	3.24	3.70	4.17
20	1.85	2.47	3.09	3.71	4.32	4.94	5.55
25	2.32	3.08	3.86	4.63	5.40	6.17	6.95
30	2.78	3.70	4.64	5.56	6.48	7.41	8.33
35	3.24	4.32	5.40	6.48	7.57	8.64	9.72
40	3.70	4.94	6.18	7.41	8.64	9.88	11.13
45	4.17	5.55	6.95	8.33	9.72	11.10	12.52
50	4.63	6.17	7.72	9.26	10.80	12.34	13.88

[a] Multiply by 2.54 to convert to cm.
[b] Multiply by 0.00823 to convert to m^3/m^2.

- The rate of water application
- The climatic patterns, particularly rainfall and temperature
- The land topography as it relates to soil and water erosion control and runoff
- The methods of soil incorporation
- The permeability of the soil (texture and structure)
- Adequate drainage both internally and externally
- The established and new crop production patterns
- The soil fertilizer program
- The mechanical land program of planting, tilling, and harvesting
- The monitoring for pollution and environmental impact

A. Application Mode for Wastewaters on Cultivated Land

The selection of a method for the application of organic wastewaters to cultivated land depends on a great number of factors and local conditions, as does the rate of loading (Tables 16 and 17). Application may be (a) made directly to the soil surface or (b) incorporated into the upper few centimeters (inches) by injection or furrow placement.

1. Surface Application (Methods)

A great variety of methods and equipment for application are available (Tables 16 and 17). Irrigation may be by (a) spray (sprinkler), (b) ridge-and-furrow spreading, (c) over-land-flow, and (d) rapid-infiltration methods (Figure 7). Truck-tank, farm-tank wagon and tractor, flexible irrigation hose with plow-furrow or disc cover, tank truck with plow-furrow cover and farm-tank wagon and tractor represent a few of the equipment options that have been used successfully for surface mode of application. General methods of surface application listed for various parts of the U.S.A. in Table 17 include all of those in Table 16 with the addition of rapid infiltration whereby water is intended to be filtered and purified by the soil and sent to the underground source as recharge. Figure 7 illustrates what is involved by these surface application approaches. Although infiltration-percolation begins

Table 16
SOME SUGGESTED SURFACE APPLICATION METHODS OF WASTEWATERS, CONTAINING ORGANIC WASTES, TO CULTIVATED LAND

Method	Characteristics	Topographical and seasonal suitability
Irrigation		
Spray (Sprinkler)	Large orifice required on nozzle; large power and lower labor requirement; wide selection of commercial equipment available; sludge must be flushed from pipes when irrigation completed.	Can be used on sloping land; can be used year-round if the pipe is drained in winter; not suitable for application to some crops during growing season; odor (aerosol) nuisance may occur.
Ridge and furrow	Land preparation needed; lower power requirements than spray.	Between 0.5 and 1.5% slope depending on percent solids; can be used between rows of crops.
Overland flow	Used on sloping ground with vegetation with no runoff permitted; suitable for emergency operation; difficult to get uniform areal application.	Can be applied from ridge roads.
Tank Truck	Capacity 1,892 to more than 7,500ℓ; larger volume trucks will require flotation tires; can use with temporary irrigation set-up; with pump discharge can spray from roadway onto field.	Tillable land; not usable with row crops or on soft ground.
Farm Tank Wagon and Tractor	Capacity, 1,900 to 12,000ℓ; larger volume will require flotation tires; can use with temporary irrigation set-up; with pump discharge can spray from roadway onto field.	Tillable land; not usable with row crops or on soft ground.
Flexible irrigation hose with plow furrow or disc cover	Use with pipeline or tank truck with pressure discharge; hose connected to manifold discharge on plow or disc.	Tillable land; not usable on wet or frozen ground.
Tank truck with plow furrow cover	1,100ℓ commercial equipment available; sludge discharged in furrow ahead of plow mounted on rear of 4-wheel-drive truck.	Tillable land; not usable on wet or frozen ground.
Farm tank wagon and tractor Plow furrow cover	Sludge discharged into furrow ahead of plow mounted on tank trailer-application of 380 to 500 wet t/ha; or sludge spread in narrow bank on ground surface and immediately plowed under — application of 110 to 300 wet t/ha.	Tillable land; not usable on wet or frozen ground.

White, R. K., Utilization of Municipal Sewage Sludge on Forested and Disturbed Land, The Pennsylvania State University Press, University Park, 1979, 285.

as a surface application, it results in a recharge method for underground water. Therefore, the organic wastes in the water and presence of hazardous pollutants and heavy metals must be limited to that amount which can be completely biodegraded in the microbially active layer(s) of soil. The soil layers between the surface and the capillary fringes of the under-

Table 17

WASTEWATER MANAGEMENT RELATING DISPOSAL LOCATION, TYPE OF ORGANIC WASTE, MODE, RATE, TIME OF APPLICATION, AND GROUND COVER

Location	Type waste	Application			Ground cover
		Mode[a]	Rate[b]	Period	
Penn State College State College, Pa.	Domestic	SI	2"/acre—wk @ 0.25"/hr	32 wk—crops 52 wk—forest	Mixed crop land and forest land
Campbell Soup Co. Paris, Tex.	Cannery	OR	2.5"/acre—wk summer, 1.25"/acre—wk winter; at 0.5"/day summer, 0.25"/day winter	52 wk	Crop land (hay)
Muskegon, Mich. (under construction)	Municipal	SI	2"/acre—wk, max. 4"/wk rain plus spray	35 wk	Crop land (corn)
Flushing Meadows Project Phoenix, Ariz.	Municipal	RI	400'/acre—yr @ 2 wk wet followed by 10 days (summer) or 20 days (winter) dry—up	52 wk	Open basin with Bermuda grass cover
Santee, Calif.	Domestic	RI	110'/acre—yr intermittent spread and dry	52 wk	3-acre spreading area
Whittier Narrows Los Angeles, Calif.	Municipal	RI	15 mgd	52 wk	—
South Lake Tahoe, Calif.	Domestic	OR	4.5"/acre—wk @ 2 days spray, 4—5 days rest	Study periods winter and summer 1963—1965	Forest land
North Lake Tahoe, Calif.	Domestic	RI	Approx. 1.25 mgd 17 acre site with 15,000 LF percolation trenches	52 wk	Volcanic cinder cone
Celotex Corporation L'Anse, Mich.	Industrial (insulation board)	SI	3.8"/acre—wk @ .035"/hr, 16 hr/day	24 wk	Crop land (Reed canary grass)
Seabrook Farms Seabrook, N.J.	Frozen vegetable	RI	10"/acre—wk @ .25"/hr, 5 days/wk	—	Original oak forest altered to marsh grasses

Table 17 (continued)
WASTEWATER MANAGEMENT RELATING DISPOSAL LOCATION, TYPE OF ORGANIC WASTE, MODE, RATE, TIME OF APPLICATION, AND GROUND COVER

Location	Type waste	Mode[a]	Application		Ground cover
			Rate[b]	Period	
Westby, Wis.	Municipal	RF	21"/acre—wk, 2 wk on, 2 wk off	52 wk	Reed canary grass (burned each spring)
Sunkist Growers, Inc. Corona, Calif.	Lemon processing	RF	6"/acre, load in 1 day rest and dry 2 wk	14 wk canning season	Bare soil during spreading season. Crops planted during 38-wk dormant season
Campbell Soup Co. Napoleon, Ohio	Cannery (tomato)	OR	7"/acre-wk @ .083"/hr, 6 hr on, 6 hr off	50-day canning season	Reed canary grass, seaside bent and red top
Campbell Soup Co. Chestertown, Md.	Cannery (poultry)	OR	Similar to Napoleon, Ohio	52 wk	Similar to Napoleon, Ohio
Howard Paper Mills Urbana, Ohio	Industrial (paper)	SI	1.3"/acre—wk @ 0.2"/hr	52 wk	Alfalfa hay cover. Underdrains for recapture
Sunapee State Park Mt. Sunapee, N.H.	Domestic	SI	2"/acre—wk @ 0.25"/hr	20 wk	Forested
Beardmore and Company Toronto, Ontario, Canada	Industrial (tannery)	SI	6"/acre—wk @ 0.2"/hr	30 wk	Reed canary, bone and timothy grass
Shoemaker's Dairies Bridgeton, N.H.	Industrial (milk processing)	SI	0.2"/acre—wk @ 0.02"/hr	52 wk	Hay and grasses
Commercial Solvents Corp. Terre Haute, Ind.	Industrial (fermentation)	SI	2"/acre in one 10-hr spray followed by 2-wk rest	52 wk	Bare ground
H. J. Heinz Co. Salem, N.J.	Cannery (tomato)	SI	0.5"/acre—wk @ 0.03"/hr	Canning season	Reed canary grass
Riegel Paper Co. Hughesville, N.J.	Industrial (paper)	SI	3.2"/acre—wk @ .2"/hr for 4 hr and 20—hr rest	52 wk	Grass
Green Valley Farms Avondale, Pa.	Cattle wastes	SI	2"/acre-wk @ 0.25"/hr	52 wk	Crop land (hay) in summer, forest in winter
Masonite Corp. Towanda, Pa.	Industrial (wall board)	SI	Approx. 0.5"/acre—wk @ .05"/hr	52 wk	Reed canary grass

Table 17 (continued)
WASTEWATER MANAGEMENT RELATING DISPOSAL LOCATION, TYPE OF ORGANIC WASTE, MODE, RATE, TIME OF APPLICATION, AND GROUND COVER

[a] Application mode
SI Spray irrigation
OR Overland runoff
RI Rapid infiltration
RF Ridge and furrow spreading

[b] Multiply "/acre (i.e. in/ac) by 6.276 to convert to cm/ha. Multiply "/acre (i.e. ft/ac) by 0.753 to convert to m/ha. Multiply mgd by 3785 to convert to m³/d.

Reed, A. D., Wildman, W. E., Seyman, W. S., Ayers, R. S., Prato, J. D., *Calif. Agric.*, 27, 6, 1973. With permission.

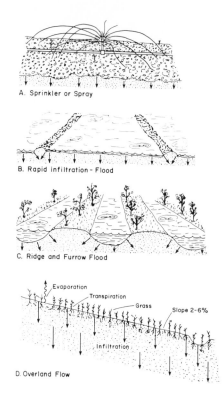

A. Sprinkler or Spray

B. Rapid infiltration - Flood

C. Ridge and Furrow Flood

D. Overland Flow

FIGURE 7.   Illustration of wastewater application
to the land for treatment purposes.

Table 18
LOADING RATES OF WATER WITH CONTAINED ORGANIC WASTES BY DIFFERENT IRRIGATION SYSTEMS TO SOILS

Rate of water addition cm/day	Irrigation system	Soil texture	Cover crop	Industrial process
12.5	Ridge and furrow	—	Yes	Dairy wastes
15.2	Ridge and furrow	—	Yes	Citrus wastes
1.3	Spray on slope	—	Yes	Food processing
13.7	Spray	—	Yes	Food processing
12.7	Ridge and furrow	Loam	Yes	Food processing
4.2	Spray	—	Woods	Vegetable cannery
1.3	Spray	Clay	Yes	Vegetable cannery
5.5	Ridge and furrow	Sandy Loam	Yes	Vegetable cannery

Norstadt, F. A., Swanson, N. P., and Sabey, B. R., *Soils for Management of Organic Wastes and Waste Waters*, American Society of Agronomy, Madison, Wis., 1977, 349. With permission.

ground water must be capable of removing (renovating) pollutants from the wastewaters applied. Thus, many kinds of equipment provide many methods of application with many levels of loading rates and no single set of management options can rigidly be set above another for all cases.

The rate of water addition as given in Table 18 again illustrates the diversity in the state of the art and the variability in the amount of water added per day, the kind of irrigation systems, the soil texture effects, kind of crop cover, and industrial wastewater being put onto the land. The rate of water application, for example, varies 11- or 12-fold.

2. Soil Incorporation

Most soils are notably deficient in soil organic matter as a natural circumstance of climate, and as a consequence, the nitrogen fertility status of soil also is low. Nitrogen is the single most limiting nutrient element in the achievement of maximum crop production. Water, of course, is the one great natural resource limiting plant growth in arid, semiarid, and subhumid lands. The ultimate means of correcting these deficiencies and perpetuating food, fiber and organic raw materials production is in the recycling of the waste materials and waters back to the land. Indeed, a great proportion of crop residues returns to the land as a natural result of agricultural operations. This is considered to be good economy and essential for maintaining maximum soil productivity and crop yields. Plant residue, food, and animal waste recycling have reached even a more sophisticated stage of compost production for land improvement as a means of conserving soil fertility and reuse of wastes.[22,24,28] Municipal sludges also are classified as biodegradable organic waste suitable for land application.

Industrial wastewaters and sludges are generated in an ever increasing volume and some create unusual problems of suitable disposal because they are noxious, odorous, and/or highly volatile. With proper management, some of them could function as potential resources and not as wastes. Approximately 4.5 billion dry kg/yr (10 billion dry lb/yr) of sludge are produced at municipal wastewater plants in the U.S. This amount is expected to double by the late 1980s.[55,56]

Data in Table 19,   reproduced from compilations by Phung et al.,[9] demonstrates the accelerated generation of wastewaters expected through 1985. There is an urgent need for viable alternatives to those that dominate present operations. One promising option is soil injection. Subsurface injection of liquid fertilizers may be considered a forerunner of the soil injection of wastewaters, waste slurries, and sludges. Subsurface applications involve the same wastes as surface applications but have the added advantage of prudent disposal of those organic wastes that pose a risk factor in potential odor nuisance, runoff and spreading beyond the confines of the disposal area, and dispersion of toxic volatiles.

Subsoil applications may be made by two main methods (a) injection ahead of a plow blade or (b) direct injection beneath the soil surface with powerful truck-tank injection probes. For example, liquid manure manufacturers can supply soil injection attachments for placing liquid wastes up to 45 cm below the land surface with minimal modification to the present tankwagons.[57] The "plow-furrow-cover" technique, another agronomic practice, consists of applying the waste in a deep furrow (15 to 30 cm) and covering the furrow by trailing a moldboard plow attachment. Several variations of this procedure depend on the kind of wastes that are being land-treated and the inventiveness of the crop producer with the equipment available. By modifying the land treatment into three operations, for example, energy can be conserved. One operation opens up a furrow, a second places the waste into the furrow, and a third covers. By combining the first and last operation, energy can be conserved. As a new furrow is opened, it covers the first after the wastewater or slurry is applied.

Soil injection now can be performed in one operation with the development of new tank trucks equipped with high pressure injection hose connections to knives capable of delivering wastewaters containing solids up to 8% to depths ranging up to 45 cm (Figure 8).   Although this new soil injection technique is less economical than surface application methods, there are certain advantages, such as:

- Solves waste disposal problems prudently
- Eliminates the nuisance of offensive odors

Table 19
ESTIMATED QUANTITIES OF INDUSTRIAL WASTEWATERS AND SLUDGES SUITABLE FOR LAND CULTIVATION

Industry	Waste type	Waste quantities[a]		
		1975	1980	1985
Food and kindred products				
Meat packing	Wastewater[b]	150	170—190	190—230
Poultry dressing plants	Wastewater[b]	87	98—110	110—140
Dairy products	Wastewater[b]	45	51—57	58—70
Fruits and vegetables	Wastewater[b]	280	320—350	360—440
Cane sugar	Wastewater[b]	68	79—87	89—110
Malt beverages	Wastewater[b]	87	98—110	110—140
Textile mill products				
Textile finishing	Secondary wastewater treatment sludge[c]	3.9×10^4	8.5×10^4	1.8×10^5
Lumber and wood products				
Wood preserving	Wastewater[b]	1.9	2.2—2.3	2.5—2.9
Paper and allied products	Primary wastewater	1.4×10^6	1.6×10^6	1.8×10^6
	treatment sludge[c]	$1.7 \times 10^{6-}$	$1.9 \times 10^{6-}$	$2.1 \times 10^{6-}$
Chemicals and allied products				
Organic fibers, noncellulosic	Secondary wastewater treatment sludge[c]	4,800	7,300	9,800
Pharmaceuticals	Waste mycelium[c]	6.3×10^4	8.1×10^4	1.0×10^5
Soap and other detergents	Wastewater[b]	12—14	17—18	20—22
Organic chemicals	Wastewater treatment sludges[c]	5.5×10^6	6.8×10^6	8.5×10^6
Petroleum refining and related industries				
Petroleum refining	Non-leaded product tank bottoms[c]	4.1×10^4	5.1×10^4	6.2×10^4
	Waste bio sludge[c]	4.0×10^4	5.0×10^4	6.2×10^4
	API separator sludge[c]	3.4×10^4	4.2×10^4	5.2×10^4
	Dissolved air flotation float[c]	2.9×10^4	3.7×10^4	4.5×10^4
	Slop oil emulsion solids[c]	1.7×10^4	2.1×10^4	2.6×10^4
	Crude tank sludge[c]	390	490	610
Leather and leather products				
Leather tanning and finishing (vegetable)	Wastewater[b]	2.5	2.3	2.2
	Secondary wastewater treatment sludge[c]	820	740	680
Total wastewater[b]		735	840—920	940—1,160
Total sludge[c]		7.2—	8.8—	10.8—
		7.5×10^6	9.1×10^6	11.1×10^6

[a] Waste quantities for 1975 based on ref. 1 to 15 and contacts with industry representatives and current researchers. Values for 1980 assume the 1975 relationship between production and waste generation, and are based on industry-specific production projections (if available).

[b] Wastewater quantities are given in units of millions of cubic meters per year.

[c] Sludge quantities are given in units of metric tons per year on a dry weight basis.

Phung, T., Barker, L., Ross, D., and Bauer, D., U.S. Environmental Protection Agency, Cincinnati, Ohio, 1978, 205.

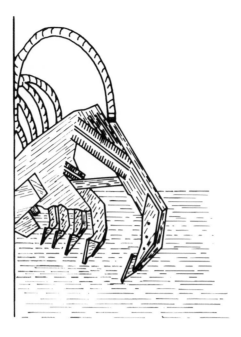

FIGURE 8. Soil injection technique for waste-waters, slurries, and semi-liquids.

- Eliminates problems with flies, vermin, and rats
- Minimizes erosion and spreading of surface runoff
- Effectively reduces loss of ammonia
- Improves the soil aeration and water intake
- Minimizes volatilization of potential hazardous vapors and gases
- Conserves water (particularly important in arid and semiarid lands)
- Places plant nutrients well within the root zone
- Places organic matter in the root zone area for better root growth and penetration

B. Wastewater Loading Rates

Wastewaters are limited in land loading not only by quantity of organic matter they contain, but by the liquid vehicle of transport itself. Therefore, most loading rates of natural organic wastes are calculated on a basis of biological oxygen demand (BOD) and amount of water discharged to the land. Some examples of industrial organic wastewater loading rates for soil as related to BOD and N, P, and K are presented in Table 20 as compiled by Norstadt et al.[58] These data may be compared to those summarized by Wallace, Table 21, for liquid organic wastes. Neither of these tables takes into account liquid fluids other than water.[59]

Organic solvents, many of which are highly toxic like trichloroethylene (TCE), cannot be loaded onto the land at the rates suggested above, nor can they be placed on the land during crop production. Highly toxic and hazardous wastes and aqueous wastes that contain toxic substances require another set of criteria for land treatment than BOD, and amount of liquid applied as related to permeability and rate of internal soil drainage. Temporarily, this problem is not being raised with much vigor since wastewater disposal on cultivated land should not contain highly toxic fluids or organic chemicals until more is known about their rate of degradation in soil, phytotoxicity, effects on microbial inhibition, and movement through the soil. Because of the wide diversity of solvents and great number of toxic organic compounds, industrial hazardous waste fluids require evaluation almost on a case-by-case basis, a very time consuming task. The end result cannot lend itself to a simple loading rate

Table 20
LOADING RATE OF ORGANIC WASTE TO SOILS

Waste description		Rates of amendments				
				Nutrients (kg ha^{-1} year^{-1})		
Kind	State	Waste[a]	BOD mg/liter	N	P	K
Cattle feedlot lagoon water	Liquid	30	1,400[b]	510	130	1,460
Cattle feedlot lagoon water	Liquid	30	—	371	73	—
Meatpacking waste water	Liquid	528	1,340	6,500	900	—
Prepared dinners	Liquid	528	1,900	2,400	1,100	—
Dressings, sauces, spreads	Liquid	528	2,600	740	580	—
Meat specialties	Liquid	528	820	2,500	580	—
Fish and vegetable	Liquid	395	387	1,820	560	—
Potato processing	Liquid	234[c]	1,442	1,357	—	—
Fruits and vegetables	Liquid	300	582[d]	510	105	—
Cheese and powdered milk	Liquid	150	1,062[d]	570	510	—

[a] Rate of waste amendments cm/year for liquids. (Some values taken from articles abstracted in U.S. Environmental Protection Agency, 1974.).
[b] This example of COD from Kreis et al. (1972).
[c] Average values for 2 years of data and COD value instead of BOD.
[d] COD value instead of BOD.

Norstadt, F. A., Swanson, N. P., and Sabey, B. R., *Soils for Management of Organic Waste and Wastewaters*, American Society of Agronomy, Madison, Wis., 1977, 349. With permission.

criterion as it can for wastewaters containing natural plant and animal wastes, such as listed in   Tables 20 and 21.   The limitation on loading of highly toxic and hazardous solvents and chemical compounds cannot be based strictly on any single soil criterion either, such as permeability alone. Land (soil) treatment of such hazardous wastes, therefore, must be considered independently of the other organics.

1. Criteria Determining Wastewater Loading

Wastewaters acceptable for disposal on cultivated land should meet certain criteria of quality. Superimposed on these criteria are those of loading rates. Assuming that the wastewaters meet the quality criteria of (a) capability of biodegradation of solids or soluble components, (b) no long-term toxicity to plants or microorganisms, (c) each migration at practical rates of application to the groundwater, (d) no adverse influence on the natural physical and chemical properties of the soil at reasonable rates of application, and (e) no long-term limitation of agricultural productivity of the land. Some kinds of industrial wastes and accompanying wastewaters that can meet these criteria originate from industries of:

- Paper pulp and paper production
- Wood preserving
- Textile finishing
- Organic chemicals production
- Leather tanning and finishing
- Food processing including meatpacking wastewater
- Petroleum refining

Table 21
SUMMARY OF HYDRAULIC AND ORGANIC LOADING RATES USED IN EXISTING LAND APPLICATION SYSTEMS FOR INDUSTRIAL WASTES[a]

Type of Waste	Hydraulic Load		Organic Load (BOD)
	gal/ac/day[a]	in/wk	lb/ac/day[b]
Biological chemicals	1,500	0.39	370
Fermentation beers	1,350	0.35	170
Vegetable tanning			
Summer	54,000	13.91	360
Winter	8,100	2.09	54
Wood distillation	6,850	1.76	310
Nylon	1,700	0.44	287
Yeast water	15,100	3.89	—
Insulation board	14,800	3.81	138
Hardboard	6,000	1.55	85
Boardmill whitewater	15,100	3.89	38
Kraft mill effluent	14,000	3.61	26
RI[c]	350,000	90.13	120
Semichemical effluent	72,000	18.54	90—210
Paperboard	7,600	1.96	13—30
Drinking	32,400	8.34	108
Poultry	40,000	10.30	100
Peas and corn			
57 day pack	49,000	12.62	238
35 day pack	34,400	8.86	2,020
Dairy			
Low value	2,500	0.64	10
High value	30,000	7.73	1,000
Soup	6,750	1.74	48
Steam peel potato	19,000	4.89	80
Instant coffee and tea	5,800	1.49	92
Citrus	3,100	0.80	51—346
Cooling water — aluminum casting (RI)	95,000	24.46	35

[a]   Multiply by 9.35×10^{-3} to convert to m^3/ha/day.
[b]   Multiply by 0.89 to convert to kg/ha/day.
[c]   RI — Rapid Infiltration.

From Wallace, A. T., in *Land Treatment and Disposal of Municipal and Industrial Wastewaters*, Sanks, R. L. and Asono, R., Eds., Ann Arbor Science Publishers, Inc., Ann Arbor, Mich., 1976, 147. With permission.

Specific factors to consider for industrial wastewater loading rates on land relate mostly to the concentration of potentially toxic and hazardous components and the nature of the environment. Briefly listed, they are;

A. Wastewater quality
   Organic matter (BOD, chemical oxidizable constituents, COD, and total organic carbon, TOC)
   Heavy metals

Total dissolved solids (TDS)
Suspended solids
Nitrogen (several forms)
Phosphorus
Sodium-absorption-ratio (SAR)
Boron
Bacteriological composition (human and animal diseases)
Intestinal parasites
Organic chemicals (toxic, hazardous, and carcinogens)
Organic solvents (volatility, aerosol drift, polar and nonpolar)
BPT constituents (best practice treatment technology)

B. Soil quality
   Texture
   Structure
   Permeability, infiltration
   Presence of confining soil barriers
   Depth to water table
   Drainage

C. Climate
   Rainfall amount and intensity factor
   Temperature
   Wind velocities and direction
   Evapotranspiration

D. Topography
   Ground slope
   Soil and water erosion potential
   Flood hazard
   Topography of watershed

E. Geologic formation
   Depth to bedrock
   Limestones

F. Groundwater
   Depth to groundwater
   Direction and rate of flow
   Perched water tables
   Location, depth, and quality of wells

These same criteria relate to site selection and will be discussed fully in that section of the text along with other site factors.

2. Wastewater Quality and Loading Rates

Organic matter in wastewaters usually is evaluated by the BOD and TOC. Certain organic synthetic compounds like those of plastic origin decompose so slowly the BOD test is not always applicable. Organics that are highly toxic to soil microorganisms and very resistant to degradation such as DDT also affect the loading rate. In these instances, the TOC (and Cl) may be used to quantify the rate of loading or aid in deciding whether or not the organic

Table 22
TYPICAL WASTE LOADING AT DESIGNATED RATES
FOR LAND TREATMENT

	Load (lb/acre)			
	SI or OR		RI	
Material	Per week[a]	Per year	Per week[b]	Per year
BOD	11.1	330	240	1800
Organic N	0.9	27	18.5	139
+ NH₄	4.4	132	91	680
− NO₂	0.0	0	0	0
− NO₃	3.8	114	76	570
Total N	9.1	273	185.5	1389
Phosphorus	4.5	135	92	700
Phenols	0.1	3	2.8	21
Refractory organics	20.0	600	415	3112
Cadmium	0.04	1.2	0.9	6.8
Chromium	0.09	2.7	1.8	13.5
Copper	0.04	1.2	0.9	6.8
Iron	0.04	1.2	0.9	6.8
Lead	0.04	1.2	0.9	6.8
Manganese	0.09	2.7	1.8	13.5
Nickel	0.09	2.7	1.8	13.5
Zinc	0.09	2.7	1.8	13.5
Boron	0.3	9	6.5	49
Chlorides	45.0	1350	900	6750
Sulfate	56.0	2520	1150	8630

Note: Values for other loading rates or application seasons can be calculated
directly. Example:
  Find BOD load at 3 in./wk for 40 weeks
    $(3/2)(11.1) = 16.7$ lb/wk
    $(16.7)(40) = 668$ lb/year
For other waste concentrations:
  Multiply mg/liter concentration by 0.45 to produce a load in pounds
  per acre at a 2 in./wk rate.

[a] Assumed for this table, not necessarily optimum for any location. At 2 in./
week, 30 week season.
[b] At 6 in./day for 14 days, 14 days rest during total 30 week season.

Reed, A. D., Wildman, W. E., Seyman, W. S., Ayers, R. S., Prato, J. D., and
Rauschkolb, S. R., *Calif. Agric.*, 27, 6, 1973. With permission.

matter is amenable to land treatment. Some organic chemicals are so highly toxic they should
not be allowed on cultivated land without some detoxification. Most of the research to
determine optimum loading rates has been done with wastewaters having their origin in plant
and animal wastes. A few data are reproduced in Tables 20 and 21.   Some of the
constraints on loading rates of those wastewaters that qualify for land treatment may profitably
be discussed in more detail.

Biodegradation of organic wastes is vitally important to both initial and subsequent land
loading. The rate of decomposition determines the frequency of application. The BOD test
(Tables 22 and 23) is often used to evaluate biodegradation rate, whereas COD and TOC
measure the amount of organic matter available. Respiration tests as represented by the rate
of CO_2-evolution from wastewater-treated soil also are used to evaluate the extent of biodegradation of the organic constituents.[53,60] The loss of carbon as CO_2 can be equated to loss

Table 23
ORGANIC LOADING RATES IN TERMS OF POUNDS BOD/ACRE/YEAR

Site location or reference	Type of waste	Wastewater flow, mgd[c]	Average BOD, mg/ℓ	Average hydraulic loading in/day[d]	lb BOD/acre/day[e]
Seabrook, New Jersey	Vegetable processing	16	—	12	—
[61]	Cannery	—	—	6.4	210
[122]	Corn processing	—	—	3.4	860
Bridgeton, New Jersey	Tomato processing	3	—	2.5	—
[63]	Kraft mill	1	1,000	1.8	420
Sumter, So. Carolina	Poultry processing	3.75	520	0.9	110
L'Anse, Michigan	Board mill	0.6	1,200	0.55	138
[4]	Citrus processing	0.82	3,300[a]	0.45	350
Acton, Canada	Tannery	1.25	800	0.45	50 (600)[b]
Firth, Idaho	Potato processing	0.63	1,200	0.3	78
Fremont, Michigan	Food processing	0.8	1,000	0.3	74
Fairmont, Minnesota	Food processing	1.5	1,500	0.1	47

[a] Estimated from COD measurements.
[b] Loading during spring.
[c] Multiply by 3785 to convert to m³/d.
[d] Multiply by 2.54 to convert to cm/d.
[e] Multiply by 0.89 to convert to kg/ha/d.

Pound, C. E. and Crite, R. W., U.S. Environmental Protection Agency, Cincinnati, Ohio, 1973, 281.

in organic matter from the waste source. The more rapid the rate of CO_2 loss, the more rapid is the rate of decomposition or biodegradation. The CO_2-evolution test also is used to detect the presence of inhibition of microbial activity due to the presence of some toxic substances or to unfavorable environmental conditions. The optimum and maximum loading rate in soils may be readily determined by comparing the CO_2 evolved for given levels of wastewater-treated soils over selected periods of time. Tables 20 and 21 provide a comparison of the organic load (BOD) of wastewaters from a number of industries. There is a limit in the amount of readily decomposable organic matter a soil can accept before it becomes inefficient and stagnates due to lack of available oxygen just as in any burning or oxidative process. Where plants are growing, anaerobosis of the soil greatly inhibits growth and finally leads to complete killing.

The salt contents of wastewaters are highly variable. The adverse effects of salty soils on crops are quite well known to the western growers. Irrigation practices take into account salinity and alkali control. In humid, temperate climates where rainfall provides abundant leaching, there has been little need for this kind of knowledge. The use of wastewaters that contain high concentrations of salts require closer attention in designating suitable loading rates. Some plants are sensitive to even low levels of salt. The USDA Salinity Laboratory provides some valuable information on the relative sensitivity of a wide range of food and fiber plants to soluble salts in soils (Table 24). Thus, soils and wastewaters vary in salt content and plants vary in sensitivity to soluble salts. All of this influences the recommendations of loading rates for wastewaters on cultivated soils.

Heavy metals and trace elements also limit the amount of irrigation water or wastewater that can be applied to cultivated land (Table 25). Page[61] shows the FWPCA (Federal Water Pollution Control Administration, 1968)[62] surface water criteria to be virtually the same as the U.S. drinking water standards. Later data (1973) suggest different water quality criteria for coarse-textured (sands) and fine-textured (clays) soils.[63] These data along with those for fine-textured soils and short-time use only are given in Table 26 for comparative purposes and to emphasize the importance of soil particle size. The FWPCA relaxed its earlier water quality recommendations because more recent research information shows that metal attenuation in soils is much greater in fine (clays) than coarse textured soils.

Since cation exchange capacity (CEC) in mineral soils is correlated with particle size distribution of soil (texture), and surface area of the soil particles per unit weight, some soil scientists relate the loading rate of sludge metals allowed on cultivated land with this property (Table 2). This scheme for loading rates also takes into consideration the differences in metal attenuation by soils. The amount that may be applied is greater for soils with larger CEC, with the metals following in allowable amounts of Pb > Zn > Cu > Ni > Cd. Although these same principles will hold for metals in other wastes and wastewaters, the specific maximum allowable amounts of each will vary with the kind and nature of the waste and projected land use. Wastewaters with high metal contamination, one of the listed criteria, may be met with a single application. With better quality waste, several or many applications may be made.

Nitrogen is another factor to be considered in wastewater loading rates when the land is to be cropped. Either too much or too little nitrogen results in poor crop production. Wastewaters with too little nitrogen (i.e., wide C/N ratio) can readily be corrected by supplementary applications of nitrogen fertilizers. Excessive nitrogen in the form of nitrate can accumulate in the groundwater as a pollutant since it is wholly soluble and mobile. Nitrates move through soil with the water and with little attenuation. Excess nitrogen in the system may require limiting the loading rate to the amount of N the plants can assimilate. Loading rates in terms of amount of N for different crops and different locations appear in Table 27.[64] Obviously the nitrogen loading rates differ markedly depending on (a) the N content of the wastewater, (b) the liquid loading rate, (c) location, and (d) plants grown (Table

Table 24
YIELD DECREMENT TO BE EXPECTED FOR FIELD AND FORAGE CROPS DUE TO SALINITY OF WATER APPLIED TO THE SOIL SURFACE

Crop	0%			10%			25%			50%			Maximum
	ECe	ECw	TDS	ECe	ECw	TDS	ECe	ECw	TDS	ECe	ECw	TDS	ECdw
Bermuda Grass	8.7	5.8	3,712	13	8.7	5,568	16	10.7	6,840	18	12	7,680	44
Tall Wheat Grass	7.3	4.9	3,136	11	7.3	4,672	15	10	6,400	18	12	7,680	44
Crested Wh. Grass	4	2.7	1,728	6	4	2,560	11	7.3	4,672	18	12	7,680	44
Tall Fescue	4.7	3.1	1,984	7	4.7	3,008	10.5	7	4,480	14.5	9.7	7,208	40
Barley (hay)	5.3	3.5	2,240	8	5.3	3,392	11	7.3	4,672	13.5	9	5,760	36
Perennial Rye	5.3	3.5	2,240	8	5.3	3,392	10	6.7	4,288	13	8.7	5,568	36
Harding Grass	5.3	3.5	2,240	8	5.3	3,392	10	6.7	4,288	13	8.7	5,568	36
Birdsfoot Trefoil	4	2.7	1,728	6	4	2,560	8	5.3	3,392	10	6.7	4,288	28
Beardless Wild Rye	2.7	1.8	1,152	4	2.7	1,728	7	4.7	3,008	11	7.3	4,672	28
Alfalfa	2	1.3	832	3	2	1,280	5	3.3	2,112	8	5.3	3,392	28
Orchard Grass	1.7	1.1	704	2.5	1.7	1,088	4.5	3	1,920	8	5.3	3,392	26
Meadow Foxtail	1.3	.9	576	2	1.3	832	3.5	2.3	1,472	6.5	4.3	2,752	24
Clover	1.3	.9	576	2	1.3	832	2.5	1.7	1,088	4	2.7	1,728	14
Barley	8	5.3	3,392	12	8	5,120	16	10.7	6,848	18	12	7,680	44
Sugarbeets	6.7	4.5	2,880	10	6.7	4,288	13	8.7	5,568	16	10.7	6,848	42
Cotton	6.7	4.5	2,880	10	6.7	4,288	12	8	5,120	16	10.7	6,848	42
Safflower	5.3	3.5	2,240	8	5.3	3,392	11	7.3	4,672	14	8	5,120	38
Wheat	4.7	3.1	1,984	7	4.7	3,008	10	6.7	4,288	14	9.3	5,952	40
Sorghum	4	2.7	1,728	6	4	2,560	9	6	3,840	12	8	5,120	36
Soybean	3.7	2.5	1,600	5	3.7	2,368	7	4.7	3,088	9	6	3,840	26
Sesbania	2.7	1.8	1,152	4	2.7	1,728	5.5	3.7	2,368	9	6	3,840	26
Rice (paddy)	3.3	2.2	1,408	5	3.3	2,112	6	4	2,560	8	5.3	3,392	24
Corn	3.3	2.2	1,408	5	3.3	2,112	6	4	2,560	7	4.7	3,008	18
Broadbean	2.3	1.5	960	3	2.3	1,472	4.5	3	1,920	6.5	4.3	2,752	18
Flax	2	1.3	832	3	2	1,280	4.5	3	1,920	6.5	4.3	2,752	18
Beans (field)	1	.7	448	1	1	640	2	1.3	832	3.5	2.3	1,472	12

Note: Conversion from ECe to ECw assumes a three-fold concentration of salinity in soil solution (ECsw) in the more active part of the root zone due to evapotranspiration. ECw × 3 = ECsw; ECsw + 2 = ECe. Tolerance during germination (beets) or early seedling stage (wheat, barley) is limited to ECe about 4 mmho/cm. (ECe means electrical conductivity of saturation extract in millimhos per centimeter (mmho/cm); ECw means electrical conductivity of irrigation water (in mmho/cm). TDS in mg/L = ECw × 640. ECdw shows maximum concentration of salts in drainage water permissible for growth. Use to calculate leaching requirement (LR = ECw/ECdw × 100 = %) to maintain needed ECe in active root area; Leaching Requirement (LR) means that fraction of the irrigation water that must be leached through the active root zone to control soil salinity at a specific level.)

Ayers, R. S., CWRCB, Sec. 1, 1, 1976. With permission.

Table 25
DRINKING AND FWPCA IRRIGATION WATER
STANDARDS FOR HEAVY AND TRACE METALS,
AND NITROGEN AND PHOSPHORUS

Substance	Drinking water (mg/liter)	Irrigation water (mg/liter)	Controlling concentration (mg/liter)
BOD	1[a]	—	1
COD	1[a]	—	1
+ NH$_4$ (as N)	—	—	—
− NO$_2$ (as N)	—	—	—
− NO$_3$ (as N)	10	—	10
P	—	—	—
Phenols	0.001	—	0.001
Cadmium	0.01	0.005	0.005
Chromium	0.05	0.005	0.005
Copper	1.0	0.2	0.2
Iron	0.3	—	0.3
Lead	0.05	5.0	0.05
Manganese	0.05	2.0	0.05
Nickel	—	0.5	0.5
Zinc	5.0	5.0	5.0
Boron	—	0.75	0.75
Chlorides	250	—	250
Sulfates	250	—	250
Suspended solids	—[b]	—	5
Color	15	—	15
Taste	Unobjectionable	—	Unobjectionable
Odor	3	—	3
Turbidity	5	—	5
Aluminum	—	1.0	1.0
Beryllium	—	0.5	0.5
Selenium	0.01	0.05	0.01
Silver	0.05	—	0.05
Vanadium	—	10.0	10.0
Arsenic	0.05	1.0	0.5
Barium	1.0	—	1.0
Cyanides	0.2	—	0.2
Cobalt	—	0.2	0.2

[a] Carbon chloroform extract to measure organic contaminants.
[b] Suspended solids should approach turbidity requirements.

Reed, A. D., Wildman, W. E., Seyman, W. S., Ayers, R. S., Prato, J. D., and Rauschkolb, R. S., *Calif. Agric.*, 27, 6, 1973. With permission.

28).   The figurative data presented in this section should not be accepted without reservations when making one's own wastewater loading rate plans. The basic principles are sound but actual values vary greatly with specific circumstances. The information provides broad guidelines, the details of which require much more specific identification of the nature of disposal waste, environment, pollutant load, etc.

Phosphorus is an element required by all living cells. Charges of accelerated entrophication of streams, lakes, water holes, and ponds have been laid at the doorstep of phosphorus. Although in some instances this may be correct, where wastes high in soluble phosphorus have reached water sources, many other constituents may accelerate entrophication. Phosphoric acid wastes originate from metal bath operations in higly industrialized manufacturing,

Table 26
SURFACE AND IRRIGATION WATER QUALITY CRITERIA FOR TRACE ELEMENTS[61]

Element	Surface Water (mg/ℓ) FWPCS[a]	Irrigation Water (mg/ℓ)		
		Continuous Use		Short-Term Use FWPCA[c] (Fine-Textured Soil)
		FWPCA (Any Soil)	NAS[b] (Coarse-Textured Soil)	
As	0.05	1.0	0.1	10
Ba	1.10	—	—	—
B	1.0	0.75	0.75	2.0
Cd	0.01	0.005	0.01	0.05
Co	—	0.2	0.05	10.0
Cr	0.05	5.0	0.1	20.0
Cu	1.0	0.2	0.2	5.0
Pb	0.05	5.0	5.0	20.0
Mn	0.05	2.0	0.2	20.0
Mo	—	0.005	0.01	0.05
Ni	—	0.5	0.2	2.0
Se	0.01	0.05	0.02	0.05
V	—	10.0	0.1	10.0
Zn	5.0	5.0	2.0	10.0
Ag	0.05	—	—	—

[a] U.S. Dept. of Interior, Federal Water Pollution Control Administration (1968). Surface water criteria are virtually the same as drinking water standards (U.S. Dept. of Health, Education, and Welfare, 1962).

[b] National Academy of Sciences (1973). Recommended maximum concentrations of trace elements in irrigation waters used for sensitive crops on soils with low capacities to retain these elements in unavailable forms.

[c] For short-term use only on fine-textured soils.

Page, A. L., U.S. Environmental Protection Agency, Cincinnati, Ohio, 1974, 97.

exemplified by automobile parts brightening and is considered a very hazardous waste. Under highly acid conditions phosphorus compounds remain soluble. Discharged into water sources in this condition, phosphorus can be a serious pollutant. Fortunately, phosphorus combines readily with a great number of elements to become insoluble. Strong phosphoric acid (2 *N* H_3PO_4) placed in soils readily precipitates and moves very short distances. Any soil (except coarse sand) readily immobilizes inorganic phosphorus forms. Organic phosphates, however, move through the soil more readily although phosphorus compounds move far less readily than nitrogen.[65-68] Much of the phosphorus wastes in the past originated from soap and detergent cleaning agents.

Bacteria and virus phathogenic to man, animals, and plants cannot be allowed to migrate through soils to underground water sources. Loading rates based on the presence of pathogenic organisms both in kind and amount vary from state to state. Also the permissible levels vary with the presence or absence of a crop and the kind of crop. For example, a higher *E. coli* count in wastewater is tolerated if the crop is not edible (cotton), rather than edible (lettuce). The regulations also differ depending on the method of land application and incorporation and the longterm cultivation program. These regulations relate to survival

Table 27
EXISTING NITROGEN LOADING RATES

Location	Seasonal liquid loading rate, in./week[a]	Annual liquid load ft/year[b]	Nitrogen loading rate, lb/acre/year[c]	Crop
St. Petersburg, Florida	10.6	45.9	1,490	Grass
Tallahassee, Florida	4.0	17.3	467	Grass, corn, millet, sorghum
Golden Gate Park, San Francisco, California	1.0	2.5	337	Grass
Irvine, California	1.5	6.5	176	Citrus, vegetables
Oceanside, California	1.3	5.6	113	Alfalfa, corn, grass
Woodland, California	1.5	2.5	72	Milo
Calabasas, California	1.9	8.2	67	Alfalfa
Abilene, Texas	3.0	2.0	65	Cotton, coastal bermuda, maize
Laguna Hills, California	0.3	1.3	18	Grass

[a]   Multiply by 2.54 to convert to cm/wk.
[b]   Multiply by 0.3048 to convert to m/yr.
[c]   Multiply by 1.12 to convert to kg/ha/yr. lb/acre/yr = concentration, mg/L × annual liquid load, ft/yr × 2.7.

Pound, C. E. and Crites, R. W., U.S. Environmental Agency, Cincinnati, Ohio, 1973, 231.

rates for pathogenic organisms in soils (Table 29). Related to this information of importance to loading rates is the data from the University of California, Berkeley, reporting the distance of travel by some pathogenic organisms and hazardous chemicals in soil and groundwater (Table 30). The movement of virus through soils also was determined in the Santee, California, project. The virus moved no more than 61 m (200 ft) of vertical travel when added to percolating water.[69]

The destiny of virus in the soil is not well known although a large number of investigations have been underway during the 1970s and 1980s. The human and animal viruses do not multiply in the environment but they may survive for extended periods of time.[70] Viruses can be isolated from groundwaters when domestic waste is applied to or released into certain coarse textured soils.[71-73] Viruses have been found redistributed from 5 to 8 m below surface sources in coarse sandy soils. However, domestic wastewater applied by spray-irrigation onto clay soils, yielded no viruses below a depth of 1.4 m.[74] In a similar study on medium-textured soil at a seven year old recharge site, viruses were not found in water samples 6 to 9 m below the surface.[75] Groundwaters at 17 m at the Arizona Flushing Meadows recharge site were virus free.[76] Virus-soil interactions are complex due to many variables. Loading rates, to insure retention of viruses by the soil, depend on the texture of the soil and extent of conventional treatment the domestic wastewaters receive. It now appears that full conventional treatment may be required to give reasonable assurance that a virologically safe drinking water is maintained in shallow soils, particularly if they are sandy and groundwater is near the surface.

Helminths (*Ascaris lumbricoides*) are intestinal parasites which have almost identical species parasitic for animals (pigs) and man. Ova of ascarid roundworms, and a nematode, are the most resistant disease-producing organism in human and animal wastes. The ova survive in the soil for many months, even years and, therefore, some remain as a possible

Table 28
ANNUAL NITROGEN, PHOSPHORUS, AND POTASSIUM UTILIZATION BY SELECTED CROPS[a]

Crop	Yield	Nitrogen	Phosphorus	Potassium
		lb per acre		
Corn	150 bu[d]	185	35	178
	180 bu	240	44	199
Corn silage	32 tons[c]	200	35	203
Soybeans	50 bu[f]	257[b]	21	100
	60 bu	336[b]	29	120
Grain sorghum	8,000 lb[g]	250	40	166
Wheat	60 bu[f]	125	22	91
	80 bu	186	24	134
Oats	100 bu[h]	150	24	125
Barley	100 bu[i]	150	24	125
Alfalfa	8 tons[c]	450[b]	35	398
Orchard grass	6 tons	300	44	311
Brome grass	5 tons	166	29	211
Tall fescue	3.5 tons	135	29	154
Bluegrass	3 tons	200	24	149
Coastal bermuda grass	8 tons	480—600	50—60	250—450

[a] Values reported above are from reports by the Potash Institute of America and are for the total above-ground portion of the plants. Where only grain is removed from the field, a significant proportion of the nutrients is left in the residues. However, since most of these nutrients are temporarily tied up in the residues, they are not readily available for crop use. Therefore, for the purpose of estimating nutrient requirements for any particular crop year, complete crop removal can be assumed.
[b] Legumes get most of their nitrogen from the air, so additional nitrogen sources are not normally needed.
[c] Multiply by 1.12 to convert to kg/ha.
[d] Multiply by 62.8 to convert to kg/ha.
[e] Multiply by 0.907 to convert to metric tons/ha.
[f] Multiply by 67.2 to convert to kg/ha.
[g] Multiply by 0.4536 to convert to kg/ha.
[h] Multiply by 35.8 to convert to kg/ha.
[i] Multiply by 53.9 to convert to kg/ha.

source of human disease. Fortunately, helminths are finally destroyed by exposure to external environments, such as the soil, just as are pathogenic bacteria, virus, and protozoa.[77] The exposure or dose of helminths and illness is complex. Inclusion of helminths with pathogenic organisms, as a part of water quality and loading rate of wastewater to cultivated land, can only serve to alert wastewater users of a potential hazard that should be considered in recommended loading rates of domestic wastewaters for food crop production or animal watering.[78]

3. Soil Properties and Loading Rates

Acceptable land loading rates and frequency of loading depend as much on soil characteristics as those of the wastewater quality. All of the soil properties described in Section I come into play. The soil functions in wastewater loading as a filter, like a chromatographic column selectively retaining certain pollutants, slowing others, and not appreciably affecting still others. In addition, the soil functions as a medium for biodegradation and chemical reactions, that can renovate wastewaters (or any other vehicle of transport) to the extent the quality of underground water is maintained, provided loading rates are not excessive.

Table 29
SURVIVAL TIMES OF PATHOGENIC ORGANISMS IN WATER, IN SOIL, AND ON VEGETATION

Organism	Media	Survival time
Anthrax bacteria	In water and sewage	19 days
Ascaris eggs	On vegetables	27—35 days
	On irrigated soil	2—3 years
	In soil	6 years
B. dysenteriae flexner	In water containing humus	160 days
B. typhosa	In water	7—30 days
	In soil	29—70 days
	On vegetables	31 days
Cholera vibrios	On spinach, lettuce	22-29 days
	On cucumbers	7 days
	On nonacid vegetables	2 days
	On onions, garlic, oranges, lemons, lentils, grapes, rice and dates	Hours to 3 days
Coliform	On grass	14 days
	On clover leaves	12—14 days
	On clover at 40—60% humidity	6 days
	On lucerne	34 days
	On vegetables (tomatoes)	35 days
	On surface of soil	38 days
	At −17°C	46—73 days
Entamoeba histolytica	On vegetables	3 days
	In water	Months
Enteroviruses	On roots of bean plants	At least 4 days
	In soil	12 days
	On tomato and pea roots	4—6 days
Hookworm larvae	In soil	6 weeks
Leptospira	In river water	8 days
	In sewage	30 days
	In drainage water	32 days
Liver fluke cysts	In dry hay	Few weeks
	In improperly dried hay	Over a year
Poliovirus	In polluted water at 20°C	20 days
Salmonella	On grass (raw sewage)	6 weeks
	On clover (settled sewage)	12 days
	On vegetables	7—40 days
	On beet leaves	3 weeks
	On grass	Over winter
	On surface of soil and potatoes	40 days
	On carrots	10 days
	On cabbage and gooseberries	5 days
	In sandy soil — sterilized	24 weeks
	In sandy soil — unsterilized	5—12 weeks
	On surface of soil (raw sewage)	46 days
	In lower layers of soil	70 days
	On surface of soils (stored sewage)	15—23 days
	In air dried, digested sludge	17 weeks
Schistosoma ova	In digestion tanks	3 months
	In sludge at 60—75°F (dry)	3 weeks
	In septic tank	2—3 weeks
Shigella	On grass (raw sewage)	6 weeks
	On vegetables	7 days
Streptococci	In soil	35—63 days
	On surface of soil	38 days
S. typhi	In water containing humus	87—104 days

Table 29 (continued)
SURVIVAL TIMES OF PATHOGENIC ORGANISMS IN WATER, IN SOIL, AND ON VEGETATION

Organism	Media	Survival time
Tubercle bacteria	On grass	10—14 days
	In soil	6 months
	In water	1—3 months
Typhoid bacilli	In loam and sand	7—17 days
	In muck	40 days
Vibrio comma	In river water	32 days
	In sewage	5 days

Pound, C. E. and Crites, R. W., U.S. Environmental Protection Agency, Cincinnati, Ohio, 1973, 251.

The different methods of wastewater application require different soils criteria to maximize wastewater loading rates. For example, the four major designs (a) low-rate irrigation, (b) overland-flow, (c) infiltration-percolation, and (d) subsurface-injection, require soils of (a) moderately permeable loamy sands to clay loams, (b) slowly permeable silt loams to clays, (c) rapidly permeable sandy loams to sands, and (d) tillable soils, respectively. The permeability as influenced by the texture and structure of the soil, therefore, controls the loading rate as much as, if not more than, any other single factor (Table 31). Land applications by spray overland flow procedures require soils of low permeability to permit the water to flow overland along a slight grade so purification may occur, primarily by contact with the surface vegetation. With low-rate irrigation (e.g., sprinkler), when soil and crop absorption occur and penetration into the goundwater is desired, purification is provided by contact with the soil surface, and a moderate rate of permeability is required. In the third case of infiltration-percolation, the design approach is for rapid recharge of relatively high quality water and a relatively rapid flow rate is desired. Liquid loading rates for the three systems are suggested as 1.3 to 10 cm, 5 to 14 cm, and 10 to 300 cm, respectively.[9]

Permeability is dependent on soil texture and structure. These physical properties influence infiltration rate, subsoil percolation rate, moisture holding capacity, (Table 31; Figure 9) and attenuation (such as adsorption reactions) of waste components. Fine-textured soils are dominated by clay and coarse-textured soils by sands. Soils of medium texture, i.e., silt, silt loam, loam, and sandy clay loam, are most suitable for land treatment. Fine-textured soils posses a larger total pore space and a higher water holding capacity than coarse-textured soils. They transmit water more slowly than medium or coarse-textured soils. Fine-textured soils restrict loading rates of wastewaters and the flow occurs along the surfaces of the soil structural aggregates and cracks rather than through the entire soil volume. Unless the loading rate on the clay soils is greatly restricted, the soil will become saturated, free water will accumulate, and anaerobic conditions will develop. Biodegradation and chemical attenuation are more limited under anaerobic conditions. Nuisance odors can develop.

When it is necessary to apply wastewater to clay soils (or clay subsoils), it becomes necessary to facilitate internal drainage, particularly under humid climatic regions. The number of drainage lines, for land receiving wastewaters, is greater than for the usual removal of rainwater and the spacing usually is closer to insure aerobic conditions. Independent soil-wastewater interaction studies at the land treatment site result in valuable data essential in determining long-term wastewater loading rates of the specific pollutant involved. Definitive values for permeability and infiltration rate of a soil for the duration of the disposal are complicated by the biological activity associated with soil clogging.

Microbial activity in soils receiving organic substances takes place through two systems, dissimilation (biodegradation) and assimilation (synthesis of organic matter, cells, tissues,

Table 30
DISTANCE OF TRAVEL OF SOME PATHOGENIC ORGANISMS AND HAZARDOUS CHEMICALS IN SOIL AND GROUNDWATER

Nature of pollution	Pollutant	Observed distance of travel[a]	Time of travel
Sewage polluted trenches intersecting groundwater	Coliform bacteria	65 ft	27 wk
	Chemicals	115 ft	—
Polluted trenches intersecting groundwater	Coliform bacteria	232 ft	—
	Uranin	450 ft	—
River water in abandoned wells	Intest, pathogens	800 ft	17 hr
	Tracer salts	800 ft	17 hr
Sewage in bored latrines intersecting groundwater	Coliform bacteria	10 ft	
	Anaerobic bacteria	50 ft	
	Chemicals	300 ft	
Sewage in bored latrines lined with fine soil	Coliform bacteria	10 ft	
Sewage in bored latrines intersecting groundwater	Coliform bacteria	35 ft	
	Chemicals	90 ft	
Sewage in bored latrines intersecting groundwater	Coliform bacteria	80 ft, regressed to 20 ft	
Coliform organisms introduced into soil	Coliform bacteria	50 m	37 days
Sewage effluents on percolation beds	Coliform bacteria	400 ft	
	Ammonia	1,400 ft	
Sewage effluents on percolation beds	Bacteria	150 ft	
Sewage polluted groundwater	Bacteria	a few meters	
Introduced bacteria	*Bacillus prodigiosus*	69 ft	9 days
Chlorinated sewage	Phenols, fungi	300 ft	
	Dye	300 ft	24 hr
Industrial wastes	Tar residues	197 ft	
	Picric acid	several miles	
Industrial wastes in cooling ponds	Mn, Fe, hardness	2,000 ft	
Chemical wastes	Miscellaneous chemicals	3—5 mi	
Industrial wastes	Chromate	1,000 ft	3 yr
	Phenol	1,800 ft	
	Phenol	150 ft	
Weed killer wastes	Chemical	20 mi	6 mo

[a]   For bacteria, the distance observed was the extent of travel; for some chemicals the observations were taken along paths of unknown extent.
Multiply by 0.3048 to convert ft to m.
Multiply by 1.609 to convert mi to km.

California State Solid Waste Management Board, Resource Recovery Program, Vol. II, Sacramento, Calif. Sec 1, 1, 1976.

gums, and slimes). These two systems influence soil infiltration differently. The level of the assimilative accumulations depends on the quality of the organic component(s) and the amount of solids in the wastewater, and the amount and frequency of wastewater applications. Laboratory soil permeability tests may be misleading and acceptable loading rates exceeded if this tendency for clogging of soil pores by microbial tissues and products is ignored. Orlob and Butler[79] who experimented with sewage spreading on five California soils, found that soil infiltration rates passed through three distinct stages (a) abrupt decrease in rate attributed to suspension of soil particles, (b) an increase in rate due to solution of trapped gases into the percolating liquid, and (c) a decrease due to accumulation of biological slimes in the soil voids. Over longer periods of time (more than a few days) continuous application may result in a fourth phase, (d) a decided decrease in infiltration rate. The latter phase

Table 31

RANGE IN AVAILABLE MOISTURE FOR SOILS HAVING DIFFERENT TEXTURES OR RATIOS OF SAND, SILT, AND CLAY

Texture range	Range of available water Inches per foot of soil[a]
Coarse Texture	
Very coarse sands	0.40—0.75
Coarse sands, fine sands, and loamy sands	0.75—1.25
Medium Texture	
Sandy loams and fine sandy loams	1.25—1.75
Very fine sandy loams, loams, and silt loams	1.50—2.30
Fine Texture	
Clay loams, silty clay loams, and sandy clay loams	1.75—2.50
Sandy clays, silty clays, and clays	1.60—2.50
Peats and mucks	2.0—3.00

[a] Multiply by 8.33 to convert to cm per m of soil.

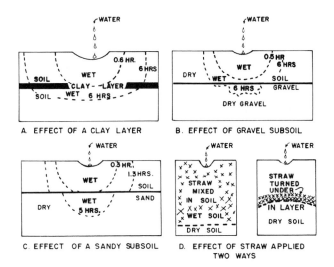

FIGURE 9.   Water movement in the soil is influenced by changes in soil texture, structure, and layering of organic waste residues, all of which are associated with nonhomogeneous soil condition. (Modified from Gardner, W. H., How water moves in the soil, Part 1, The basic concept, *Crops and Soils*, 15, 1, 1962.)

appears to be slow but continuous with decreasing infiltration rates and renovating (attenuation) capacity (Pound and Crites, 1973; Bouwer, 1970, 1972).[73,76,78] This led to the initiation of the cyclic-loading-systems which allows for application periods to be interrupted by periods of rest or no application. The length of resting period depends on the drying rate

Table 32
SOME HYDRAULIC LOADING CYCLES

Location	Loading objective	Application period	Resting period
Flushing Meadows, Arizona			
Maximum infiltration	Increase ammonium adsorption capacity	2 days	5 days
Summer	Maximize nitrogen removal	2 wk	10 days
Winter	Maximize nitrogen removal	2 wk	20 days
Hyperion, California	Maximize infiltration rates	Continuous[a]	—
		0.5 hr	23.5 hr
Lake George, New York			
Summer	Maximize infiltration rates	9 hr	4—5 days
Winter	Maximize infiltration rates	9 hr	5—10 days
Tel Aviv, Israel	Maximize renovation	5—6 days	10—12 days
Vineland, New Jersey	Maximize infiltration rates	1—2 days	7—10 days
Westby, Wisconsin	Maximize infiltration rates	2 wk	2 wk
Whittier Narrows, California	Maximize infiltration rates	9 hr	15 hr

[a] Abandoned after 6 months.

Pound, C. E. and Crites, R. W., U.S. Environmental Protection Agency, Cincinnati, Ohio, 1973, 231.

of the soil. Drying allows shrinking to occur and cracking of crust established by algae, bacteria, fungi, and organic matter. A more rapid rate of liquid infiltration is restored upon soil dehydration. There appears to be no single length of resting period for all sites or even for a given wastewater quality. Table 32 illustrates this point. Resting periods range from a few hours to two weeks. The most advantageous resting period varies with time required to dry the soil sufficiently to permit shrinking and cracking. Many factors of climate, soil, and vegetative cover determine the rate of drying. The optimum loading cycle will therefore depend on the objectives of the wastewater application. Maximizing infiltration rate, for example, may minimize the renovation. Denitrification of nitrates progresses more rapidly under anaerobic than aerobic conditions. Removal efficiencies vary with every site regardless of maintenance of optimum loadings as related to site characterization (Table 33). The difficulty to fully characterize all the parameters that entered into the selection of the specific loading rate, such as (a) soil variability, (b) climatic variations from the means, and (c) wastewater seasonal variations, no doubt, makes for these differences in removal efficiencies and ability to predict optimum loading. Where tillage is permissible with disk, harrow, Rototiller®, or ripper equipment, satisfactory permeability can be attained.

The first soil characteristics to consider in loading rate decisions are (a) slope, (b) general soil permeability, (c) depth of soils, and (d) depth to groundwater. Some of these appear in soil survey publications and maps. Soil survey maps cover more detailed features if the land is cultivated or is amenable to cultivation. Permeability classes are available on a soil series basis. General permeability classes such as in Table 34 are presented.[4] Related to these are classes of permeability or percolation rates for saturated subsoils (Table 36).[81] Most often permeability is qualitatively described as (a) slowly permeable silts and clays, (b) moderately permeable clay loams, loams, and sandy loams, and (c) rapidly permeable sandy loams, loamy sands, and sands. Another system[80] compares permeability data of selected porous media including rock (Table 35).

Soil characteristics at the second level of interest for loading rate may include the following:

- Soil type
- Size of the area of a given type

Table 33
REMOVAL EFFICIENCY AT SELECTED INFILTRATION-PERCOLATION SITES[64]

Location	Depth samples, ft[d]	Removal Efficiency, % BOD	SS	N	P	E. coli
Flushing Meadows, Arizona	30	98	100	30—80	60—70	100
Hyperion, California	7	90	—	—	—	98
Lake George, New York	10	96	100	43—51	8—61	99.3
Santee, California	200[a]	88	—	50	93	99
Westby, Wisconsin	3	88	—	70	93	—
Whittier Narrows, California	8	—	—	0[b]	96	0[c]

ᵃ Lateral flow.
ᵇ Short frequent loading promotes nitrification, but not denitrification.
ᶜ Coliforms regrow in the soil.
ᵈ Multiply by 0.3048 to convert to m.

Pound, C. E. and Crites, R. W., U.S. Environmental Protection Agency, Cincinnati, Ohio, 1973, 251. With permission.

Table 34
CLASSES OF PERMEABILITY OR PERCOLATION RATES FOR SATURATED SUBSOILS

Class	Hydraulic conductivity or percolation rate cm/hr	Comments
Extremely slow	<0.003	So nearly impervious that leaching process is insignificant. Unsuitable for wastewater renovation under most circumstances.
Very slow	0.003 to 0.025	Poor drainage results in staining; too slow for artificial drainage. Wastewater renovation possible under restricted conditions.
Slow	0.025 to 0.25	Too slow for favorably airwater relations and for deep root development. Usable under controlled conditions; drainage facilities may be required; runoff likely to be a problem. Good nitrate removal possible.
Moderate	0.25 to 2.5	Adequate permeability (conductivity). Ideal for most irrigation systems.
Rapid	2.5 to 25.4	Excellent water-holding relations and permeability (conductivity). Ideal for most irrigation systems. Application rates may have to be reduced to ensure renovation.
Very rapid	>25.4	Associated with poor waterholding conditions. Infiltration and drainage may be too rapid to achieve complete renovation. Extreme caution required.

SCS Engineers, U.S. Army Engin. Waterway Expt. Sta. Vicksburg, Miss., 1977, 210.

Table 35
PERMEABILITY DATA OF SELECTED
POROUS MEDIA[80]

Material	Permeability (gpd per sq ft @ 60°F)[a]		
Granite	0.0000009	to	0.000005
Slate	0.000001	to	0.000003
Dolmite	0.00009	to	0.0002
Hematite	0.000002	to	0.009
Limestone	0.00001	to	0.002
Gneiss	0.0005	to	0.05
Basalt	0.00004	to	1
Tuff	0.0003	to	10
Sandstone	0.003	to	30
Till	0.003	to	0.5
Loess	1	to	30
Beach sand	100	to	400
Dune sand	200	to	600
Alluvium	(see individual materials below)		
Clay	0.001	to	1
Silt	1	to	10
Very fine sand	10	to	100
Fine sand	100	to	1000
Medium sand	1000	to	4500
Coarse sand	4500	to	6500
Very coarse sand	6500	to	8000
Very fine gravel	8000	to	11000
Fine gravel	11000	to	16000
Medium gravel	16000	to	22000
Coarse gravel	22000	to	30000
Very coarse gravel	30000	to	40000
Cobbles	Over 40000		

[a]   Multiply by 4.716×10^{-5} to convert to cm/s.

- Depth to bedrock
- Subsurface geology
- Depth of individual soil horizons
- Permeability of individual soil horizons
- Flooding frequency
- Erosion class

Finally those specific soil properties that are of interest and are readily measureable include:

- pH
- Electrical conductivity
- Soil texture (sand, silt, and clay)
- Soil structure
- Clay mineral type
- Hydraulic conductivity
- Infiltration rate

Table 36
LOADING RATES VS. SOIL TYPE AND CROP

Location	Loading rate in./week[c]	Soil type	Method of Application	Crop
Infiltration-percolation				
Eglin AFB, Florida	11.2[a]	Sand	Spraying	Forest
Quincy, Washington	7.2[a]	Silty sand	Flooding	Corn, wheat
High irrigation[b]				
Hanford, California	4.2	Sandy loam	Flooding	Corn, oats, cotton
Tallahassee, Florida	4.0[a]	Sand	Spraying	Corn, millet, sorghum, grass
San Bernardino, California	3.7[a]	Sand	Spraying	Grass
Hillsboro, Oregon	3.1[a]	Sandy loam	Spraying	Grass, forest
Moderate irrigation[b]				
Abilene, Texas	3.0	Clay loam	Flooding	Cotton, maize, coastal bermuda grass
Alamogordo, New Mexico	2.5	Silty clay loam	Flooding	Corn, oats, sorghum, alfalfa
Pleasanton, California	2.2[a]	Loam	Spraying	Grass
Light irrigation[b]				
Rawlins, Wyoming	1.5	Gravel	Flooding, spraying	Alfalfa
Bakersfield, California	1.2	Clay loam	Ridge and furrow, flooding	Cotton, corn, barley, alfalfa
Ely, Nevada	0.3	Sandy loam	Flooding	Alfalfa

[a] Year-round rate.

[b] High, 3 to 4 in./wk; moderate, 2 to 3 in./wk; light, <2 in./wk.

[c] Multiply by 2.54 to convert to cm/wk.

Pound, C. E. and Crites, R. W., U.S. Environmental Protection Agency, Cincinnati, Ohio, 1973, 251.

- Bulk density
- Cation exchange capacity
- Hydrous iron oxide
- Organic matter
- Drainable porosity

Loading rates do not always require measurement of all of these characteristics to predict optimum and maximum loading rates. Analyses should be made on a basis of need.

4. Climatic Characteristics and Loading Rates

Since the macroclimate cannot be changed by present technology, the system of disposal, including loading rates, must be adjusted to the climate at the specific loading.[82,96] Climatic characteristics that have a distinct effect on loading include:

- Temperature (air and soil) (length of growing season)
- Precipitation (intensity, annual amount, and seasonal distribution)
- Wind velocity and prevailing direction
- Humidity (relative humidity)
- Evapotranspiration

Temperature of the air and soil is controlled by the energy received from the sun. In winter, the day length and the sun's angle decrease away from the equator. The total receipt of energy decreases rapidly with concurrent colder conditions prevailing. In summer, day length increases away from the equator whereas the angle of the sun's rays still decreases. The total receipts of energy are considerably less variable latitudinally in summer than winter. The air and soil temperatures respond to these energy relationships. Climatic constraints on land treatment and loading rate respond strongly to changes in geographic latitudinal differences. Temperature controls the length of the growing season for crop plants, and concurrently, the length of the land treatment with wastewaters. Irrigation and overland-flow systems are restricted during freezing weather. Application of wastewater to snow cover isn't a desirable or effective practice for many real reasons. Storage capacity must be provided in regions of freezing winters. Overland-flow systems, dependent on plant life, must be restricted. Vegetative cover is a major factor in the continual success of most wastewater land treatment systems. The accumulation of ice and freezing of water lines, spray nozzles, and other delivery systems, make winter applications of wastewaters unsuccessful. In arid and semiarid land of the warmer climates, for example in the southwestern U.S., wastewater disposal can be a year round practice. Climatic influences on loading rates relate primarily to temperature and moisture relationships as they influence (a) freezing and thawing, (b) soil moisture, (c) intensity of precipitation and annual distribution, (d) temperatures favoring optimum biodegradation, (e) evapotranspiration, and (f) vegetative growth and kind of crops (Table 36).[64] The importance of temperature and precipitation are readily demonstrated by data in Table 37 that provides tabulation of average monthly temperature and precipitation values of representative stations in each of the five climatic zones.[64]

Often overlooked is the effect of climate on the activity of the soil microorganisms. Temperatures below about 6°C (42°F), for example, reduce nitrification of ammonium to nitrite and nitrite to nitrate. As a general observation, microbiological activity rates double with every 10°C rise in temperature until some limiting temperature is reached.[82] The upper limiting temperature for the ubiquitous mesophillic organisms, which is about 46°C, is seldom reached in soils except for dry soil surfaces. Dry surfaces are more common in arid and semiarid regions and occasionally on other surfaces lacking vegetation.

Precipitation (soil moisture) constraints also may limit the best functioning of land treatment systems. The moisture broadly affects loading rates as follows:

Table 37
AVERAGE MONTHLY TEMPERATURE AND PRECIPITATION VALUES AT REPRESENTATIVE STATIONS IN EACH OF THE FIVE CLIMATIC ZONES

Zone	Location		Jan	Feb	Mar	Apr	May	June	July	Aug	Sept	Oct	Nov	Dec	Annual
A	Sacramento, Calif.	T[a]	46.2	50.2	54.3	59.9	66.0	72.5	77.4	76.1	73.6	64.9	54.3	47.5	61.9
		P[b]	3.19	2.99	2.36	1.42	0.59	0.12	Trace	0.04	0.20	0.79	1.46	3.23	16.30
A	Los Angeles, Calif.	T	55.8	57.0	59.4	61.9	64.8	68.0	73.0	73.0	72.0	67.5	62.8	58.3	64.4
		P	3.07	3.35	2.24	1.18	0.16	0.08	Trace	0.04	0.24	0.39	1.06	2.87	14.69
B	Midland, Tex.	T	44.1	48.4	55.2	64.9	73.4	81.1	82.9	82.2	75.4	65.8	52.5	45.9	64.2
		P	0.79	0.59	0.35	0.83	2.09	1.61	1.89	1.50	1.77	1.65	0.47	0.67	14.25
B	Phoenix, Ariz.	T	50.7	54.5	60.4	68.7	77.0	85.6	91.2	89.1	84.4	72.1	59.2	52.5	70.5
		P	0.75	0.87	0.67	0.31	0.12	0.08	0.79	1.10	0.75	0.47	0.47	0.87	7.20
C	Macon, Ga.	T	49.3	51.1	56.8	65.7	73.9	80.8	81.9	81.3	76.5	66.2	55.4	48.9	65.7
		P	3.39	4.25	4.92	3.74	3.31	3.35	5.63	4.17	2.76	2.01	2.44	4.06	44.09
C	Little Rock, Ark.	T	40.6	44.4	51.8	62.4	70.5	79.0	81.9	81.3	74.3	63.1	49.5	41.9	61.7
		P	5.24	4.33	4.80	4.92	5.28	3.62	3.35	2.83	3.23	2.87	4.13	4.09	48.66
C	Portland, Ore.	T	38.5	42.1	46.0	51.8	57.4	62.1	67.3	66.6	62.2	54.1	45.1	41.4	52.9
		P	5.35	4.21	3.82	2.09	2.01	1.65	0.39	0.67	1.61	3.62	5.31	6.38	37.17
D	Columbus, Ohio	T	31.6	32.7	40.5	52.0	63.1	72.7	76.5	74.8	67.8	56.5	43.3	33.4	53.8
		P	2.91	2.20	2.87	3.35	3.46	3.78	3.66	2.80	2.32	1.89	2.24	2.24	33.74
D	Omaha, Neb.	T	22.3	26.4	36.9	51.6	63.0	73.0	78.4	76.3	66.9	55.8	38.8	28.2	51.4
		P	0.83	0.94	1.46	2.56	3.46	4.53	3.39	3.98	2.64	1.73	2.24	2.24	27.56
D	Spokane, Wash.	T	25.3	30.0	38.1	47.3	55.8	61.3	69.4	67.5	59.9	48.6	35.2	29.7	47.3
		P	2.44	1.85	1.50	0.91	1.22	1.50	0.39	0.39	0.75	1.57	2.24	2.44	17.20
E	Albany, N.Y.	T	25.7	26.8	35.8	48.4	59.9	69.1	73.9	71.8	63.7	53.1	41.7	29.5	49.8
		P	2.52	2.24	2.87	2.91	3.62	3.74	4.29	3.31	4.02	2.83	2.91	2.72	37.95
E	Huron, S.D.	T	12.6	16.5	28.8	45.0	57.6	67.6	75.0	72.9	61.9	48.7	31.3	18.9	44.8
		P	0.47	0.59	1.10	1.85	2.36	3.15	1.81	2.09	1.54	1.14	0.67	0.55	17.32

[a] Temperature, deg. F.
[b] Precipitation, in.
Multiply by 25.4 to convert to mm.

Pound, C. E. and Crites, R. W., U.S. Environmental Protection Agency, Cincinnati, Ohio, 1973, 251.

- Moisture is essential to vegetative cover and microbial activity. Levels below about 30% moisture and above normal field capacity have a limiting effect on aerobic microflora biodegradation.
- Moisture cools the soil surfaces during hot weather by evaporation, providing a more desirable habitat for living organisms.
- Moisture is related to evapotranspiration and water loss which effect loading rates.
- Moisture from precipitation adds to the burden of loading rates. High intensity and prolonged rainstorms may overload the soil causing erosion and requiring the establishment of auxiliary storage facilities.
- Moisture from prolonged precipitation periods in some humid temperate climates, particularly in spring, may cause waterlogging and contribute to unfavorably prolonged anaerobic conditions, aggravated by soluble organic wastes of high BOD, with accompanying malodors, reduced biodegradation, and difficulty in accessibility for land treatment operations.
- Excessive precipitation in humid climatic regions requires more land to accommodate the wet soil conditions and requires the establishment of elaborate and/or expensive drainage systems.
- Precipitation in arid and semiarid regions usually does not limit loading rates but will extend loading rates, relative to humid regions.

*5. Time and Loading Rates**

Reports on the effect of time on loading rates are limited in number. Even with use of the best quality industrial wastewaters over long periods of time, there is a hazard of overloading the soil with organic residues resulting in damage to the ecosystem and ineffectual biodegradation rates.[64] The soil mantle must be recognized as having some limited capacity to absorb and degrade fully, organic matter in wastewaters. One solution to this is reducing the loading rate but extending the time the land is used. Frequency of loading also may be reduced when it appears organic (BOD) loading is becoming limited. In general, organic loading of wastewaters does not occur unless the system involves very high BOD wastewaters, e.g., BOD greater than 10,000 mg/ℓ.[64] Using a spray irrigation distribution system and close control where low volumes of cannery wastewater were applied, one site was effective for at least 10 years.

Although a relatively large number of successful land treatment systems for wastewater disposal are in progress, specific questions with respect to loading rates and length of time the soil can effectively purify the water, remain to be answered. An objective in one of the few long-term studies (30 years) was to determine the long-term ability of the soil to treat municipal wastewater.[83] This was achieved by calculating the mass of selected constituents retained in the soil profile and determining quality of underlying ground water. The location was Hollister, Calif., and the system was a fully managed groundwater recharge program for renovating primary municipal effluent by surface spreading at high rates. The basins received 30 cm of wastewater each at intervals of 14 to 21 days. Infiltration rates averaged about 10 cm/d. Disking between applications was necessary to maintain this favorable rate of infiltration.

Organic matter accumulated only slightly during the 30 year land treatment. The pH of the sandy loam, originally about 8.1, declined in the 0 to 15 cm layer to about 6.5 and native calcium carbonate was depleted to a depth of 30 cm. The total soluble salts, total dissolved solids (TDS), and boron and phosphorus content increased throughout the 300 cm depth. Although levels of heavy metals (Ni, Co, Zn, Cd, and Cu) accumulated within the 300 cm soil depth, the level of soil DTPA-extractable metals was not sufficiently elevated,

---

*   Additional references for loading rates are provided in State and Federal Publications, 97, 98, 99.

within crop plant root zones, to adversely affect future agricultural potential of the Hollister site soil. Groundwaters were not measurably polluted with heavy metals or organics.

REFERENCES

1. U.S. Environmental Protection Agency, *Hazardous Waste Management System; Permitting Requirement for Land Disposal Facilties,* CFR 47 (143), RCRA P.L. 94-580, Washington, D.C., 1982, 32324
2. **Elliott, L. B. and Stevenson, J. B., Eds.,** *Soils for Management of Organic Wastes and Waste Waters,* Am. Soc. Agron., Madison, Wis., 1977, 1.
3. **Kreis, R. D. and Shuyler, L. R.,** *Beef Cattle Feedlot Site Selection for Environmental Protection,* EPA-R2-72-129, U.S. Environmental Protection Agency, RSKERL, Ada, Okla., 1972, 39.
4. **Loehr, R. C., Jewell, W. J., Novak, J. D., Clarkson, W. W., and Friedman, G. S.,** *Land Application of Wastes,* Vol. 7, Environmental Engineering Series, Van Nostrand Reinhold, New York, 1979, 308.
5. **Overcash, M. R. and Pal, D.,** *Design and Land Treatment Systems for Industrial Wastes — Theory and Practice,* Ann Arbor Sci. Pub. Inc., Ann Arbor, Mich., 1979, 684.
6. **Meeks, B., Chesnin, L., Fuller, W., Miller, R., and Turner, D.,** Guidelines for manure utilization in the Western Region USA. Rept. WRRC, W-124 Comm. Washington State University, Col. Agric. Res. Cntr., Bull., 1975, 1.
7. **Kreis, R. D., and Shuyler, L. R.,** *Beef Cattle Feedlot Site Selection for Environmental Protection,* EPA-R-2-72-129, U.S. Environmental Protection Agency, RSKERL, Ada, Okla., 1972, 39.
8. **Durlak, E. R.,** Solid Waste Disposal by Landspreading Techniques, TN No. N-1491, Naval Construction Battalion Center, Port Hueneme, Calif., 1977, 15.
9. **Phung, T., Barker, L., Ross, D., and Bauer, D.,** Land Cultivation of Industrial Wastes and Municipal Solid Wastes: State-of-the-Art Study, Vol. 1, EPA 600/12-78-140a, PB 278-080/AS, U.S. Environmental Protection Agency, MERL, Cincinnati, Ohio, 1978, 205.
10. **Byrne, T. G., Davis, W. B., Booher, L. J., and Werenfels, L. F.,** Vertical mulching for improvement of old golf greens, *Calif. Agric.* 19, 12, 1965.
11. **Fuller, W. H.,** *Management of Soils of the Southwest Desert,* University of Arizona Press, Tucson, 1975, 195.
12. **Cress, F.,** Petroleum mulches, *Agrochem. West,* Vol. 21, 1967.
13. **Mays, D. A.,** Special problems and opportunities in use of waste heat for soil warming, in *Soils for Management of Organic Wastes and Waste Waters,* Elliott, L. F. and Stevenson, F. J. Eds., Amer. Soc. Agron., Madison, Wis., 1977, 511.
14. **Unger, M., Fuller, W. H., Watson, J., and Pepper, I.,** A comparison between municipal sewage sludge from an industrialized and non-industrialized cities: II. Impact of heavy metals on growth of barley, 1984 (accepted for publication in *Plant and Soil).*
15. **Watson, J., Pepper, I., Unger, M., and Fuller, W. H.,** Yield and leaf element composition of cotton growth on sludge amended soil, *J. Environ. Qual.,* 1984, (accepted for publication).
16. **Fuller, W. H. and Tucker, T. C.,** Utilization and disposal of organic wastes in arid regions, in *Soils for Management of Organic Waste and Waste Waters,* Elliott, L. F. and Stevenson, F. J., Eds., Amer. Soc. Agron., Madison, Wis., 1977, 471.
17. **Cottrell, N. Mc.,** Disposal of Municipal Wastes on Sandy Soil: Effect of Plant Nutrient Uptake, M.S. thesis, Oregon State Univ. Corvallis, 1975, 201.
18. **Halvorson, G. A.,** Movement of Elemental Constituents in Sagehill-Loamy Sand Treated with Municipal Waste, M.A. thesis, Oregon State Univ. Corvallis, 1975, 1.
19. **Hart, S. A., Flocker, W. J., and York, G. K.,** Refuse stabilization in the land, *Compost Sci.,* 11, 4, 1969.
20. **Volk, V. V.,** Application of trash and garbage to land, in *Land Application of Waste Materials, Soil Conservation Society of America,* Ankeny, Ind., 1976, 154.
21. **Polson, R. L.,** Refractory metal processing waste utilization on Dayton silty clay loam, M.S. thesis, Oregon State Univ., Corvallis, 1976, 145.
22. **Golueke, C. G.,** *Composting, A Study of the Process and Its Principles,* Rodale Press, Emmanus, Pa, 1975, 110.

23. **Gotaas, H. B.,** *Composting: Sanitary Disposal and Reclamation of Organic Wastes,* World Health Organization, Palais des Nations, Geneva, Switzerland, 1956, 205.

24. **Huag, R. T.,** *Composting Engineering: Principles and Practices,* Ann Arbor Sci. Pub. Inc., Ann Arbor, Mich., 1980, 11.

25. **King, L. D. and Morris, H. D.,** Municipal compost for crop production, *Georgia Agric. Res.,* 10, 4, 1969.

26. **Mays, D. A., Termon, G. L., and Duggan, J. C.,** Municipal compost: Effects on crop yields and soil properties, *J. Environ. Qual.,* 2, 87, 1973.

27. **Tiejen, C. and Hart, S. A.,** Compost for agricultural land, *J. Sanit, Engn. Div., Proc., Amer. Soc. Civil Engn.,* 95, 269, 1968.

28. **Fuller, W. H.,** New organic pelleted compost, *Compost Sci.,* 6, 30, 1966.

29. **Kincannon, B. C.,** Oily Waste Disposal by Soil Cultivation Process, EPA-R2-72-110, Office of Resources and Monitoring, U.S. Environmental Protective Agency, Washington, D.C., 1972, 45.

30. **Huddleston, R. L.,** Treatment of Oily Wastes by land farming, in *Disposal of Industrial and Oily Sludges by Land Cultivation,* Resource Systems and Management Assoc., Northfield, N.J., 1980, 231.

31. U.S. Environmental Protection Agency, Proceedings of the Open Forum on Management of Petroleum Refinery Wastewater, Tulsa, Okla., and R. S. Kerr Environ. Res. Lab., U.S. Environmental Protection Agency, Ada, Okla., 1976, 512, 273.

32. **Dolar, S. G., Boyle, J. R., and Keeney, D. R.,** Paper mill sludge disposal on soils: effect on the yield and mineral concentration of oats (Avena sativa), *J. Environ. Qual.,* 1, 405, 1972.

33. **Aspitarte, T. R., Rosenfeld, A. S., Samle, C. B., and Amberg, H. R.,** Pulp and paper mill sludge disposal and crop production, *Tappi,* Tech. Assoc. Pulp Paper Ind., Wis., 56, 140, 1973.

34. **Bollen, W. B. and Lu, K. C.,** Effect of Douglas-fir sawdust mulches and incorporation on soil microbial activities and plant growth, *Soil Sci. Soc. Amer. Proc.,* 21, 35, 1957.

35. **Briedenbach, W.,** *Composting of Municipal Solid Waste in the United States,* U.S. Environmental Protection Agency, SW-47R, Washington, D.C., 1971, 1.

36. **Kirsch, R. K.,** Effects of sawdust mulches, Oregon State University, Corvallis, Tech. Bull., 49, 16, 1959.

37. **Fuller, W. H.,** Unpublished data, 1983.

38. **Hunt, P. G. and Peele, T. C.,** Organic matter removal from peach wastes by percolation through soil and interrelations with plant growth and soil properties, *Agron. J.,* 60, 321, 1968.

39. **Reed, A. D., Wildman, W. E., Seyman, W. S., Ayers, R. S., Prato, J. D., and Rauschkolb, R. S.,** Soil recycling of cannery wastes, *Calif. Agric.,* 27, 6, 1973.

40. **Loehr, R. C.,** *Agricultural Waste Management: Problems, Processes and Approaches,* Academic Press, New York, NY, 1974, 51.

41. **Fuller, W. H.,** *Management of Southwestern Desert Soils,* Univ. Arizona Press, 1975, 95.

42. **Fuller, W. H.,** Reclamation of saline and alkali soils, *Plant Food Rev.,* Fall, 7, 1962.

43. **Fuller, W. H. and Ray, H.,** Gypsum and sulfur-bearing amendments for Arizona soils, Univ. Arizona Bull., A-27, 1, 1963.

44. **Fuller, W. H. and Lanspa, L.,** Uptake of iron and copper by sorghum from mine tailings, *J. Environ, Qual.,* 4, 417, 1975.

45. **Miyamoto, S. and Stroehlein, J. L.,** Sulfuric acid for increasing water penetration into some Arizona soils, *Prog. Agric. Ariz.,* 27, 13, 1975.

46. **Ryan, J., Miyamoto, S., and Stroehlein, J.,** Solubility of manganese, iron, and zinc, as affected by applications of sulfuric acid to calcareous soils, *Plant Soil,* 40, 421, 1974.

47. **Wallace, A. T.,** Massive sulfur applications to highly calcareous agricultural soil as a sink for waste sulfur, *Research,* 2, 263, 1977.

48. **Jastrow, J. D., Zimmerman, C. A., Dvorak, A. J., and Hinchman, R. R.,** Comparison of lime and flyash as amendments to acidic coal mine refuse: Growth response and trace-element uptake of two grasses, Argonne Natl. Lab. ANL/LRP-6, NTIS, U.S. D.C., Springfield, Va, 1979, 36.

49. **Martens, D. C.,** Availability of plant nutrients in fly ash, *Compost Sci.,* 12, 15, 1971.

50. **Chang, A. C., Lund, L. J., Page, A. L., and Warneke, J. E.,** Physical Properties of Fly Ash — amended Soils, *J. Environ. Qual.,* 6, 267, 1977.

51. **Jones, C. C. and Amos, D. F.,** Physical Changes in Virginia Soils Resulting From the Addition of High Rates of Fly Ash, in Proc. 4th Cong., MERC/SP-76/4 (conf.-760322), ERDA Morgantown Energy Research Center, Morgantown, WV, 1976, 624.

52. **Fuller, W. H. and Paget, C.,** The Effects of Discing, Rototilling, and Water Action on the Structure of Some Calcareous Soils, Univ. Ariz., Agric. Expt. Sta. Tech. Bull., 134, 26, 1958.

53. **Fuller, W. H. and Giraud, C.,** The influence of soil aggregate stabilizers on the biological activity of the soil, *Soil Sci. Soc. Amer. Proc.,* 18, 33, 1954.

54. **Fuller, W. H., Gomness, N., and Sherwood, L.,** The Influence of Soil Stabilizers on Stand, Composition, and Yield of Crops on Calcareous Soils of Southern Arizona, Univ. Ariz., Agric. Expt. Sta. Tech. Bull. 129, 261, 1953.

55. **Bastian, R. K.,** Municipal sludge management: EPA construction grants program, in *Land as a Waste Management Alternative,* Loehr, R., Ed., Ann Arbor Sci. Pub. Co. In., Ann Arbor, Mich., 1977, 32, 86.

56. **Pahren, H. R.,** An Appraisal of the Relative Health Risk Associated with Land Application of Municipal Sludge, 50th Ann. Conf. Water Poll. Control Fed., Philadelphia, Pa, 1977, 210.

57. **Minor, J. R. and Hazen, T. E.,** Transportation and application of organic waste to land, in *Soil for Management of Organic Wastes and Waste Waters,* Elliott, L. F. and Stevenson, F. J., Eds., American Society of Agronomy, Madison, Wis., 1977, 379.

58. **Norstadt, F. A., Swanson N. P., and Sabey, B. R.,** Site design and management for utilization and disposal of organic wastes, in *Soils for Management of Organic Wastes and Waste Waters,* Elliott, L. F. and Stevenson, F. J., Eds., American Society of Agronomy, Madison, Wis., 1977, 349.

59. **Wallace, A. T.,** Land disposal of liquid industrial wastes, in *Land Treatment and Disposal of Municipal and Industrial Wastewater,* Sanks, R. L. and Asavo, T., Eds., Ann Arbor Sci. Pub. Inc., Ann Arbor, Mich., 1976, 147.

60. **Fuller, W. H. and Clark, K. G.,** Microbiological studies on ureaformaldehyde preparations, *Soil Sci. Soc. Amer.,* 2, 198, 1947.

61. **Page, A. L.,** Fate and Effect of Trace Elements in Sewage Sludge when Applied to Agricultural Land, EPA-670/2-74-005, U.S. Environmental Protection Agency, Cincinnati, Ohio, 1974, 97.

62. U.S.D.I., Federation Water Pollution Control Administration, Surface Water Standards of Quality, Washington, D.C., 1968, 55.

62. National Academy of Sciences, Maximum Recommended Concentration of Trace Elements in Irrigation Waters, Washington, D.C., 1973, 1.

64. **Pound, C. E. and Crites, R. W.,** Wastewater Treatment and Reuse by Land Application, EPA-660/2-73-006a, U.S. Environmental Protection Agency, Cincinnati, Ohio, 1973, 251.

65. **Hannapel, R. J., Fuller, W. H., Bosma, S., and Bullock, J. S.,** Phosphorus movement in a calcareous soil: I. Predominance of organic forms of phosphorus in phosphorus movement, *Soil Sci.,* 97, 350, 1964.

66. **Hannapel, R. J., Fuller, W. H., and Fox, R. H.,** Phosphorus movement in a calcareous soil: II. Soil microbial activity and organic phosphorus movement, *Soil Sci.,* 97, 421, 1964.

67. **Fuller, W. H.,** Phosphorus: Element and geochemistry, in *The Encyclopedia of Geochemistry and Environmental Sciences,* Encycl. Earth Sci., Ser. IV A, Van Nostrand Rheinhold, Co., New York, NY, 1972a, 942.

68. **Fuller, W. H.,** Phosphorus: Phosphorus Cycle, in *The Encyclopedia of Geochemistry and Environmental Sciences,* Encycl. Earth Sci., Ser. IV A, Van Nostrand Rheinhold, Co., New York, NY, 1972b, 946.

69. **Merrell, J. C., Jr., Katko, A., and Pantler, H. E.,** The Sante Recreation Project, Sante, Calif, Summary Rept. 1963-64, Pub. Health Serv. Pub. No. 999-WP-27, Washington, D.C., 1965, 1.

70. **Weaver, R. W., Dronen, N. O., Foster, B. F., Heck, F. C., and Fehrmann, R. C.,** Sewage Disposal on Agricultural Soils: Chemical and Microbiological Implications, Vol. II, Microbiological Implications, EPA-600/2-78-131b, U.S. Environmental Protection Agency, ORD, RSKERL, Ada, Okla., 1978, 93.

71. **Mack, W. N., Lu, Y-S, and Coohan, D. B.,** Isolation of Poliomyelitis Virus and Wastewater Systems, *Health Serv. Rept.* 87, 271, 1972.

72. **Wellings, F. M., Lewis, A. L., and Mountain, C. W.,** Demonstration of solid-associated virus in wastewater and sludge, *Appl. Environ. Microbiol.,* 31, 354, 1976.

73. **Vaughn, J. M.,** The Fate of Human Viruses in Groundwater Recharge Systems Utilizing Tertiary Treated Effluents, Sept. 1977, Interium Rept. EPA Grant No. R-804776, Cincinnati, Ohio, 1977, Sec. 2.

74. **Sagik, B. P. and Sorber, C. A.,** Human Enteric Virus Survival in Soil Following Irrigation with Sewage Plant Effluents, Ann. Rept. U.S. Environmental Protection Agency Grant No. R-803844, Cincinnati, Ohio, 1977, 232.

75. **Gilbert, R. G., Rice, R. C., Bower, H., Gerba, C. P., Wallis, C., and Melnick, J. L.,** Wastewater renovation and reuse: Virus removal by soil filtration, *Science,* 192, 1004, 1976.

76. **Bouwer, H., Rice, R. C., Escorega, E. D., and Riggs, M. S.,** Renovating secondary sewage by groundwater recharge with infiltration basins, U.S. Environmental Protection Agency, Water Pollut. Control Res. Ser. Proj. No. 16060 DRU, Washington, D.C., 1972, 101.

77. **Fitzgerald, R. P.,** Helminth and Heavy Metal Transmission From Aerobically Digested Sewage Sludge, EPA-600/52-81-024, U.S. Environmental Protection Agency, Cincinnati, Ohio, 1981, 101.

78. **Reimers, R. S., Little, M. D., Englande, A. J., Leftwich, D. B., Bowman, D. D., and Wilkinson, R. F.,** Parasites in Southern Sludges and Disinfection by Standard Sludge Treatment, EPA-600/52-81-166, U.S. Environmental Protection Agency, Cincinnati, Ohio, 1981, 160.

79. **Orlob, G. T. and Butler, R. G.,** An Investigation of Sewage Spreading on Fine California Soils, SERL, Univ. Calif., Tech. Bull. No. 12, I.E.R. Ser. 37, Berkley, 1955, 30.

80. **Wang, C. H.** Disposal of Environmentally Hazardous Wastes, Task Force Rep. to Oregon State Department of Environmental Quality, December, 1974, Oregon State University, Corvallis, 1974, 210.

81. SCS Engineers, Feasibility of Inland Disposal of Dredged Material: A Literature Review, Environmental Effects Lab., U.S. Army Engin. Waterway Expt. Sta., Vilksburg, Miss., 1977, 210.

82. **Metcalf and Eddy, Inc.,** *Waste Water Engineering,* McGraw and Hill Book, Co., New York, N.Y., 1972, 9.

83. **Pound, C. E., Crites, R. W., and Olson, J. V.,** Long-term Effects of Land Application of Domestic Wastewater, Hollister, Calif., Rapid Infiltration Site, EPA-60/2-78-084, U.S. Environmental Protection Agency, R. S. Kerr Environ. Res. Lab., Ada, Okla., 1978, 150.

84. **Fuller, W. H.,** Investigation of Landfill Leachate Pollutant Attenuation by Soils, EPA-600/2-78-158, U.S. Environmental Protection Agency, MERL, Cincinnati, Ohio, 1978, 219.

85. U.S. Environmental Protection Agency, Third Report to Congress: Resource Recovery and Waste Reduction, Solid Waste Management Series Publication, SW-161, 1975, 10.

86. U.S. Environmental Protection Agency, Use of Water Balance Method for Predicting Leachate Generation from Solid Waste Disposal Sites, SW-168, U.S. Environmental Protection Agency, OSWMP, Washington, D.C., 1975, 40.

87. U.S. Environmental Protection Agency, Composting of Municipal Solid Waste in the United States, SW-47P, U.S. Environmental Protection Agency, Stock No. 5502-0033, 1971, 4.

88. **McCalla, T. M., Peterson, J. R., and Lue-Lunin, C.,** Properties of agricultural and municipal waste, in *Soils for Management of Organic Wastes and Wastewaters,* Elliott, L. F. and Stevenson, F. J., Eds., Amer. Soc. Agron., Madison, Wis., 1977, 11.

89. **Snyder, H. J., Rice, G. B., and Skujins, J. J.,** Disposal of waste oil re-refining residue by land farming, in *Residual Management by Land Disposal,* Fuller, W. H., Ed., EPA-600-9-76-015, U.S. Environmental Protection Agency, MERL, Cincinnati, Ohio, 1976, 195.

90. **Allison, F. E.,** Decomposition of wood and bark sawdust in soils, nitrogen requirements and effects on plants, USDA-ARS Tech. Bull., Sec. 1, 1332, 1965.

91. **White, R. K.,** Methods and techniques for sludge application on the land, in *Utilization of Municipal Sewage Sludge on Forested and Disturbed Land,* Sopper, W. E. and Kerr, S. N., Eds., The Pennsylvania State University Press, University Park, 1979, 285.

92. **Ayers, R. S.,** Water quality criteria for Agriculture, UC-Committee of Consultants, CWRCB, 1973, Sec. 1, 1.

93. California State Solid Waste Management Board, Resource Recovery Program, Vol. II, Sacramento, Calif., Sec. 1, 1, 1976.

94. **Sommers, L. E.,** Chemical Composition of sewage sludge and analyses of their potential use as a fertilizer, *J. Environ. Qual.,* 6, 225, 1977.

95. **Lindsay, W. L.,** *Chemical Equilibrium in Soils,* John Wiley and Sons, New York, 1979, 1.

96. **Page, A. L. and Chang, A. C.,** Trace elements and plant nutrient constraints of recycled sewages on agricultural land, in *Second Natl. Conf. on Water Reuse: Water's Interface with Energy, Air, and Solids,* Inst. Chem. Engin., and U.S. Environmental Protection Agency, Chicago, IL, Washington, D.C., 1975, 36.

97. **Page, A. L., Gleason, T. L., Smith, J. E. Jr., Iskandar, I. K., and Sommers, L. E.,** Utilization of wastewater and sludge on land, in Procedures Workshop, University of California, Riverside, 1983, 480.

98. U.S. Department of Agriculture Soil Conservation Service, *Natl. Soils Handbo.,* 430, 606, 1983.

99. U.S. Environmental Protection Agency, Hazardous Waste Management Systems; permitting requirements for disposal facilities, *Federal Register,* 143, 47, 1982.

Chapter 3

WASTE UTILIZATION ON NONCULTIVATED LAND

I. RATIONALE AND SCOPE

The usefulness of noncultivated lands as sites for waste disposal is not new. Historically, people and animals have sought the woods, bush, and grass to hide all kinds of wastes. Using the forest for solid waste and wastewater as a resource is a more modern concept. Many years of sewage farming in Europe and the United States demonstrate this practice to be a viable alternate to conventional municipal sewage treatment programs.[1,2] For this discussion, noncultivated lands may be described as being all land not tilled for commercial agricultural food, fiber, or forest products, as well as home gardens and urban landscaping. They include forestlands and woodlands (which constitute about one-third of the land area of the United States) rangeland and grassland used for animal grazing, wetlands, strip-mine lands, mine spoils, and wastelands. The potential advantages of these lands for waste utilization over other lands and forms of disposal, as a part of the land treatment option, are manifold.[3-6]

A. Advantages and Disadvantages

Noncultivated lands compared to cultivated lands are

- Usually removed from human food producing areas.
- Usually located in areas less sensitive to public objections related to health or other concerns of social significance.
- More easily managed, often requiring a minimum of attention.
- Relatively low in plant nutrients, particularly N, P, and K, which may be readily absorbed by plants from wastes containing them.
- Characterized by fewer interruptions in treatment.
- More readily adaptable to manufacturing operations that require close access to large acreages.
- Adaptable to the dual purpose of waste disposal and waste utilization systems.
- Particularly well suited for recharge of groundwater.
- Associated with relatively long cropping periods.
- More capable of being upgraded in quality and land uses.
- Less sensitive to excesses of disposal since many noncultured lands are not expected to return high level income per acre basis and, therefore, disposals.
- Less susceptible to the demands of cost-effective land returns.

Noncultivated lands also may have limitations and disadvantages:

- Some may be so remote from manufacturing and municipalities that they cannot be used.
- Much of wasteland is on rough terrain and steep slopes, or is stony, rocky and near clean streams, lakes, and rivers.
- Environmental impact on wildlife, fish, and game may be so unfavorable as to cause some areas to be unsuited for certain disposals.
- Vegetation (forests, shrubs, brouse, and grass lands) on nonagricultural land has growth and survival limits just as does that of cultivated land.
- Some soils also have severe loading limitations for leaking pollutants to shallow underground water sources.

- Noncultivated land may be used for grazing, thus, establishing a link directly into the food chain. This may require more restrictive disposal management of wastes and more stringent monitoring of animal food products than otherwise would be necessary.
- Soil and water erosion can be more serious on steep and rough topography, making it essential to limit the size and extent of the operation to prevent sediments from reaching surface waters.
- Forest land and other virgin wildernesses, and even wasteland, evoke a pristine image that may make waste disposal unacceptable to some for fear that such use would mar its unspoiled appearance.

Some soils clearly are not well suited for disposals, and the risks are great if they fail to meet minimum requirements, as for example:

- Shallow soils — Less than 60 cm (2 ft) to bedrock
- Steep Slopes — In excess of 12%
- Excessive rate of permeability — Sands, gravels, and stones
- High water table — Less than 60 cm (2 ft)
- Subject to flooding, ponding, and erosion
- Restrictive layers of soil or rock impervious to water
- Locations over active aquifers.

Fortunately, not all noncultivated lands share such disadvantages; and there are large acreages available free of these limitations.

B. Kinds of Wastes Adapted to Uncultivated Land

The kinds of wastes acceptable for disposal on noncultivated lands may be enlarged and loading rates extended somewhat as compared with cultivated land treatment. Yet, except for small areas, both support vegetation. Therefore, both lands must share in similar constraints as to acceptable waste sources. The cultural and human health restraints against the application of certain wastes to crops on cultivated land often are relaxed on noncultivated land. For example, certain insecticide, pharmaceutical, and organic chemicals may find suitable disposal on forest, range, and grasslands at modest loading rates. Odorous and offensive wastes unacceptable for disposal near urban metropolitan areas may be well adapted to rural lands further removed from population centers and suburban homes.

The differences in kinds of wastes acceptable for noncultivated lands are associated with great diversity in types of (a) vegetation, (b) soils, (c) equipment traffic, (d) interruptions of land cultural practices, (e) game and fish, (f) topography, (g) micro-climate, and (h) remoteness from human habitation. A brief summary of some types of industrial effluents applied to land is given in Table 1. In addition, Table 2 lists some possible solid wastes amenable to land utilization. Before a complete list of recommended industrial wastes acceptable for utilization on noncultivated and disturbed lands can be developed, a great number of details on long-term effects on the ecosystems and environment must be obtained. Certainly the capacity of the soil to alter the water quality over long periods of time must be ascertained and a quantitative restraint developed. Efforts in this direction have been made by personnel at the School of Forest Resources, the Institute for Research on Land and Waste Resources, The Pennsylvania State University, University Park, Pa., and the U.S. EPA, R. S. Kerr Environmental Research Laboratory, Ada, Okla. This chapter relates to those wastes from which the vegetation and soils derive some benefit, i.e., waste as a resource.

C. Land Types Most Suited

The land types most suitable for waste utilization are those having only slight slopes and moderate soil textures (such as loams). Excessively sandy or clayey soils should be avoided,

Table 1
SUMMARY OF TYPES OF INDUSTRIAL EFFLUENTS DISPOSED OF ON LAND

Industry	Industry, continued
Food processing	Pulp and paper
Canning, frozen and convenience foods	Sulfite
Vegetables, including baby food	Kraft (sulfate)
Soup	Semichemical
Fruit, except citrus	Strawboard
Instant coffee and tea	Groundwood
Instant coffee and tea	Hardboard and insulation
Dairy products	Boxboard and paperboard
Milk plants	Deinking
Cheese (whey)	
Meat products	Other
Miscellaneous	Tanning
Potato processing	Textiles
Sugar beets	Pharmaceuticals
Starch plants	Biological chemicals
Wine	Explosives
Distilling (spirits)	Wood distillation
	Rope and hemp
Petroleum wastes	Coal—tar chemicals
	Cooling water from aluminum casting

Table 2
SELECTED SOLID INDUSTRIAL WASTES APPARENTLY AMENABLE TO NON-CULTIVATED LAND TREATMENT/UTILIZATION

Petroleum industry wastes	Organic solvents/residues
oils, greases, asphalts	degreasers
salty waters	laundry cleaning
	certain hospital waste
Acidic and caustic wastes	paint industry
metal brightening acids	
mining acid wastes	Municipal solid wastes (MSW)
electroplating acids	shredded MSW
caustic residues	composted MSW
	sewage sludges
Electronic industry wastes	
acidic and caustic	Woody and paper wastes
cyanide wastes	sawdust
noble metal residues	bark
	shredded paper
Agricultural wastes	waste pulp
animal manures	urban pruning waste
pruning	grass clippings
crop residue	
composted residues	Solid canning wastes
	fruit
Fly ash	vegetable
	meat, poultry, and fish scraps

except for specialized uses where these soils fit into a proven plan of disposal. Lands suitable for solid wastes, for example, are not necessarily suitable for wastewaters. Wastewater disposal often is less demanding for suitable sites. Since our discussion has turned from cultivated to noncultivated land, the restrictions of tillability assumes less importance. Certainly the forest and woodlands limit the use of soil tillage equipment for incorporation. It should not be implied, however, that applications on noncultivated land are restricted to the surfaces only. Some of the most suitable noncultivated soils for waste application and resource recovery are those too marginal in quality, i.e., are sloping, small, irregular, saline (soil reclamation), droughty, gravelly, stony, or too wet for commercial agricultural production. Should these soils be sufficiently deep to accept tillage equipment, soil incorporation methods provide certain real advantages for solid waste disposal. Fortunately, the use of soil incorporation techniques are not necessary for most wastewater disposal on noncultivated land.

The acceptance of noncultivated land for waste utilization will depend largely on the specific nature of the waste and desired loading rate of the waste being applied. Methods of application of both solid waste and wastewater will depend on the nature of the specific terrain together with the soil and underlying layers as will be explained later in this chapter under headings of loading rates.

The establishment and maintenance of precise loading rates on noncultivated land well within the limits of groundwater pollution, may not appear to be as necessary in remote areas as on land near population centers and country homes. Yet, remoteness is an out-dated concept. There are few regions of the United States so remote that some habitation does not exist. Should these few exceptions become polluted, another generation will be facing the same problems some of us now face — of habitation migration. Another difference in land treatment of cultivated and noncultivated land centers on the concept that those wastes having the greatest value as resources for soil improvement should be placed on the most valuable land or land that will return the greatest immediate profit. This concept permeate the literature, and indeed may seem to do so in this discussion, although not intended.

II. SOIL MANAGEMENT OF SOLID WASTES

The estimated solid wastes of industrial origin constitute only 10% of the total waste generated. The bulk is liquid. Solid waste application to noncultivated land, except manures, is in its infancy as a generally acceptable practice, specifically for resource recovery. The potential appears to be great, but experience is lacking and knowledge of the best techniques on ways to use untilled land is lacking. Except in a few instances, experience has been confined primarily to uncontrolled spreading rather than with controlled and monitored designs. In those few instances of more quantitative research, concentration of effort has centered on municipal sewage sludge, animal wastes, and composts from sludge, some woody residues and municipal solid waste. More recently, petroleum and fossil hydrocarbon/oily wastes have received controlled experimental attention, although the greatest effort involves landfarming demonstrations as opposed to controlled quantitative experimentation.

A. Methods of Application

The lack of suitable roads, bridges, and utilities limit access into much of the noncultivated lands. The transportation and application of industrial wastes to these areas are seriously limited. Solid waste disposals, therefore, have at times improvised with available farm equipment such as manure spreaders, trucks, wagons and bulldozers. Sophisticated application methods and equipment for solid waste loading remains yet to be developed for noncultivated lands. The technique used for land application of manure, municipal sludge, and food processing wastes, whether solids or slurries, appears to be much the same. Solid wastes are defined here as those that do not flow hydraulically in contrast to liquid wastes that behave hydraulically, in much the same manner as water.[7] Large-bore irrigation nozzles have been adapted to apply heavy slurries. They can deliver 350 to 1500 ℓ/min (100 to 400 gal/min) over an area of 0.2 to 1 hr (0.5 to 2.2 a) per setting. Shredded municipal refuse and sludge have been applied by several types of methods including dumping and spreading with a dozer blade. Compost and compost-sludge mixtures have also employed truck, dozer, and blade spreading on noncultivated land in a number of experiments throughout the United States. Application of solid wastes to western U.S. rangeland offers great opportunity for new innovations in techniques and equipment because of large areas of available land, favorable topography, many suitable soils, and remoteness from population densities.

B. Placement or Design

Solid-waste placements are made either by spreading on the surface or by incorporation into soil. Surface placement is the least expensive method of disposal but has some possible disadvantages: (a) susceptibility to wind and water erosion, (b) attraction to flies, insects, vermin, and rodents, (c) slower biodegradation, (d) less acceptability by society, and (e) less apt to be attenuated by the soil.

Pretreatment of solid wastes may be necessary to eliminate, or at least modify, some of these disadvantages. For example, solid municipal refuse requires the pretreatment of shredding and grinding accompanied by air elution or classification to remove some heavy particles of metal and glass. Composting is another pretreatment option after removal of large pieces of glass and metal and other undesirables. Surface placement cannot successfully accommodate MSW solid waste without pretreatment of shredding and size reduction. Because of its diverse nature, subsurface incorporation is much more aesthetically acceptable. Other pretreatment possibilities desirable for solid wastes before placement on the land or into the soil include (a) size and volume reduction by changing its physical form for ease of handling, (b) storing for a short period to eliminate obnoxious odors, gases, and aerosols through aeration and composting, (c) storing to make it available at a more convenient climatic period, and (d) mixing with other wastes that favor disposal of the final product over separate disposals, e.g., sewage sludge with municipal solid waste.

Some suggested pretreatments for more successful placement are listed in summary:

- Sorting
- Shredding and grinding
- Air classification
- Incineration
- Storage
- Composting
- Drying
- Mixing with fertilization and/or soil amendments (e.g., lime, gypsum)

For those soils and landslopes that permit subsurface placement, a number of attractive options are available. The "Plow-furrow-cover" method deposits solid waste at depths of 15 to 30 cm (6 to 12 in.). The "subsod injection" equipment places slurry waste just below the soil surface with minimal sod disturbance. Placement on pastures, rangeland, and mine spoil favors the subsurface technique.

Trench incorporation of solid organic wastes combines the possibility of disposal of relatively large amounts of waste with the desirable features of subsurface placement.[8] Trench incorporation appears to be well suited for solid waste application to marginal agricultural and disturbed land. For example, municipal sewage sludge rates of 800 and 1200 t/ha dry solids were placed in trenches 60 × 60 cm and 60 cm apart and others 60 cm wide by 120 cm deep by 120 cm apart successfully on marginal agricultural land.[8] Movement of heavy metals away from the entrenchment was not detected and fecal coliform and salmonella were confined to within a few centimeters of soil after 19 months of placement.

C. Solid Waste Loading Rates

Loading rates of solid residues on noncultivated land can be as great, if not greater than those suggested for cultivated soils for the same material, with all other factors being

Table 3
ANNUAL MANURE APPLICATION RATES
NEEDED TO INSURE 200 kg OF AVAILABLE
N/ha WHICH COMPENSATES FOR
RESIDUAL EFFECTS

Nitrogen in manure %	Year of application					
	1st	2nd	3rd	5th	10th	20th
	metric tons/ha					
3.5	7.6	7.2	7.0	6.8	6.6	6.2
2.5	20.0	14.6	13.8	12.6	11.0	9.4
1.5	38.2	29.6	26.2	24.2	20	16.8
1.0	100.0	71.4	62.6	52.6	38.6	27.8

Powers, W. L., Wallingford, G., Murphy, L. S., Whitney, D. A., Manges, H. C., and Jones, H. E., Coop. Ext. Serv. Bull., Kansas State University, Manhattan 1974, 11. With permission.

reasonably similar. Characteristically noncultivated lands may not be tilled for agricultural crops for good reason. They may have excessive slope; erodable, stony or shallow soil; wet soils with high water tables, and numerous other easily recognizable problems. Such landscape and soil limitations understandably influence the loading rates of wastes. Subsurface applications may be made at considerably greater loadings in contrast to surface loadings since there is much more control over the nuisance of flies, odors, and insects, and over pollution spreading through surface water runoff soil erosion, and wind action. As described by Reed,[9] for example, dairy manure was loaded successfully at rates of 380 to 500 t/ha (170 to 225 T/a) to a medium fine-textured soil by the plow-furrow-cover method.

Loading rates are influenced by the water content of the solid waste thus pretreatment such as composting or dewatering may prove cost saving. For example, in small municipal units, anaerobically digested sludge may be dewatered in sand drying beds to a desirable consistency for surface application with a manure spreader, thus saving cost of subsurface disposal. Loading rates on nonagricultural land must balance the benefits derived in terms of soil and vegetation quality as a resource with the cost of disposal of a waste solely to be eliminated.

Length of time during which applications can be made to one piece of land is also a consideration in loading rates, particularly as it affects available N. Nitrogen is both a necessary plant nutrient and a pollutant of groundwater for drinking (in excess of 10 ppm). Thus, loading rates should be made to balance out these two opposing effects over a long period of time as illustrated in Table 3.[10] The residual N in the waste is shown to vary both with time and original N concentration. The lower the N content of the material, the greater the acceptable loading rate, but greater also is the percentage decrease each succeeding year as compared with material higher in N.

Another factor in planning long time annual loading rates of organic wastes (resources) to land is the rate of decomposition or rate of biodegradation. Rate of biodegradation can be determined and plotted fairly accurately with the help of a few parameters of the system and a theoretical model.[11] A model output for annual (June) additions of municipal sewage sludge at four different sites where the ratio of carbon (C) accumulation to one annual carbon addition is plotted as a function of time (Figure 1). Ratios after 50 years of sludge addition for the four sites are 5.9, 4.3, 7.7, and 7.9 from site 1 to 4, respectively. The characteristic shape of the curve is parabolic and flattens to a near plateau as time progresses. Site 2 represents the nearest to optimal soil moisture and temperature conditions for quick bio-

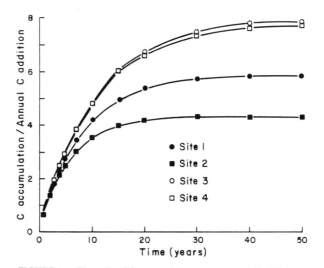

FIGURE 1. The ratio of C accumulation to one annual C addition as a function of time for yearly additions of sludge. (From Gilmour, J. T. and Gilmour, C. M., Simulation model for sludge decomposition in soil, *J. Environ. Qual.*, 9, 198, 1980. By permission of the Soil Science Society of America.)

degradation. Sites 3 and 4 represent suboptimal conditions and the largest accumulations. Using the equations below, Gilmour and Gilmour computed residue accumulated after many annual additions of a decomposable material to a soil:[11]

$$\log f_1 = (-0.30103 \ t)/t_{1/2} \tag{1}$$

$$r = C_o \left[1/(1 - f_1) \right] - C_o \tag{2}$$

where f_1 is the fraction left after 12 months, t is 12 months, $t_{1/2}$ is the half-life in months, r is the accumulated residue prior to an annual addition, and C_o is the initial concentration. For r to be the ratio used in Figure 1, C_o was set equal to 1. The ratios computed using Equations (1) and (2) for a June addition of sludge were 5.7, 4.3, 7.1, and 7.1 for sites 1, 2, 3, and 4, respectively. They conclude that:[11]

"First, time of addition had only a small effect on overall decomposition suggesting that no large advantage with respect to decomposition would be gained by amending soil with sludge during a given month. Second, estimates of sludge decomposition based on yearly average data could lead to errors in predicting sludge decomposition. Third, laboratory studies conducted under "optimum" conditions should be considered over estimates of field decomposition rate when field soil temperatures and water contents diverge from the "optimum". And, fourth, computations of sludge accumulation due to many annual additions should be evaluated with care where suboptimal soil conditions for the decomposition process exist. These results suggest that in many cases detailed information on the soil environment may be needed in addition to the detailed information usually collected on the sludge if accurate estimates of decomposition of sludge are to be obtained."

Precise loading rates cannot be provided as in a general recommendation for all lands. The variations in land and soil type, in macro- and microclimate, geology and hydrology, and quality of even a single waste source and moisture level, are so diverse, that loading rates must be determined on a case-by-case basis. However, we can give some rates as a guide toward the development of a design for disposal at a specific site and for a specific type of waste.

1. Shredded Municipal Solid Waste

Shredded municipal solid waste (MSW) loading rates near Odessa, Texas, ranged from 134 to 278 t/ha (60 to 124 T/a) in a single application. The purpose was to improve the soil fertility and organic matter content and reduce waste disposal cost.[12] About 60% of the ferrous metal had been removed. First year results failed to demonstrate any adverse effects on the native vegetation when the waste was applied to the nearly level rangeland surface. In fact, vegetative grass cover was improved.

In an adjacent area, loading rates of 78 to 278 t/ha (35 to 124 T/a) were tested with a predominant loading rate of 112 t/ha (50 T/a). The MSW on this rangeland was rototilled to a depth of about 256 cm (100 in.) and rolled to compact. Applications were made throughout the year. Except for slight increases in the B, Pb, Na, and Mn content of the wild buckwheat, the vegetation was not altered appreciably the first few years and apparently was suitable for cattle feed. Because of the dry climate, biodegradation of some waste was slow. Cardboard, plastic, leather, and metals presented the greatest problem.

According to Halvorson[13] and Cattrell,[14] both of Oregon, loading rates of municipal solid waste should not exceed 448 t/ha (200 T/a) on a sandy soil under their climates.

2. Sewage Sludge and Septic Tank Waste

Solid wastes in the form of municipal sewage and septic tank slurry vary in rates of loading on land, depending in part, on whether or not they are mixed with other organic solid wastes such as shredded MSW, wood chips, compost, or strawy manure which effects their physical and metal content (Table 4). Composted sewage sludge will be discussed under Compost.

When used alone on idle land in Michigan, municipal sludge was loaded at rates ranging from 7.6 to 22 t/ha (dry weight) (or 3 to 10 T/a) for the purpose of improving soil quality and eliminating odors in a disposal program. The topography was flat to rolling hillsides with natural grass sod being the predominate vegetation.[15] These rates appeared suitable for sod production under Michigan climatic conditions. This same type of sludge was applied to aspen and poplar on a tree farm also with apparent success. According to Phung[12] experimental loadings ranging from 7 to 54 t/ha (3 to 24 T/a) on a dry weight basis were tested for long-term effects by the U.S. Forest Service. The vegetation was not adversely affected the first year.

Sandy soils offer some loading problems because of their rapid permeability, acidic character in humid regions, low attenuating capacity, low organic matter content, and limited water-holding capacity. They require precautionary management practices for land loading wastes to avoid groundwater pollution and site degradation.[16] Conversely, sandy sites free of water table problems and not associated with underground aquifers have certain desirable features; contact with the traffic of developing communities may be less in remote areas of low productivity, sludge and other wastes can be loaded throughout the year because sand supports heavy equipment during wet periods better than clayey soils, and improvement of the soil fertility and organic matter content is a more realistic goal than in finer textured soils. Over 40 municipalities in Michigan, alone, use sandy soils for sludge and effluent disposals. In a study by Harris,[16] sludge loadings on loamy sand were made at rates of 7.5, 15, and 23 dry t/ha on one aspen stand and 11.5, 22.3, and 46 dry t/ha (5.1, 10.3, and

Table 4
ANALYSES OF COMPOSTED REFUSE AND FRESHLY SHREDDED REFUSE

Constituent	Japan[a]	Florida[b]	Tennessee[c]	Washington[d]	Ontario[e]
			% dry-weight basis		
Total N	1.0	1.2	1.3	0.78	0.57
P	0.24	0.26	0.26	0.16	0.08
K	0.98	0.38	0.97	0.27	0.314
Ca	5.1	1.30	4.6	1.97	0.850
Mg		0.07	0.5	0.11	0.209
Na			0.6	0.26	0.187
Fe				0.51	
S			0.5		
Total C	28		27	48	
Organic C					37
			g/g dry-weight basis		
Mn		130		242	250
B		25		58-62	
Cu		125	100	96	28
Zn		250	1,500	715	400
Pb					200
Cd					7
Cr					31

[a] Egawa, 1974. Composted refuse.
[b] Hortenstine and Rothwell, 1968. Composted refuse.
[c] Terman et al., 1973. Compost may contain up to 20% sewage sludge.
[d] Volk and Ullery, 1972 (unpublished report). Freshly shredded refuse >2-mm fraction.
[e] King et al., 1974. Freshly shredded unsorted refuse.

Elliot, L. F. and Stevenson, F. J., *Soils for Management of Organic Wastes and Waste Waters*, American Society of Agronomy, Madison, Wis., 1977, 650. With permission.

20.5 T/a) on another. Soil data as reproduced in Table 5 shows that the nutrient content of the 0 to 5-cm soil layer was enhanced by the sludge loadings. Seasonal loadings of sludge and effluent differ primarily as a result of differences in seasonal rainfall distribution patterns during the early biodigestion period.

Nitrogen content of sludge frequently is used to determine total sludge loading rates on nonagricultural land. Nitrogen as nitrite and nitrate move with the soil water since the soil retains them poorly. The less mobile NH_4^+ ion is retained better against rapid migration but is transformed to nitrate readily in aerobic soils. Loading rates of 92 kg/ha to 117 kg/ha of total N are considered to be modest for sewage sludge on a dry weight basis. Unless the soil is very sandy and the water table is near the soil surface, groundwater pollution should not occur within the 10 ppm limit for drinking water standards[17] at these loading rates for municipal sewage sludge. Density and type of vegetation affects the nitrate in the percolating water concentration. For example, before clean-cutting all trees and removing them, the nitrate-nitrogen concentration at the 120 cm (47 in.) soil depth was 28.2 mg/ℓ in 1969. After "clean-cutting", undergrowth vegetation increased and the nitrate-nitrogen decreased 1 and 2 years later to 8.3 and 2.9 mg/ℓ, respectively.[18]

In general, municipal sewage sludge application to noncultivated, uninhabited land offers few serious problems except where nitrates and heavy metals in industrial waste are more highly concentrated.

Table 5

AVERAGE NUTRIENT CONCENTRATIONS IN THE 0- TO 5-cm SOIL LAYER AT ASPEN SITE WITH SLUDGE FERTILIZATION

Plot	Application Rate (t/ha)	Kjeldahl N (%)	Total P (ppm)	ECE (meq/100 g)	Base Saturation (%)	Organic Matter[a] (%)	Humus (t/ha)	Humus[b] pH
Control	0.0	0.095	182	11.83	31	4.2	171	5.7
Sludge applied fall 1975	7.5	0.110	233	11.92	29	4.4	100	5.7
	15.0	0.097	230	10.89	32	3.7	108	5.8
	23.0	0.081	166	8.75	33	3.4	140	5.7
Sludge applied spring 1976	11.5	0.115	248	13.00	37	4.7	87	6.2
	23.0	0.100	244	12.33	45	5.0	90	6.2
	46.0	0.136	325	14.73	41	5.1	110	6.4

[a]   Loss on ignition.
[b]   Above soil surface.

Harris, A. R., *Utilization of Municipal Sewage Effluent and Sludge on Forested and Disturbed Land*, The Pennsylvania State University Press, University Park, 1979, 155. With permission.

Table 6
COMPOSITION OF RAW AND DIGESTED SLUDGES FROM THE WASHINGTON, D.C., BLUE PLAINS WASTEWATER TREATMENT PLANT, AND THEIR RESPECTIVE COMPOSTS PROCESSED AT THE USDA COMPOSTING FACILITY, BELTSVILLE, MD.

Component	Raw Sludge	Raw Sludge Compost	Digested Sludge	Digested Sludge Compost
pH	5.7	6.8	6.5	6.8
Water, %	78	35	76	35
Organic carbon, %	31	23	24	13
Total N, %	3.8	1.6	2.3	0.9
$NH_4^+ - N$, ppm	1540	235	1210	190
Phosphorus, %	1.5	1.0	2.2	1.1
Potassium, %	0.2	0.2	0.2	0.1
Calcium, %	1.4	1.4	2.0	2.0
Zinc, ppm	980	770	1760	1000
Copper, ppm	420	300	725	250
Cadmium, ppm	10	8	19	9
Nickel, ppm	85	55	—	—
Lead, ppm	425	290	575	320
PCB,[a] ppm	0.24	0.17	0.24	0.25
BHC,[b] ppm	1.22	0.10	0.13	0.05
DDE,[c] ppm	0.01	0.01	—	0.008
DDT, ppm	0.06	0.02	—	0.06

[a]   Polychlorinated biphenyls as Arochlor 1254.
[b]   The gamma isomer of benzene hexachloride is also called lindane.
[c]   DDE results from the dehydrochlorination of DDT.

Epstein, E. and Parr, J. P., *Waste Management Technology and Resource and Energy Recovery*, 1977, 321.

3. Compost

Composts are made from many wastes of plant, animal, and human origin. The loading rate of compost to noncultivated land is almost unlimited except where the landscape and soils offer serious problems. Rates may range from 37 dry t/ha[19] to 188 t/ha[20] per year without causing serious soil productivity. When municipal compost containing 20% sewage sludge was loaded at rates varying from 23 to 327 t/ha over two years, nitrogen was still needed for seeded grass (*Cynodon dactylon* L.).[21] Municipal compost has been effective in controlling erosion and runoff under western European climatic conditions.[22]

Municipal solid waste compost developed through research by the Tennessee Valley Authority applied to a pine woods on sandy soil in Florida at a loading rate of 45 t/ha, showed no adverse or beneficial tree growth responses.[23] Metal concentration (metal iron removed) in pine foliage and ground vegetation was little changed by municipal compost loading rates of 224 and 488 wet t/ha on a poorly drained Florida *Spodosol*.[5] Filtered sewage sludge, composted by Epstein and Wilson[24] in a USDA research program at Beltsville, Md., was found to be highly suitable for cultivated as well as noncultivated land at rates up to 500 t/ha. Considerable modification in the sludges takes place as a result of composting (Table 6).[25]

4. Wood and Paper Mill Wastes

Wood wastes free of certain toxic preservatives, may be applied to noncultivated lands with a greater range of rates than on cultivated lands; although loading rates of woody

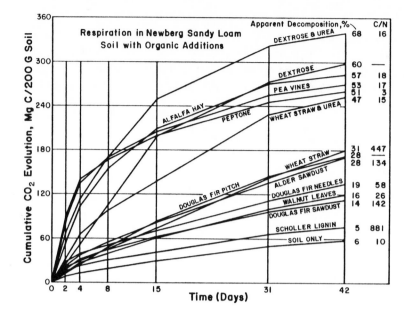

FIGURE 2.   Respiration in Newbery sandy loam soil with organic additions. (From Bollen, W. L. and Lu, K. C., Effects of Douglas-fir sawdust mulches and incorporation on soil microbial activity and plant growth, *Soil Sci. Soc. Amer. Proc.*, 21, 35, 1957. By permission of the Soil Science Society of America.)

products, once considered to be reasonably unlimited, have limitations even in some remote forested areas. Disposal of large quantities of bark, sawdust, and forest waste in lagoons and ponds may release undesirable wood turpentines (pitches), tannic acid, and lignin substances to the waters. When such water was allowed to infiltrate to the ground water, surrounding domestic wells became polluted with foul tastes and odors, thus being quite socially unacceptable. Bark waste in ponds can create undesirable tastes and colors, e.g., pale yellow to brown, depending on the tree source. Ponding, lagooning, and piling of natural woody wastes at high point-source concentrations should be discouraged where the soil is porous, underground water sources are shallow, and domestic wells are located near the disposal area. Other than these unusual nuisances, loading rates may be heavier than for cultivated land without fear of heavy metal, pathogens, salts, or chemical pollution affecting the underground water or the food chain.

Cotton gin-mill waste is woody in nature and somewhat similar in rate of biodegradation to some tree wastes. Gin-mill waste has been applied to infiltration-percolation basins as a mulch to aid in maintaining infiltration rates with some success in California and Arizona.

Douglas fir sawdust applied at a depth of 12.7 cm (5 in.) to forest soils without addition of nitrogen depressed growth of seeded grasses.[26] With added nitrogen, however, growth was enhanced. The depression was due to the wide C/N ratio and not to the presence of toxic compounds. The respiration data in Figure 2 appears to confirm this conclusion.[26]

Paper mill sludges, although variable in composition, and plant nutrients as discussed in Chapter 2, generally are low in nutrients, particularly nitrogen. A C/N ratio of 150 is not uncommon. In one study, not until the C/N ratio was brought to 12 was there maximum decomposition and release of nutrients.[27] Loading rates on noncultivated land, therefore, depend more on BOD or TOC than any other factor.

5. Tannery Wastes

Some residues from leather tanning and finishing industry are considered to be potentially hazardous. This is largely due to the use of certain heavy metals (chromium, lead, and zinc)

Table 7
ELEMENTAL COMPOSITION OF THE
TANNERY WASTE AT TWO
SAMPLING DATES: PENNSYLVANIA
SITE[a]

	Sampling Date	
Element	**August 1975[b]**	**September 1977[c]**
	%	
N	—	6.75
Organic C	—	84.8
Al	0.80	—
Si	0.41	—
Ba	0.63	—
Cr	4.76	2.25
Ti	1.60	—
Fe	0.31	—
	μg/g	
Ca	530	—
Cu	13	53.1
Cd	—	2.12
Co	—	1.41
Ni	—	5.0
Pb	180	133
Mn	51	13.5
Zn	700	159
Se	—	0.056
S	—	983
Sr	82	—

[a]   Concentrations expressed on dry weight basis.
[b]   Unpublished data, SCS Engineers. Results are from spectrographic analyses of a grab sample of the buffing dust scrubber slurry.
[c]   A grab sample of the dewatered sludge taken from the field.

Phung, T., Barker, L., Ross, D., Bauer, D., U.S. Environmental Protection Agency, Cincinnati, Ohio, 1978, 205.

in the final leather processing. These metals, which are usually in relatively high concentrations, limit the usefulness and rate of application of some tannery wastes. One site in eastern Pennsylvania described as a 12-ha (30 a) plot of 1 to 5% slopes and having a moderately permeable, well-drained silt loam, has been in land treatment operation for an extended period. The solid waste (scrubbing slurry, buffering dust) was applied between 1961 to 67 with the waste waters at a loading rate of 3.78×10^4 ℓ/d (1000 g/d). The composition is given in Table 7. Between 1967 to 1972 only scrubber slurries and buffering dusts were applied. Since 1972, only buffering dust has been land treated. Placement was on the surface with some shallow soil mixing using a disc. Neither the 0.1 N HCl — or aqueous — extractable heavy metals were significantly greater in the surface layer of the soil than in untreated soil. The loading rates were probably not sufficiently high enough to permit accumulation of significantly soluble heavy metals despite the low soil pH (5.3 to 6.4).

The leather tanning and finishing industrial solid wastes also produce (a) solvent-based finished residues, (b) blue trim and shavings, and (c) leather trimmings. The solvent is recovered from solvent-based finishing residues leaving a finishing residue of small proportions. Blue trim and shavings has been applied onto fruit orchard soil surfaces for its fertilizer value, principally nitrogen, and for soil conditioning. Leather trimmings do not usually reach land treatment sites, but are recovered for other leather uses. Loading rates of leather materials are relatively light since the quantities of waste generated are not large.

6. Cannery Wastes

The loading rate of cannery wastes to noncultivated land does not lend itself to general figures. The quality and composition varies from cannery to cannery and produce to produce. Loading rates probably should be calculated on BOD, COD, or TOC level per unit volume and the soil and topography characteristics of the noncultivated land site. Most cannery wastes may be expected to contain more N-P-K plant nutrients than paper mill sludge, a narrower C/N ratio, less total salts and a low heavy-metal content. Most of the cannery wastes are liquids and are adapt to irrigation procedures that relate to hydraulic loading rate criteria, even in forest land.[5] Generally, solid cannery loading rates are relatively high compared with municipal sludges.

7. Acid Wastes

Disposal of strong acids (and caustics) on noncultivated land as a land treatment (utilization) system has a greater chance of acceptability than on cultivated land. Because of the high hydrogen ion activity of acids, however, loading rates necessarily must be light and closely monitored. Most acid waste streams, if they are inorganic, contain significant quantities of potentially hazardous pollutants like heavy metals. If the acid is organic, the soluble carbon is high and readily biodegradable at favorable pH values. Therefore, it harbors a serious BOD problem, as is exemplified by acetic acid. Pretreatment to neutralize the acidity before application to the land is required because the high hydrogen ion activity (acidity) is not tolerated by living systems and is destructive to the soil. Lime ($CaCO_3$) neutralizes acids most effectively because the anion rises as a harmless gas into the atmosphere and calcium is tolerated by plants better than any other cation. In certain cases, calcium hydroxide ($Ca(OH)_2$) may be preferred to lime for reasons of solubility. Calcium also forms slowly soluble salts with acids as well as, if not better than, most common cations in the soil. Loading rates will, therefore, depend on the kind of acid and residual constituents formed. Neutralized acids leave salts that differ markedly in their pollution potential. For example, the calcium salts of nitric, sulfuric, phosphoric, and acetic acid have reactions ranging from virtually nonpolluting (e.g., gypsum) to highly polluting (e.g., calcium nitrate). Loading rates, therefore, must be developed on a highly individual basis. Only very small levels of calcium nitrate can be applied to the land because of the tendency of nitrates to move with water. Neutralized sulfuric acid, gypsum, is a common, harmless constituent in most soils of arid lands including productive soils. Calcium acetate has a high BOD loading rate factor and, therefore, should be applied to noncultivated land on a basis of BOD or hydraulic loading rate, whichever is more limiting.

8. Oils and Oily Wastes

The composition of petroleum refinery waste, like other industrial waste streams, is highly variable (Table 8).   According to some definitions, refinery wastes are classed as solids although most have some fluid properties. Water and soil (sand, silt, and clay) are commonly associated with refinery sludges as are heavy metals and a variety of organic compounds. The oil content ranges from 1 to 82% by weight (Table 8).   Some wastes are highly alkaline and some acid. Not all petroleum refinery wastes are suitable for land utilization, although most of them may be disposed of by land treatment if loading rates are restricted to kind and amount of toxic metal present in the waste. Waste biosludge, tank bottom, and API separator sludge respond well to land spreading.[12]

Table 8

RANGES OF CONCENTRATIONS AND TOTAL QUANTITIES FOR REFINERY SOLID WASTE SOURCES (ALL VALUES IN MILLIGRAM PER KILOGRAM EXCEPT WHERE NOTED)

Parameters	Sludge from Clarified Once Through Cooling Water	Exchange Bundle Clearing Sludge	Slop Oil Emulsion Solids	Cooling Tower Sludge	API/Primary Clarifier-Separator Bottom	Dissolved Air Flotation Float	Kerosene Filter Clays	Lube Oil Filter Clays	Waste Biosludge
Phenols	0.0—2.1	8—18.5	5.7—68	0.6—7.0	3.8—156.7	3.0—210	2.0—25.2	0.05—6.4	1.7—10.2
Cyanide	0.01—0.74	0.0004—3.3	0—4.6	0—14	0—43.8	0.01—1.1	Trace	0.01—0.22	0—19.5
Selenium (Se)	0.1—1.7	2.4—52	0.1—6.7	0—2.4	0—7.6	0.1—4.2	0.01—26.1	0.1—2.1	0.01—5.4
Arsenic (As)	0.1—18	10.2—11	2.5—23.5	0.7—21	0.1—32	0.1—10.5	0.09—14	0.05—1.4	1.0—0.6
Mercury (Hg)	0.42—1.34	0.14—3.6	0—12.2	0—0.1	0.04—7.2	0.07—0.89	0—0.05	0.04—0.33	0—1.28
Beryllium (Be)	0.013—0.63	0.05—0.34	0—0.5	Trace	0—0.43	0—0.25	0.025—0.35	0.025—0.5	Trace
Vanadium (V)	15—57	0.7—50	0.12—75	0.12—42	0.5—48.5	0.05—0.1	13.2—42	0.5—65	0.12—5
Chromium (Cr)	16.6—103	310—311	0.1—1325	181—1750	0.1—6790	2.8—260	0.9—25.8	1.3—45	0.05—475
Cobalt (Co)	5.5—11.2	0.2—30	0.1—82.5	0.38—7	0.1—26.2	0.13—85.2	0.4—2.3	1.3—5	0.05—1.4
Nickel (Ni)	20.5—39	61—170	2.5—288	0.25—50	0.25—150.4	0.025—15	0.025—15	0.25—22	0.013—11.3
Copper (Cu)	56—180	67—75	8.5—111.5	49—363	2.5—550	0.05—21.3	0.4—12,328	0.5—8.0	1.5—11.5
Zinc (Zn)	93—233	91—297	60—656	118—1,100	25—6,596	10—1,825	6.6—35	0.5—115	3.3—225
Silver (Ag)	0.84—1.3	Trace	0—20.1	0.01—1.6	0.05—3	0—2.8	0.02—0.7	0.013—1.0	0.1—0.5
Cadmium (Cd)	0—1.0	1.0—1.5	0.025—0.19	0.06—0.6	0.024—2.0	0—0.5	0.19—0.4	0.025—1.5	0.16—0.54
Lead (Pb)	17.2—138	0.5—155	0.25—380	1.2—89	0.25—83	2.3—1,320	4.25—12	0.25—2.3	1.2—17
Molybdenum (Mo)	0.5—33	1.0—12	0.25—30	0.25—2.5	0.25—60	0.025—2.5	0.012—8.8	0.025—0.05	0.25—2.5
Ammonium Salts (as NH$_4^+$)	0.01—13	5—11	0—44	0.07—14	0.05—24	8.7—5.2	~0.01	2—4	28—30
Benz-a-pyrene	0—1.8	0.7—3.6	0—0.01	0—0.8	0—3.7	0—1.75	1.7—1.8	0.02—0.2	Trace
Oil (wt. %)	0.24—17.0	8—13	23—62	0.07—4.0	3.0—51.3	2.4—16.9	0.7—5.6	~3.9	0.01—0.53
Total Weight Metric Tons/yr.	9.7—18.0	0.4—1.0	1.4—29.2	0.1—0.13	0.3—45	13.6—31.0	0.79—127	102—682	1.8—38.5

Table 8 (continued)

RANGES OF CONCENTRATIONS AND TOTAL QUANTITIES FOR REFINERY SOLID WASTE SOURCES
(ALL VALUES IN MILLIGRAM PER KILOGRAM EXCEPT WHERE NOTED)

Parameters	Coke Fines	Silt from Storm Water Runoff	Leaded Tank Bottoms	Non-Leaded Product Tank Bottoms	Neutralized HF Alkylation Sludge (CaF$_2$)	Crude Tank Bottoms	Spent Line from Boiler Feedwater Treatment	Fluid Catalytic Cracker Catalyst Fines
Phenols	0.4—2.7	6.3—13.3	2.1—250	1.7—1.8	3.2—15.4	6.1—37.8	0.05—3.6	0.3—10.5
Cyanide	Trace	0.48—0.95	Trace	0—14.7	0.21—4.6	0.01—0.04	0—1.28	0.01—1.44
Selenium (Se)	0.01—1.6	1.1—2.2	0.1—3.1	1.5—22.4	0.1—1.7	5.8—53	0.01—9.2	0.01—1.4
Arsenic (As)	0.2—10.8	1.0—10	63—455	Trace	0.05—4.5	5.8—53	0.05—2.3	0.05—4.0
Mercury (Hg)	0—0.2	0.23—0.36	0.11—0.94	0.41—0.04	0.05—0.09	0.07—1.53	0—0.5	0—0.16
Beryllium (Be)	0—0.2	Trace	Trace	0.025—0.49	0.012—0.13	Trace	Trace	0.025—1.4
Vanidum (V)	400—3.500	25—112	1.0—9.8	9.1—34.6	0.25—5	0.5—62	0—31.6	74.4—1,724
Chromium (Cr)	0.02—7.5	32.5—644	9.0—13.7	12.7—13.1	0.75—5	1.9—75	0.025—27.9	12.3—190
Cobalt (Co)	0.2—9.2	11.0—11.3	26.5—71	5.9—8.2	0.3—0.7	3.8—37	0—1.3	0.25—37
Nickel (Ni)	350—2,200	30—129	235—392	12.4—41	7.4—103	12.8—125	0.13—26.2	47.5—950
Copper (Cu)	3.5—5.0	14.8—41.8	110—172	6.2—164	2.5—26	18.5—194	0.22—63.2	4.1—336
Zinc (Zn)	0.2—20	60—396	1190—17,000	29.7—541	7.5—8.6	22.8—425	2.0—70	19—170
Silver (Ag)	0.10—3.0	0.4—0.6	0.05—1.7	0.5—0.7	0.12—0.25	0.02—1.3	0.05—0.7	0.5—2.0
Cadmium (Cd)	0.015—2	0.1—0.4	4.5—8.1	0.25—0.4	0.012—0.12	0.025—0.42	0—1.3	0—0.5
Lead (Pb)	0.5—29	20.5—86	158—1,100	12.1—37.3	4.5—9.6	10.9—258	0.01—7.3	10—274
Molybdenum (Mo)	0.1—2.5	6.3—7.5	0.5—118	0.25—18.2	Trace	0.025—95	0—0.05	0.5—21
Ammonium Salts (as NH$_4^+$)	No value	1.0	No value	0.2	Trace	2.0	Trace	No value
Benz-a-pyrene	Trace	0.03—2.5	0.02—0.4	0.3—0.9	No value	0—0.6	Trace	0—1.0
Oil (wt. %)	0—1.3	2.2—5.5	18.9—21	45.1—83.2	6.7—7.1	21—83.6	0.04—0.5	0.01—0.8
Total Weight Metric Tons	0.06—4.2	2.7	0.2—1.3	34.7—77	28—67	0.14—0.26	28.5—214.7	0.65—23.6

Stewart, W. S., U.S. Environmental Protection Agency, Cincinnati, Ohio, 76, 291, 1978.

For the most part, refinery waste has yet to be proved a resource rather than a waste. There is little hope to apply refinery waste in significant amounts to land during periods of crop growth. For this same reason, application of petroleum wastes to noncultivated land should be done with consideration for vegetation that may be present. Loading rates, of necessity, must be very light and disposal must be under carefully controlled conditions to prevent contact with vegetation. On noncultivated land free of vegetation, or where the presence of vegetation is of little concern, recommended loading rates may be at almost any level providing biodegradation can be maintained at a maximum or optimum rate. With addition of supplemental nitrogen a decomposition rate of 24,400 kg/ha/mo (1.0 lb/ft^3/mo or 21,800 lb/a/mo) may be expected for crude oil, bunker oil, or waxy oil.[28] The general concensus is that with tillage, added fertilizer, and optimum moisture, the upper optimum loading rate is equivalent to about 10% by weight of soil at one application a year. Single applications above 10%/W present difficulties in management. Rates of 5.0, 7.5, and 10% have all proved successful where incorporation into the soil and tilling is possible. Lesser loading rates are associated on lands where mixing with the soil is not possible, or optimum moisture and nutrient levels are not maintained. Economics may dictate a lower loading rate if land is abundant and inexpensive and a large area can compensate for the lower loading rate. This latter case is more practical for noncultivated than cultivated lands. An additional factor to consider in setting realistic loading rates is the concentrations and kinds of potentially hazardous polluting constituents carried in the oils, such as heavy metals, organic compounds, salinity, acidity, and alkalinity. The contaminants may be one or more of these constituents along with the oil itself (Table 8).

Land treatment/utilization for some oily wastes appears to be an effective, safe, and economical method of disposal. As with most land utilization of industrial wastes the development of additional quantitative information is required before finite long-term loading rates can be established as fixed guidelines.

Alaskan soils (Palmer) were found to be well supplied with hydrocarbon-oxidizing microorganisms.[29] Prudhoe Bay crude oil at 3 and 6% (by weight) loading rates evolved 19 and 12% of the carbon as CO_2 in less than a year at 10°C. Total utilization exceeded these values since some of the soil microorganisms are capable of retaining oil in their cells for delayed oxidation. Both nitrogen and phosphorus in small amounts stimulated rate of biodegradation as did stirring of the soil. The results confirm observations that oil spills in this highly pristine, predominantly cold environment can be biodegraded even at a relatively high (6%) soil-loading rate.[30]

The universality of hydrocarbon-oxidizing soil microorganisms is demonstrated further by a land utilization experiment in a semiarid, cold desert area near Ogden, Utah[31]. Loadings of waste oil (about 7.5% by weight) from a re-refining plant were biodegraded on nitrogen fertilized plots to an extent approaching 80% in one year. The climate of the area is characterized by cold wet winters and hot dry summers.

Land utilization of oils, oily waste residues, and petroleum refinery wastes is a common and successful practice by a large number of oil companies.[28,29,32] For example, Kerr McKee, Inc., has been experimenting with land utilization in Oklahoma for over 15 years (personal communication). Various methods and rates of application have been studied on a small acreage of company-owned noncultivated property. Subsurface injections of oily sludges are made by various companies when the waste is noxious, somewhat volatile, or odorous. Thus, are largely eliminated. The potential objections of surface distribution in land utilization disposals.

9. Flue-Gas Cleaning Wastes

The amount of flue-gas cleaning wastes increased by several orders of magnitude during the last two decades (1960 to 1980).[33] In 1977, 4.9×10^6 m^3/yr of flue-gas cleaning sludge

Table 9
TYPICAL CONCENTRATIONS OF TRACE ELEMENTS IN SEVERAL FLUE GAS CLEANING SLUDGES AND IN A VARIETY OF COALS

Element	Conc. range in sludges (ppm)	Median conc. in sludges (ppm)	Conc. range in coal (ppm)
Arsenic	3.4—63	33.0	3—60
Beryllium	0.62—11	3.2	0.08—20
Cadmium	0.7—350	4.0	—
Chromium	3.5—34	16.0	2.5—100
Copper	1.4—47	14.0	1—100
Lead	1.0—55	14.0	3.35
Manganese	11—120	63.0	—
Mercury	0.02—6.0	1.0	0.01—30
Nickel	6.7—27	17.0	—
Selenium	< 0.2—19	7.0	0.5—30
Zinc	9.8—118	57.0	0.9—600

— means no analysis available.

U.S. Army Engineer Waterways Experiment Station Environmental Laboratory, Cincinnati, Ohio, 1979, 117.

(FGC) was produced. By 1985 there will be 7.4×10^7 m³/yr produced. The waste increases in direct relationship to the use of coal for energy production and the tightening of air pollution restrictions. The recent installation of "throw-away" or nonregenerative systems requires new land disposal methods. The product of the flue-gas cleaning is either (q) a fine grained slurry of high water content, (b) flue-gas desulfurization sludge (FGD) or (c) flue gas cleaning sludge (FGC). The FGD usually refers to SO×-reaction products, whereas FGC refers to a general mixture of fly ash and scrubber products.[34] Three kinds of FGC wastes are being produced now:

- Wet slurry of limestone ($CaCO_3$)
- Wet slurry of hydrated lime ($CaCOH_2$)
- Double alkali (clear solution of Na_2SO_3)

Chemical reactions are as follows:

Limestone — $CaCO_3 + SO_2 + \frac{1}{2} H_2O \rightarrow CaSO_3 \cdot \frac{1}{2} H_2O + CO_2 \uparrow$

Hydrated Lime — $Ca(OH)_2 + SO_2 \rightarrow CaSO_3 \cdot \frac{1}{2} H_2O + \frac{1}{2} H_2O$

Double Alkali — $Na_2SO_3 + SO_2 + H_2O \rightarrow 2NaHSO_3$

$2NaHSO_3 + Ca(OH)_2 \rightarrow CaSO_3 \cdot \frac{1}{2} H_2O + Na_2SO_3 + \frac{3}{2} H_2O.$

All produce $CaSO_3 \cdot \frac{1}{2} H_2O$ which can oxidize to gypsum:

$2[CaSO_3 \cdot \frac{1}{2} H_2O] + O_2 + 2H_2O \rightarrow 2[CaSO_4 \cdot 2H_2O].$

The slurry contains various proportions of gypsum (sulfate) and calcium sulfite depending on the oxygen available during scrubbing. In addition, there will be sodium hydrogen sulfite from the last waste and some unreacted limestone ($CaCO_3$). The fly ash and heavy metal components will vary with the various industries and quality of coal used. In the limestone process, $CaS_2O_3 \cdot 6H_2O$ and CaS_3O_{10} will occur. Data in Table 9 presents typical concentrations of trace elements in several FGC sludges and in various coals.

The present method of disposal is lagooning and mixing with fly ash.[35,36] Land treatment or utilization of these lime-gypsum wastes is suggested at loading rates far less concentrated than occurs with the lagooning method. The retention of heavy metal of the lagoon constituents in clay soil basins is encouraging for land treatment of these wastes. Land treatment or utilization should be limited to arid and semiarid lands where rainfall is light, and soils are clayey and not acid, until further experimentation is undertaken. An advantage of land treatment is the escape from a point-source disposal.

10. Others

Application and loading rates for wastes discussed in this section are only sketchy, partly because of a lack of quantitative research and partly because land treatment or utilization has not been undertaken in any measurable extent.

Some refractory-metal-processing wastes can be well adapted to disposal on noncultivated land. Those materials with a neutral-to-alkaline pH reaction have the advantage of aiding in counteracting low pH values in acid soils, and adding calcium. Raising the pH of acid soils contributes to better retention of certain heavy metals. In Oregon, refractory-metals-processing waste eliminated the need for liming the soil for perennial ryegrass when applied at a rate of 56 t/ha (25 T/a). At an application rate of 224 t/ha/yr, Cu, Ni, Pb, and Zn migrated very little in an acid silty clay loam soil. Other metals in the waste, notably Hf and Zr offered no migration problem. If it can be demonstrated by quantitative soil analyses that heavy metals do not migrate below about 30 cm (1 ft), the loading rate may be relaxed to accommodate handling and management of the solid waste. Due consideration of the native vegetation is always a requirement.

Lime sludge waste production is increasing rapidly and is beginning to produce problems that require immediate decisions on disposal design. Some lime sludges low in heavy metals have been in demand for agricultural liming. Loading rates of lime sludges are controlled by the soil pH, particle size of the sludge solid, and $CaCO_3$ (lime) content. Lime sludges are also produced by stack-gas pollution emission control. These sludges will be high in heavy metals, polysulfides, and gypsum along with variable levels of fine flyash particulates. Gypsum ($CaSO_4 \cdot 2H_2O$) and unspent lime ($CaCO_3$) predominate.

Lime sludges also originate from primary waste water treatment. Some are high in cellulosic fibers and, therefore, are much lower in residue lime and deficient in certain nutrients, though not nitrogen. The loading rates probably should be similar to paper pulp sludge.

From industrial organic chemicals are derived some wastes that may benefit the soil of noncultivated as well as cultivated land. Some of the most useful resource-oriented wastes originate from (a) the pharmaceutical, (b) soap and other detergents, (c) synthetic noncellulosic fibers (nylon, polyester), and (d) nonspecific organic chemicals (aldehydes, ketones, and alcohols) industries. Most of these may be classed as land or soil resources due to the plant nutrients and organic matter (or presoil organic matter) they contain. Pharmaceutical residues generate fermented residues of microbial tissues (fungal and bacterial) with lime. Synthetic noncellulosic fibers degrade to yield some residual nitrogen. Soap and other detergents yield phosphorus and sometimes nitrogen in addition to organic matter. The waste streams from soap and detergents are generally high in COD, BOD, TDS, acidity, oils and grease.[37] Nonspecific organic chemicals may contain no metals and have very low plant nutrient levels. Others vary in certain plant nutrients depending on the manufacturer's mixes. Some have excessive levels of heavy metals in addition to the plant nutrients.

Loading rates of these industrial organic chemicals have yet to be determined for each material but during the early 1980s research of this nature had well begun with help from industries.

Pesticide loading rates can only be implied from agricultural application programs for crop production and a few forest land experiences. Generally, at the low rates of application reported, pesticides are readily degraded on or near the surface of the soil or they concentrate in the surface layer if insoluble and very slowly biodegradable. The relative mobility of selected pesticides as determined by soil, thin layer chromatography provides some help in establishing loading rates.[38] The rate of movement of pesticides, like the rate of biodegradation, depends on the chemical formulation and water solubility, Loading rates may be established by calculating the acceptable rates for each pesticide and adopting the most restrictive value. Input data for the establishment of these rates at this time are limited for lack of quantitative data.

The textile industry generates some wastes that may be land treated or land utilized. Here again land treatment and utilization is just developing for these wastes and loading rates have not been established.

The plastic and rubber industry produces some of the most hazardous waste streams containing oils, grease, highly acidic and caustic residues and having a high BOD and TOC. Heavy metals, phenols, and cyanides, usually occur in less than 0.1 mg/ℓ concentrations.[39] Organic solvent waste generally characterizes plastic industrial wastes.

The paint industry waste, like the plastic and rubber industry waste, contains highly toxic substances of a wide variety. Metal pigments, asbestos, cyanide, halogenated hydrocarbons, and organic pesticides are all well represented. In unaltered or untreated forms these have little, if any, place in land utilization with our present state of knowledge. Fortunately, the waste stream is relatively small since much of the residues are reworked and recovered for reuse.[40]

III. SOIL MANAGEMENT OF WASTEWATERS

Ninety percent by weight of industrial wastes are produced as liquids.[41] The U.S. EPA personnel estimates further that 40% of the liquids are inorganic and 60% organic.[42]

Information available in the literature on the land utilization of wastewaters on noncultivated land well exceeds that of solid waste utilization.[43-48] Since about 1970, a succession of symposia and conferences have centered on land management of wastes and wastewaters. Usually the discussions conclude with far more comment on the liquid rather than solid wastes. This is understandable since (a) wastewaters dominate the disposal waste world in volume and weight, (b) dewatering waste streams is costly, (c) wastewater disposal offers more perplexing disposal problems, (d) water is becoming increasingly costly and short in supply, (e) reuse of some wastewater offers cost savings opportunities, and (f) landowners, agriculturalists, recreation managers of parks, golf courses, and other public playgrounds are quick to take advantage of an inexpensive source of water. Thus, the water itself in waste streams is an additional resource, always in high demand, particularly in the arid west where short supply and irrigation is a way of life.

The treatment of unavoidable wastes of man on the land is not new, but today the success or failure of the methods adopted depends on proper design and management. This is true for all waste and wastewater applications to the soil. Today we are in a position to develop better systems. More information is available about the soil environment, the physical, chemical, and biological characteristics and reactions for pollutant attenuation, filtration, and biodegradation. The purpose of soil management of wastewaters on noncultivated land is to use this newer knowledge to develop systems for reuse, reclamation, renovation, and revegetation, all associated and compatible with the concept of land utilization. This objective includes the methods of application. The method and rate of application influences the acceptability of a wastewater for land utilization. Wastewaters have the potential of causing secondary pollution problems. These problems are no less important on noncultivated than on cultivated land. Land application can be an acceptable engineering practice, if such problems as salt buildup, nitrogen and phosphorus excesses in the soil solution and run off, and nitrogen movement to the groundwater are controlled.

In a special American Society of Agronomy publication,[49] it was stated that ''The materials we refer to as organic wastes are merely those which are not put to use in our existing technological system. Once we begin to use them, they will no longer be called wastes, and if they are in demand, we may even seek to increase their production.''

We may add to this, that the water in wastewater is a resource and that we should strive to use it, renovate it, or store it more effectively. Wastewater must be managed in such a way that it does not pollute the environment; neither the air, soil, or water resources, nor introduce toxic or pathogenic components into the groundwater or food chain. In this section, the emphasis is on the wastewater as a resource for noncultivated land. The solids and soluble constituents contained in the wastewater should also receive attention inasmuch as these constituents must be dealt with either to make certain they are eliminated as a threat to society or converted into resources. The soil is the medium for the stabilization of the waste components or their use as a resource.

The quality constraints for wastewater application to noncultivated land are similar to those of cultivated land. Nitrogen, toxic metals, organic solvents, and pathogens head the list of concerns with the usual wastewaters. Other waters containing hazardous and toxic constituents must be evaluated on an individual basis of source, kind and concentration of pollutant, and phytotoxicity if vegetation is involved. Much of the increase in interest in land utilization and treatment during the 1970s and 1980s is due to recent successes of industry, and long-term experimentation by the Universities of California, Minnesota, and Pennsylvania with the soil layer acting as a biodegradation, attenuation, and filter system. Spray irrigation is the dominant method of distribution used.

Use of secondary municipal sewage wastewater on recreational areas, parks, and golf courses is popular in the West where water is in very short supply. Some recreation parks have been using municipal sewage water for irrigation for 30 years with no serious deterioration of the soil.

A. Methods of Application

The dominant method of wastewater application to noncultivated land is by spray or sprinkler irrigation. Alternative methods can be classified into three conventional groups and three experimental groups:[50]

- Conventional: (1) Irrigation — (a) spray, (b) surface flood; (2) Overland flow or spray-runoff; and (3) Infiltration-percolation — (a) spray, (b) surface flood
- Experimental: (1) Wetland disposal (e.g., cypress dome ponds); (2) Peat filter beds; and (3) Subsurface injection

The three conventional methods are illustrated in Figure 3.[50] In forested areas spray irrigation dominates over other methods for municipal sewage waters and liquid sewage sludge. In north central Florida, disposal of secondary sewage into cypress swamps with some success has been suggested as an inexpensive method of disposing of wastewater.[51]

1. Spray Irrigation

Spray irrigation as used in this text refers to controlled spraying of liquid on land, at a rate measured in centimeters of liquid per week (cm/wk), with a flow path allowing infiltration and percolation within the boundaries of the disposal site.[52] Spray Irrigation (SI) has been adapted to forest areas for disposal of a number of different wastewaters. Forested areas are highly efficient for removal of phosphorus, nitrogen, and organic carbon constituents provided loading rates are not excessive.[53]

Ideally, spray irrigation allows all the wastewater to enter the soil, and what is not used in evapotranspiration, moves down through the profile, and merge with the groundwater without deteriorating the quality. Also ideally, surface runoff is held to a minimum, if allowed at all.

One of the advantages of surface irrigation is its versatility to operate on nearly any kind of topography, including steep slopes. It is, therefore, the dominant form of wastewater

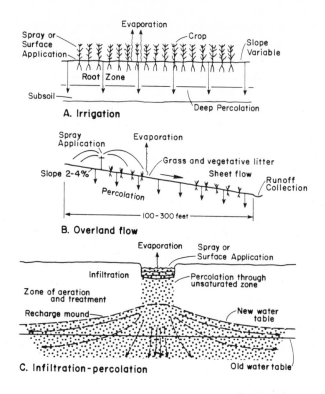

FIGURE 3.   Illustration of three general methods of wastewater application. (From Thomas, P. R., Ohio Guide for Land Application of Sewage Sludge: Report to the Task force on Land Application of Sewage Sludge, Ohio Coop. Ext. Serv. Bull., 598, 1975, 1-12.)

application on noncultivated land. In California, slopes in excess of 20% receive wastewater through sprinkling methods, care being taken to prevent erosion. Also, soils suitable for leaching fields and/or seepage beds are potentially suitable for spray irrigation.

Most tree species exhibited favorable responses to spray irrigation. Those that did not respond or were killed were on waterlogged soils subjected to severe anaerobic conditions. Soil textures described by the Unified Soil Classification most suited to each of the three wastewater disposal modes are given in Table 10.[52]   These data represent guidelines only and should not be taken literally since the textural classes represent surface soil. It should be noted, however, that there are real consistencies in the data; rapid-infiltration methods are assigned to soils of good internal drainage, and overland runoff to the more impervious soils. Subsoils may be compacted, stratified with nearly impervious clay, or excessively open and porous as with sand and gravel. Spray systems are the most expensive to install of the three methods but they are the most efficient. In some vegetation such as wooded, forest, and brush lands; sprinkler application is the only suitable method of achieving reasonable uniformity in wastewater distribution over the land. An advantage in selecting the most uniform distribution system is that a higher loading rate can be used. Generally, loading rates for the year can be greater on forest than cultivated land because application can be over a longer season. Land and soil operations of cultivation, harvesting, and plowing also do not interfere with application schedules.

Public health precautions are necessary with spray irrigation. Bacterial travel from spray irrigation depends on ultraviolet radiation intensity, humidity, and wind velocity. The travel downwind of spray was found to increase approximately 85 ft (25.9 m) for every 2.25 mph increase in wind velocity.[52] Also, there is evidence that no health hazard exists beyond the

Table 10
METHOD OF LAND DISPOSAL AS RELATED TO SOIL TEXTURES ACCORDING TO THE UNIFIED SOIL CLASSIFICATION

Symbol	Type	Drainage characteristics	Potential land disposal mode
GW	Well graded gravels or gravel-sands, little or no fines	Excellent	RI
GP	Poorly graded gravels, gravel-sands, little or no fines	Excellent	RI
GM-d	Silty gravels, gravel-sand-silt mixtures	Fair to poor	SI
GM-u	Silty gravels, gravel-sand-silt mixtures	Poor to practically impervious	OR
GC	Clayey gravels, gravel-sand-clay mixtures	Poor to practically impervious	OR
SW	Well graded sands or gravelly sands, little or no fines	Excellent	RI
SP	Poorly graded sands or gravelly sands, little or no fines	Excellent	RI
SM—d	Silty sands, sand-silt mix	Fair to poor	SI
SM—u	Silty sands, sand-silt mix	Poor to practically impervious	OR
SC	Clayed sands, sand-clay mixtures	Poor to practically impervious	OR
ML	Inorganic silts, very fine sands, silty or clayey sands	Fair to poor	SI
CL	Inorganic clays, gravelly clays, sandy clays, silty clays, lean clays	Practically impervious	OR
OL	Organic silts, organic silt clays of low plasticity	Poor	SI—OR
MH	Inorganic silts, micaceous or diatomaceous fine sandy or silty soils	Fair to poor	SI
CH	Inorganic clays of high plasticity, fat clays	Practically impervious	OR
OH	Organic clays of medium to high plasticity, organic silts	Practically impervious	OR
PT	Peat and other highly organic soils	Fair to poor	SI

SI — Sprinkler irrigation; RI — rapid infiltration; and OR — overland runoff.

From Reed, C. H., in *Managing Livestock Waste*, ASAE Pub., PROC-275, Urbana, Ill., 1975, 444. With permission.

spray or mist zone. The mist zone at 5 to 10 mph wind extends for about 46 m (150 ft) downwind from a 9 m spray radius.

2. *Ridge and Furrow and Flood Systems*

Two irrigation methods used less frequently on noncultivated lands are the ridge and furrow and the flooding systems. Ridge and furrow requires nearly flat land and accurate soil manipulation to prepare the ridge and furrow to insure uniform grading necessary for uniform infiltration.

Flood irrigation represents an historic means of spreading water over land for relief during temporary periods of drought or during periods of flood for deep storage. Originally, spreading was accomplished by inundating the level or gently sloping land with several centimeters of water by diverting runoff water from natural drainageways or streams. Several techniques are used but their purposes are similar.[53] The natural vegetation, if any, must withstand short periods of soil saturation and partial inundation.

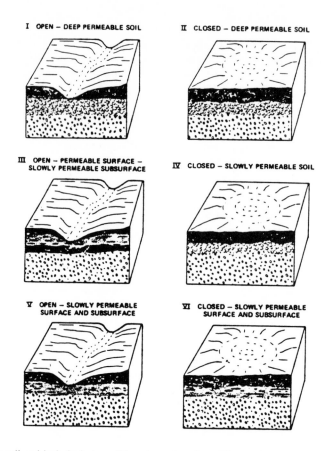

FIGURE 4.   The soil and hydrological conditions best adopted to different land application techniques. (From Larson, W. E. and Schuman, G. E., Problems and needs of high utilization rates of organic wastes, in *Soils for Management of Organic Wastes and Waste Waters,* Elliott, L. F. and Stevenson, F. J., Eds., American Society of Agronomy, Madison, Wisconsin, 1977, 589. By permission of the American Society of Agronomy.)

3. Overland Flow

Land best suited for overland flow (or runoff) allows only minimum infiltration to force the water to flow along the surface and through vegetation as a means of clarification, purification, and renovation for surface or groundwater recharge. More precisely, Overland runoff (OR) is defined as the controlled discharge of liquid onto land by spraying or other means with the flow path being downslope sheet flow.[52]

Hydrological and soil conditions best adapted to overland flow techniques are illustrated in Figure 4   by V, open — slowly permeable surface and subsurface, and IV closed — slowly permeable surface and subsurface.[54] Overland flow technique has been used in a number of ways to reclaim wastewater, although the technique is reported more frequently on cultivated (such as pastures and hay producing land) than on noncultivated land. For short flow distances, the system is feasible for partially reclaiming wastewater prior to infiltration where further renovation before final groundwater recharge will occur. The overland flow pilot experiment at South Lake Tahoe, Nev. although not fully successful, demonstrated unique possibilities for sewage water renovation. The treatment area had a slope averaging about 10% and supported a thin stand of pine. A diversion ditch located at the foot of the slope collected the runoff. The water quality with respect to nitrogen did not improve enough, however, to continue the renovation the second year. In continuing operations during the winter months over snow, ice, and frozen soil, the excessive rate of

Table 11
ESTIMATED PERCENTAGE RENOVATION EFFECTIVENESS OF
LAND DISPOSAL METHODS OF WASTEWATER APPLICATIONS
AND LOADING RATES

	Spray Irrigation[a]	Overland Runoff[a]	Rapid Infiltration Ponds
BOD	99	80	99
SS	99+	80	99
N	70—90	60—90	30—80
P	95—99	60—80	50—90
Trace metals	95—99	60—80	50—90
Organic toxicants	99	60—80	90
Viruses and bacteria	99+	90	99+
Cations	50—75	30—50	30—75
Anions	0—50	0—10	0—50

Modified by Spyridakis and Welch, 1976, and from Driver et al., 1972.
Renovation effectiveness is expressed as percentage of the original effluent constituent content that is retained in the soil, utilized by vegetation or volatilized.

[a] Removal by harvest crop is included.

application of liquid during rainy weather, and lack of fully aerobic conditions, contributed to the failure of the overland-flow systems to perform as well as expected.[52]

Overland flow requires sufficient retention time of the wastewater on the land to accomplish the desired renovation. Several variables; (a) the quality of the wastewater, (b) the length and pitch of the slope, (c) the kind and density of vegetation, and (d) the susceptibility to soil erosion, determines the final quality of the recovered water. Cannery wastewater, for example, was found to be most suited to the overland flow technique best with a slope of 6% maximum and 2% minimum and a minimum length of 53.3 m (175 ft).[55] The rates of application at the Paris, Texas, site were 6.35 cm/wk (2.5 in./wk) in the summer and 3.18 cm/wk (1.25 in./wk) in winter. Reed[52] suggests that the land used for overland runoff should have at least 15 to 20 cm (6 to 8 in.) of good top soil, with a slope distance of about 91.4 m (300 ft), and with grades ranging between 2% minimum and 8% maximum.

The estimated percentage renovation effectiveness of different land disposal techniques, as reported by Driver et al.,[56] is given in Table 11 for comparison. The overland runoff is suggested as being the least effective of the three major land treatment techniques in removal of organic toxicants. Virus and bacteria are removed less effectively by the overland runoff technique than by spray irrigation and rapid infiltration. The upper layer of soil is the most effective filter for containing pathogenic microorganisms.

Overland runoff is also the least effective of the three major land application methods for removal of heavy metals, total dissolved solids, salts, and phosphates. The procedures that involve the soil in removing wastewater constituents are, therefore, more effective than those that depend primarily on vegetation.

A comparison of the three major land application methods (Table 12) as reported by Ryan and Loehr[57] substantiate the findings presented above in Table 11 except for nitrogen. Rapid infiltration techniques were reported to be less efficient according to Ryan and Loehr than overland flow (runoff). Slow rate methods generally have the highest treatment efficiencies. Overland flow can result in a system more than a point source discharge than nonpoint discharge. This is particularly the case when the wastewater has a high BOD and the soil is coarse.

Table 12
REPORTED TREATMENT EFFICIENCIES FOR LAND TREATMENT

Treatment process	Design removal efficiency (%)				Effluent quality (mg/ℓ)			
Land application systems[a]	BOD	SS	P	N	BOD	SS	P	N
Slow rate	98+	98+	80—99+	85+	4	5	2	6
Overland flow	92+	92+	40—80	70—90	18	18	2—7	3—9
Rapid infiltration	85—99	98+	60—95	0—50	30	5	4	15—30

[a]   It is assumed that the wastewater to these systems receives preliminary treatment.

Table 13
THE REMOVAL EFFICIENCY AT SELECTED INFILTRATION-PERCOLATION SITES

Location	Depth sampled, ft[d]	Removal efficiency, %					Reference
		BOD	SS	N	P	E. coli	
Flushing Meadows, Arizona	30	98	100	30—80	60—70	100	15
Hyperion, California	7	90	—	—	—	98	58
Lake George, New York	10	96	100	43—51	8—61	99.3	6
Santee, California	200[a]	88	—	50	93	99	77
Westby, Wisconsin	3	88	—	70	93	—	10
Whittier Narrows, California	8	—	—	0[b]	96	0[c]	69

[a]   Lateral flow.
[b]   Short frequent loading promotes nitrification but not denitrification.
[c]   Coliforms regrow in the soil.
[d]   Multiply by 0.3048 to convert to m.

Pound, C. E., and Crites, R. W., U.S. Environmental Protection Agency, Sec. VI, Washington, D.C. 1973.

3. Rapid Infiltration

Rapid infiltration has often been described as infiltration-percolation or simply as ground-water recharge.[50,53] In rapid infiltration the main portion applied wastewater enters the surface and moves through the soil, the vadose (unsaturated) zone, and on to the capillary fringes of the groundwater. The wastewater quality is expected to improve because the soil attenuates pollutants as the wastewater infiltrates. Application of the water can be made either by sprinkling, spreading, or flooding techniques, but flooding is most common. A number of well-organized and documented recharge programs have demonstrated the feasibility of rapid infiltration as a viable option for wastewater treatment.

Some of the more classic experimentations in rapid infiltration are those at Phoenix and Flushing Meadows, Ariz., Helmet,[59] Whittier Narrows,[60] and Santee, Calif.,[61] and Lake George, N.Y.[53] Data in Table 13 compares removal efficiencies at these selected sites.

Modification of this system includes sprinkler and spreading applications to a variety of vegetation including forests,[6] grasslands,[53] and pastures.[5] In these cases plant nutrients,

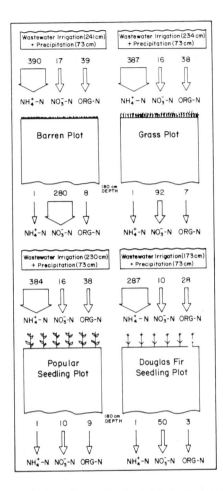

FIGURE 5. The nitrogen flux (kg/ha) during the second year of wastewater irrigation: input to field plots from wastewater plus precipitation; output from plots in soil solution passing 180 cm soil depth. (From Breuer, D. W., Cole, D. W., and Scheiss, P., Nitrogen transformation and leaching association with wastewater irrigation in Douglas fir, poplar, grass, and unvegetated systems, in *Utilization of Municipal Sewage Effluent and Sludge on Forest and Disturbed Land*, Sopper, W. E. and Kerr, S. N., Eds., The Pennsylvania State University Press, University Park, Penna., 1979, 19. With permission.)

particularly nitrogen (Figure 5)[4] and phosphorus, are removed in addition to the attenuation of metals by the soil. Rapid infiltration may be designed with the objectives of (a) recharging groundwater, (b) recovering renovated water, using wells or underdrains, and (c) intercepting renovated water by a water body.

Soil characteristics and their management control the success or failure of the application methods to meet their objectives. Surface and subsoil conditions most suited for rapid infiltration on uncultivated soil are shown in Figure 4 as I and II. Soils should be selected that allow rapid infiltration, good internal drainage, and uninterrupted percolation to the groundwater. Coarse-textured soils perform the best, in contrast with the finer textures preferred, for spray irrigation and overflow runoff. Textures most suitable include sands, sandy loams, loams, loamy sands, and some fine gravels. Textures least acceptable are

Table 3.14
ACCEPTABLE OR DESIRABLE RANGES OF SIGNIFICANT SOIL
PROPERTIES FOR LAND TREATMENT PROCESSES

Land treatment process		Acceptable ranges of property		
		Minimum	Desirable	Maximum
Slow rate	Depth (m)	0.6	≥1.	none
	Slope (%)	0	<15	35
	Permeability (cm/h)	0.50	1.5—5	50
Overland Flow	Depth (m)	0.3	≥0.9	none
	Slope (%)	0	2—8	15
	Permeability (cm/h)	none	<0.5	1.5
Rapid Infiltration	Depth (m)	1.5	≥3	none
	Slope (%)	none	<5	15
	Permeability (cm/h)	1.5	>5	50

U.S. Environmental Protection Agency, U.S. Army Corps of Engineers, Washington, D.C. 1977, 102.

clays, clay loams, silty clay loams at the fine end, and coarse sand and gravel at the coarse end of the spectrum. Of course some exceptions can be cited where ideal or near ideal conditions exist. For example, in Wesby, Wis., a silt loam successfully accepted wastewater at an annual loading rate of up to 14 m (10.4 in./wk).[62]

All soils should be well drained. Best results occur in those having a depth of 4.6 m (15 ft) or more. Three meters is considered as an absolute minimum.[50] Should a groundwater recharge mound develop, it should not be permitted to rise nearer to the soil surface than 1.3 m.[63] A further requirement is that the subsurface soil be sufficiently permeable to match or exceed the surface infiltration rate. Soil stratifications of different textures consisting of clay lenses, compacted layers, or silt barriers, cause water to accumulate as underground puddles and perched water tables and permit lateral flow that may intersect with soil surfaces. Soil heterogeneity should be identified and logged by deep soil borings prior to the establishment of a rapid infiltration system.

The hydrologic conditions of wastewater disposal locations generally receive too little attention, especially for rapid infiltration systems where recharge water adds significantly to the natural rainfall. Hydrologic features most essential to examine include groundwater movement, depth, and quality. Aquifers need to be identified and characterized. Water regulations usually require the original integrity of the quality to be maintained. In Western U.S.A., instances of groundwater improvement are not uncommon because of a variability in quality of subsurface waters. Groundwater recharge may (a) add to the storage of underground water resources, (b) improve groundwater quality, (c) limit salt water intrusion (e.g., Los Angeles) or (d) dispose of wastewater in an efficient and inexpensive land treatment program. In Figure 4. conditions I or II, deep permeable soil and permeable material, respectively, are the most suitable soil characteristics for rapid infiltration of wastewaters in land utilization. Acceptable and desirable ranges of some soil properties for different methods of application appear in Table 14 as adapted from U.S. EPA for comparative purposes.[64] Interpretation of data in this table is subject to alteration depending on different climatic conditions of temperature and rainfall.

The long-term rapid infiltration effects of land applications of primary municipal effluent at Hollister, Calif., deserve special attention because of a history of disposal extending over 33 years.[65] The more recent daily flow was reported to be about 43.8 ℓ/s (1.0 M/gal/d). An annual rate of 15.4 m was applied to 20 infiltration basins intermittently flooded for 1 to 2

Table 15
CONSTITUENT LOADING RATE FOR
WASTEWATER AT HALLISTER, CALIF.,
USING THE RAPID INFILTRATION
METHOD[a]

Constituent	Daily (kg/ha)	Annual (kg/ha)
COD	304	111,000
BOD	95	34,500
TOC	107	39,000
Total nitrogen	17.3	6,310
Total phosphorus	5.3	1,950

[a] Based on 438 1/s applied to 8.8 ha. kg/ha × 0.89 = lb/acre.

Pound, C. E., Crites, R. W., Olson, J. V., and Metcalf and Eddie, Inc. U.S. Environmental Protection Agency, Ada, Okla., 1978, 150.

days every 14 to 21 days. A maximum long-term infiltration rate in this Typic xerorthent sandy loam, was 64.6 m/yr (212 ft/yr). The constituent loading rates for the wastewater were higher than for most sites, Table 15. Clogging of the surface soil caused the main infiltration problem. Scarification of the surface by discing returned infiltration readily to the original rate. The soil profile infiltration did not change significantly during the many years of land application. The original soil pH which was slightly alkaline with some free calcium carbonate (lime). The pH decreased and free lime disappeared through the top 3 m during the 33 + years of wastewater treatment. Despite surface clogging of pores, organic matter accumulated only to a limited extent. Biodegradation apparently oxidized the organic residues at a rate nearly equivalent to that of application. Nitrogen accumulated in the surface soil, but only sufficiently to account for 2% of the total additions over the many years of treatment. Groundwater analyses indicate that 93% of the total nitrogen in the wastewater was removed from the infiltrating soil solution.[65] The phosphates also offered no problem. About 68% was retained in the 0 to 16 cm soil layer. Boron also accumulated predominantly in the surface soil. A total increase of about fourfold occurred; yet remained below toxic limits. The accumulation of organic matter in the soil profile significantly raised the cation exchange capacity, especially in the surface soil. Copper accumulated in the 0 to 16 cm depth, cadmium concentrated in the 0 to 30 cm depth, and nickel and zinc moved to about the 3 m depth. The presence of lime throughout most of the land treatment period must have exerted a pronounced effect on the differential heavy metal movement through this sandy loam. Perhaps the most surprising and gratifying finding is that a soil so low in fine particulates, and so high in sand and gravel, was capable of renovating primary municipal effluents for such a long period of time.

A similar experience in rapid infiltration is reported at the Phoenix, Ariz. (Flushing Meadows) research site with municipal effluent.[66] Both scarification of the soil surface and an intermittent period of drying were necessary to maximize infiltration rates of the municipal secondary effluent on a sandy loam soil. Renovation of the wastewater improved the quality such that the recharge water was better than the natural groundwater. Nitrates, phosphates, and heavy metals did not prove to be a problem in the quality of recharge. Of the five kinds of surfaces; (a) bare soil, (b) giant bermudagrass, (c) sudangrass, (d) rice, and (e) 6 inches of pea gravel, the vegetated surfaces were the most effective for wastewater renovation. The bare soil proved to be a better renovater than pea gravel. The vegetation advantages included (a) protection of the soil from rainfall impaction, (b) promotion of dentrification by microorganisms, and (c) removal of additional nutrients by harvesting. Disadvantages are associated with interruptions and delays in application to establish, maintain, and harvest the plants. Pea gravel did not allow for scarification of the surface as did bare soil.

The loading rates varied depending on the surface treatment. Rates of about 12,200 kg/ha/year of organic loading, and 122 m/year of liquid loading are reported to be within the range of practicability.

The influence of pea gravel mulch cannot be considered settled in light of the studies at Whittier Narrows, Calif.[60] A 6-inch layer of pea gravel over bare ground was considered to eliminate plant growth and decrease maintenance without altering infiltration. Yet, in agreement with the Flushing Meadows experiment, pea gravel on soil at another site, Helmet, Calif., was found not to be effective in maintaining infiltration rates.[67]

4. Subsurface Injection

Application of wastewater by subsurface injection to noncultivated lands is practiced only to a limited extent. The main reasons are the generally rougher terrain and the presence of a permanent vegetation of shrubs, trees, and forest. The areas most suitable for subsurface injection, therefore, are limited.[68] Such areas as grazing land, bare lake beds, and playa offer some possibilities. For example, campground sludge application by subsoil injection was studied on a native meadow.[68] Bacteria and virus pollution did not occur in the well drained, isolated soil, three meters in depth. Subsurface injection with its advantages of (a) immediate covered disposal, (b) elimination of odor, flies, and vermin, (c) placement in root zone for plants, (d) control of surface runoff and loss by soil and water erosion, and (e) complete containment of pathogenic microorganisms, can readily hold pollution to a minimum. As mentioned above, however, the method is not lacking in disadvantages. Additional factors included (a) the need for complex management, (b) requirement of special equipment with concomitant cost factor, and (c) application problems in wet soils with present equipment.

5. Wetlands Application

Wetlands method of waste utilization on noncultivated soil also is not a common practice although more recently it has grown in popularity. Natural marshlands have been investigated as purifiers of effluent from municipal treatment plants.[69,51,70] Both the influent and effluent quality of marshes have been monitored in these experiments. The vegetative removal of nitrogen and phosphorus and other constituents by harvest were followed, along with water quality analyses for BOD, COD, P, coliform, and total suspended, and dissolved solids. The conclusion is that the marshland technique of wastewater utilization is feasible and the quality of the associated water is acceptable for certain land applications and agronomic uses. Much research is needed to refine further some of these methods of wastewater utilization.[69]

Peat bed infiltration is another wetland technique that shows some promise for wastewater renovation.[71] Site characteristics for wetland treatment (utilization) compared with other processes are suggested in Table 16. The slope usually is less than 5%, soil permeability is slow to moderate, and depth to groundwater is not a critical issue. Because wetlands tend to overflow or be overflowed, storage facilities should be made available, particularly during the wet and winter months if the location is within the temperate-humid climatic region.

The utilization of fresh water wetlands for nutrient removal from secondary-treated sewage effluent may prove feasible in certain special instances.[70] Aquatic plants near the discharge

Table 16
COMPARISON OF SITE CHARACTERISTICS FOR LAND TREATMENT PROCESSES

Characteristics	Principal Processes			Other Processes	
	Slow rate	Rapid Infiltration	Overland flow	Wetlands	Subsurface
Slope	Less than 20% on cultivated land; less than 40% on noncultivated land	Not critical; excessive slopes require much earthwork	Finished slopes 2 to 8%	Usually less than 5%	Not critical
Soil permeability	Moderately low to moderately rapid	Rapid (sands, loamy sands)	Slow (clays, silts, and soils with impermeable barriers)	Slow to moderate	Slow to rapid
Depth to ground	0.6 to 1 m (minimum)	3 meters (lesser depths are acceptable where underdrainage is provided)	Not critical	Not critical	Not critical
Climatic	Storage often needed for cold weather and precipitation	None (possible modify operation in cold weather)	Storage often needed for cold weather	Storage may be needed for cold weather	None

U.S. Environmental Protection Agency, Municipal Sludge Management, Washington, D.C. 1977, 31.

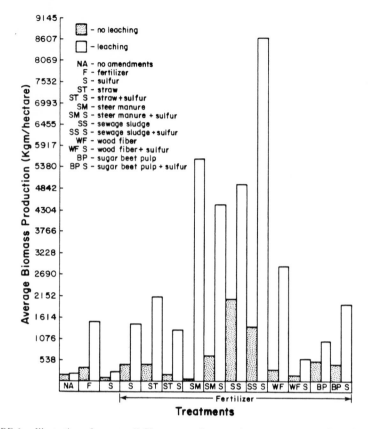

FIGURE 6. Illustration of a peat-soil filter system for secondary sewage-water. (From Farnham, R. S. and Brown, J. L., Advanced Wastewater Treatment Using Organic and Inorganic Materials: 1. Use of Peat and Peat-sand Filtration Media, Proc. 4th Int. Peat Congr., Helsinki, 4, 271, 1972.)

areas were larger and higher in phosphorus than those farther away. Nitrate was not a serious problem since the aquatic plants readily utilized nitrates in greater growth volume. The overland flow of wastewater on wetlands was shown to be a viable alternative to other acceptable methods of disposal under certain conditions.[70] The conclusions of this research, that aquatic plants of wetlands may be useful in renovating secondary sewage wastewater, are supported by Lakshman.[72] Data presented shows that aquatic plants have considerable potential in the purification of municipal wastewaters within surprisingly short periods of time. Vegetation can act as an effective sink for nitrates, phosphates, and heavy metals when cut and removed.

The peat filter bed method in conjunction with small aeration plants for tertiary treatment of sewage wastewater is not new to the U.S. Department of Agriculture (USDA) Forest Service. The renovation is based on the capacity of peat to adsorb plant nutrients and heavy metals.[73] A schematic peat soil filter system is presented in Figure 6 to illustrate the basic principles of the system.[74] In northern climates, as represented by the north central states, this biological system has special advantages over the usual spray systems. Phosphates, BOD, and coliform bacteria may be readily reduced to acceptable levels for discharge even at rates of 20 to 37 cm/h.

The cypress swamps in northcentral Florida, as mentioned also, offer an unusual means for disposal and reuse of secondarily treated sewage water for small communities on a limited budget. Cypress trees grow best in water and tolerate fluctuating water levels. Cypress strands and domes characterize certain Florida ecosystems. The strands represent slowly moving bodies of water that meander over a large area and long distances before emptying into larger bodies of water. Cypress domes are isolated ponds dispersed among pine flatwoods. They are called domes because the trees along the edges of the ponds are shorter than those in the center (more permanent deeper) water. The basic principle of the domes is to allow vegetation to absorb the plant nutrients, heavy metals, and organic constituents, and pathogenic microorganisms of secondarily treated sewage. This process permits the renovated water, that is equivalent in quality to tertiarily treated water, to filter into the main surrounding water.[51,75] A 4 year experimental field program demonstrated the feasibility of disposal and water renovation using the cypress dome technique. The duckweed on the ponds absorbed most of the nutrients and then in turn became organic sediments at the bottom of the pond. The cypress trees absorb about 10% of the nutrients into woody and more permanent tissues. The blue-gray clay and organic sediments at the bottom of the ponds attenuated any remaining pollutants of heavy metals and pathogenic bacteria and virus. Relatively pure water filtered out of the pond, characteristically, to beneficially recharge the groundwater. The lifetime of these cypress domes is predicted to be about 10 years as a minimum.[51]

B. Disturbed Lands

Some of the greatest benefits from solid waste and wastewater disposal can be realized by their utilization on disturbed land. Land reclamation is required for most mine-disturbed land since, more often than not, such land lacks vegetation, top soil, plant nutrients, and topographical features necessary for a return to the original natural status. Organic matter and plant nutrients must be replenished before natural soil regeneration processes can begin at any significant rate. Soil and water erosion process must be controlled. Land application of liquid and solid wastes to disturbed lands is an effective way of disposing of wastes and at the same time improving these environments. Many of the reported waste revegetated and reclaimed disturbed lands have received municipal sewage and wastewater.[76] Food processing wastes also have been effective for reclamation purposes. Organic waste application rates may need to be greater than for other lands and, therefore, the use of solid forms are favored. Where hauling distances are long, dewatering cost may be offset in part by smaller load weight and with consequent lower hauling cost. Moreover, some of the mine areas may be remote, lacking good roads and otherwise unsuited, or poorly suited for the routing sprinkler application systems. Soil management equipment, in any event, must be capable of application on flat benches and steep peripheral slopes. Incorporation of the solid waste concerned with the spoil results in raising the pH, encouraging better root penetration and growth, and improving the soil moisture relationships.[77] Mine spoil covers a wide range of post-mine materials. Many are acid. These are the ones requiring the most immediate attention.

Constraints are evident at some sites. To avoid nuisance conditions of odor, flies, and other pests, subsurface applications have been initiated. One of the main constraints is salinity build up in the soil, if the applications are extended on a long-term basis, especially in arid, semiarid, and possibly in subhumid climatic regions.

Soil-management practices to improve topography and infiltration, and to encourage water conservation may be necessary if the reclamation and revegetation processes are to be successful. Furrow grading is considered by Riley[39] to be a key to successful reclamation in Pennsylvania. Any mechanical practice which allows greater application rates, such as deep incorporation, has many advantages including improved infiltration and raising the pH of the acid spoil. For example, incorporation of municipal sludge to a certain mine-spoil of pH 2.5 at a rate of 659 t/ha (294 T/a) raised the soil pH to 5.8 and greatly improved attenuation and precipitation of heavy metals.[77] Mixing wastes to a depth of at least 30 cm is suggested for best results.[78]

The application of solid and liquid municipal wastes, food processing and agricultural wastes, and composts to disturbed lands for reclamation and revegetation is fast becoming a well established practice.[76] The kinds of disturbed lands receiving most attention include:

- Surface-mined areas
- Strip-mine spoil
- Coal-mine spoil
- Coal-mine land storage
- Uranium mine waste
- Orphan mined lands

The two transport surfaces, (a) the landscape or visible external surface and (b) the soil or internal surfaces must be characterized, modified, and treated to be most effective. These two surfaces in relation to the nature of the transport systems (usually aqueous) also should receive attention in determining the soil modification procedures necessary to make waste disposal more resource oriented.

1. Strip-Mine Spoil

Of all the disturbed land experimentation for reclamation and revegetation, strip-mine spoil has received the most attention.[79-84] Natural reclamation and revegetation have failed on many strip mine spoils because of:

- Excessive acidity (low pH)
- High level of toxic elements in the spoil
- Low permeability
- High surface temperature and drying in the summer
- Insufficient levels of plant nutrients (N, P, K, and Ca)
- Soil and water erosion

These are the specific factors, therefore, that must be given foremost consideration for soil modification in land utilization to bring the area back to its natural condition. Accumulation of organic matter as a forerunner for the development of a fertile topsoil is one of the most effective means of accomplishing the necessary rehabilitation to make these land areas productive again. A number of successful reclamation and revegetation examples are available in the literature demonstrating the usefulness of certain wastes as valuable resources when associated with specific soil manipulation and management techniques. Several instances of using sewage effluent and liquid digested sludge as aids to revegetation of strip mine spoil are reported from coal mining regions of the U.S. In Pennsylvania, alone, approximately 150,000 ha (370,000 a) have been disturbed by bituminous mining.[80] Anthracite refuse accounts for another 5000 ha (12,300 a). Nationally, over 4.4 million acres were disturbed by coal mining by 1975 and the area has expanded rapidly into the eighties.[85] Irrigation of both bituminous spoil and anthracite refuse with sewage effluent and/or liquid digested sludge can permit immediate vegetative covering of an agronomic nature or with trees.[80] Permanent reestablishment of vegetation required continued applications of the wastes. One short period of land treatment is not enough to maintain stabilized plant growth. Soluble aluminum and manganese are the two most toxic constituents occurring in the highly acid coal spoil and refuse. The solubility of these metals is pH-dependent and phytotoxic at low pH levels.[86] Permanent vegetation, therefore, will not become established until the pH level of the soil matrix in the root zone has been raised sufficiently (to at least 5.0) to lower the solubility of these two metals and probably iron.

In another example, a burned anthracite coal refuse bank, bare of vegetation yet moderate in pH (5.0), received heat dried sewage sludge as a revegetative measure.[85] Loading rates of 0, 40, 75 and 150 dry t/ha were mixed into the upper few centimeters of the soil with a conventional farm cultivator after surface spreading. Revegetation and reclamation of the spoil area for grass, legumes, and trees appeared permanent after 3 years treatment. The one-shot treatment of dry sludge received supplemental sewage water irrigation at a rate of 5 cm biweekly beginning the last of May and continuing on until the middle of August. The seedling hardwood trees survived better than conifers. The 150-t/ha rate produced superior herbaceous vegetation but the 40 to 75 t/ha rate produced the best seedling tree response.

At an Illinois strip mine (Palzo), the shale waste contains exposure of pyrites (FeS) that oxidize to sulfuric acid (H_2SO_4) and ferrous (Fe^{++}) salts causing highly acid spoil with pH values of 2.5 to 3.0. A field experiment of 15 spoil-plots ranging in size from 1.19 to 2.94 ha, received secondary municipal sludge of 11 to 15% solids at rates ranging from 426 to 997 dry t/ha which was incorporated in the soil surface.[87] Grasses became established the first year of treatment. Water quality improved markedly with heavy metal reductions in the seepage. In the continuation of this same experimentation,[88] at the end of the first year, those plots where the pH was above about 5.2 supported good growth of grass or legume vegetation. Limiting pH values for switchgrass, orchardgrass, and tall fescue were 3.75, 4.31, and 4.45, respectively. The heavy metal concentrations appearing in the vegetation were much greater at low pH levels of spoil than levels nearer neutral. Zinc was the best indicator element in explaining yield variation in rye. Cunningham, et al.[89] observed that for plants grown in soils receiving sludge high in heavy metals, yield reductions occurred when the concentrations of elements Cd, Cr, Cu, Mn, Ni, and Zn in plant tissue exceeded 4.8, 5.5, 23.3, 345, 163, and 289 ppm, respectively. Addition of municipal sewage sludge to spoil of the Palzo mine also benefited the survival and height growth rates of selected trees.[90] Both hardwoods and conifers attained maximum survival when planted in association with herbaceous plants. Accumulation of heavy metals occurred in selected woody plants on sludge-treated strip mine spoils at the same Palzo location.[91] Woody plants enjoy certain advantages from their ability to accumulate large biomasses, retain absorbed heavy metals for extended periods of time, and produce a minimum amount of tissues attractive for consumption by terrestrial animals. Due to the large waste applications necessary to neutralize the acids and other phytotoxic constituents of the strip mine spoil, a relatively large accumulation of heavy metals was found in the woody species in these experiments as compared with cultivated crops. This occurred despite the relatively low content of the metals, except Cd, contained in the municipal sludge applied, as compared with most sludges considered acceptable and used on agricultural lands. The most practical features of the indepth experimentation for strip mine restoration at Palzo are

- Organic waste (resource) such as municipal sewage sludge is capable of reclaiming and revegetating another waste to restore it to productivity. Both wastes become resources.
- Organic waste (resource) can reclaim some of the worst and most toxic spoil in the United States.
- Organic waste (resource) can alter pH to make it favorable for a large number of plants (grasses, legumes, and trees) to become established and survive.
- The renovation and revegetation of barren spoil research make it possible to differentiate and identify those plants most and least favored for use in the revegetation of spoil land.
- Organic waste (resource) by altering pH and supporting plant growth effectively reduced the concentration of toxic metal leaching in underground and surface waters.
- Unwanted sewage sludge is a valuable resource amendment to be considered in any strip mine reclamation and revegetation.

Another classic example of reclamation and revegetation of strip mined land is the use of composed sewage sludge as reported by researchers from the Agricultural Environmental

Quality Institute (USDA) located at Beltsville, Md.[79] The strip mine site at Frostburg, Md., received composted municipal sewage sludge at 56, 112, and 224 t/ha/yr as a mulch application and as incorporation (5 to 8 cm) material. Revegetation occurred at all loading rates. The best combination for forage production was 224 t/ha composted sludge plus 11 t/ha dolomitic limestone. The conclusion was that, composted sewage sludge with a low metal content; (a) is a valuable resource for reclamation of acid mine spoil, (b) supplies valuable plant nutrients for plant growth, (c) raises the soil pH which is both favorable for metal attenuation by the formed top soil and for plant growth with minimal heavy metal content, (d) is useful as a mulch and surface incorporated material promotes germination of seeds, and protects seedling establishment, (e) increases water retention and infiltration, and (f) ameliorates the phytotoxicity of constituents both in the sludge and in the spoil.

Lime and fly ash have received experimental attention as amendments to acid coal mine refuse (gob) primarily because of their acid neutralizing properties. Fly ash generally is less effective in raising the soil, spoil, or gob pH than lime. Fly ash mixed with gob at a rate of 134 t/ha and lime (85% $CaCO_3$ equivalent) at a rate of 67 t/ha were found to benefit the establishment and growth of tall fescue and smooth brome grass.[92]

2. Iron Mined Lands

Orphan mined lands originate from mining operations for coal and other minerals of copper, lead, zinc, phosphate, soil particulates, underground mine waste, etc.[93,94] They are referred to as "Orphan" because the miner cannot be identified or is not available. Orphan waste land varies widely in composition from one location to another and in size. Since these lands typically appear on steep slopes, are irregular, and often topographically unsuited for revegetation, they require regrading and resurfacing to adjust slopes for minimal susceptibility to soil and water erosion. Other soil management necessary includes addition of fertilizers, lime, and soil amendments, and mulch. Grass, legumes, and trees must be planted and nursed during seedling establishment. Municipal sewage sludges, compost, woody and forest residues, manures, and other non-toxic and organic agricultural residues aid in improving the soil physical properties. Sewage sludge loading rates up to 750 to 1000 dry t/ha (300 to 400 t/a) are considered feasible.[94] The material may be applied either as a liquid or solid. Requirements for such high rates of dry matter application, however, may eliminate the practicability of liquid application.

One example of an orphan mine land is provided by the Contrary Creek project, located in Louisa County, Va., that for 50 years has remained virtually barren of vegetation. The three pyrite mines border Contrary Creek which does not support fish or other aquatic life due to sulfuric acid seepage into the waters along the creek borders. The pH of the creek drops to values of 2.6 in this stretch of several miles.[94] Steps in the management of this highly toxic waste were as follows:[94] clear waste areas, grade and smooth, construct diversions and waterways, salvage and spread topsoil, excavate channels, place loose rock riprap, spread wastewater sludge, seed and plant, and provide erosion control. Municipal sludge was spread onto the leveled mine waste surface to a depth of about 10 cm and disced into the top few centimeters. After seeding, chopped straw was mulched over the surface at a rate of 4.9 t/ha (2 T/a). Emulsified asphalt was sprayed onto the straw to hold it in place. The results are remarkable; the restoration area is vegetated with established grass, acid mine mine drainage pollution is reduced, and the revegetation is pleasing to the eye. Biological hazard was assessed by analyzing water from an infiltration ditch. Coliform organisms 6 to 8 months after sludge application were recorded at 230 per 100 mℓ. This fell to 75 per 100 mℓ, an acceptable level for wildlife range, 6 months later.[94]

3. Spent Oil Shale

Spent oil shale originates from extracting oil from shale by retorting at high temperatures. Large quantities of spent oil shale may be expected to accumulate as a result of the critical world energy situation. The spent shale contains unspent carbon and, therefore, appears as a decidedly black residue. Due to its high salinity, revegetation takes place very slowly, if at all. Unlike coal spoil, spent shale is alkaline (pH 8.3) and saline (salty) with a sodium-sorption-ratio (SAR) of about 15. The SAR is a ratio for soil extracts and irrigation waters used to express the relative activity of sodium ions in exchange reactions with soil:

$$SAR = \frac{NA+}{\sqrt{Ca^{++} + Mg^{++})/2}}$$

where the ionic concentrations are expressed in milliequivalents per liter.

In an extensive greenhouse pot experiment, Williams and Packer[95] demonstrate the value of a number of organic wastes for reclaiming spent shale. The fourteen waste amendment treatments for growing tall wheat grass on spent oil shale were:[95]

- Control (no treatment)
- Fertilizer (N-P-K)
- Sulfur
- Fertilizer and sulfur
- Fertilizer and sewage sludge
- Fertilizer, sulfur, and sewage sludge
- Fertilizer and wood fiber
- Fertilizer, sulfur, and wood fiber
- Fertilizer and straw
- Fertilizer, sulfur, and straw
- Fertilizer and steer manure
- Fertilizer, sulfur, and steer manure
- Fertilizer and beet pulp
- Fertilizer, sulfur, and beet pulp

All organic materials were added at a rate of 5% of the shale by volume. A leaching treatment was imposed on duplicates of all treatments.

The data for biomass production on leached and unleached spent oil shale may be seen in Figure 7. The excessive salt problem in the spent oil shale is demonstrated in Figure 7. All of the amendments were more effective in supporting wheatgrass growth and yield as a result of removing some of the excess salt by leaching. Some of the adverse salt effect, however, was overcome by sewage sludge, manure, and beet pulp alone. Neither sulfur nor fertilizer alone significantly increased wheatgrass production. In fact, sulfur alone appeared to worsen the salt effect in the absence of leaching. These greenhouse studies provide a good screening method for selecting field treatments in a practical land utilization program.

4. Overburden, Tailings, Sand Dunes, and Playa

Overburden and tailings from mining operations occupy large areas of land in the United States.[96-100] Many of these waste lands are highly susceptible to soil, water, and wind erosion. In the western U.S wind erosion is especially a problem for the copper mine wastes, due to dryness of the powdered tailings. Waste lands from mining operations that cause pollution problems have been grouped into two main categories, (a) waste resulting from strip and surface mining consisting of soil materials, the reclamation and revegetation of which may be accomplished by the usual agricultural farming practices, and (b) waste land composed of tailings, the reclamation and revegetation of which requires specialized cultural practices. These practices are costly. On the other hand, a study that shows the effects of four soil materials (desert soil, overburden, overburden and tailings, and tailings) on growth and

General Schematic of Peat Filtration System

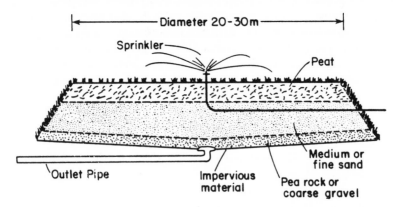

FIGURE 7. The average biomass of tall wheatgrass on leached and unleached spent oil shale subjected to various fertilizer and amendment treatments. (From Williams, B. D. and Packer, P. E., Sewage sludge and other organic materials as amendments for revegetation of spent oil shale, in *Utilization of Municipal Sewage Effluent and Sludge on Forested and Disturbed Land*, Sopper, W. E. and Kerr, S. N., Eds., Pennsylvania State University Press, University Park, 1979, 353. With permission.)

quality of grain from barley (*Hordeum vulgare*) was favorable.[101] Immediately after planting on a 1:5:1 tailing berm slope, barley straw at a rate of 11,200 kg/ha was applied to the experimental area, followed by 19 kg/ha N in 1 cm of irrigation water. During each subsequent irrigation 50 kg/ha N was applied.[101]

The barley height, heads per unit area, seeds per head, seed weight, and grain yields were greatest from desert soil, followed by barley grown in overburden, overburden and tailings, and tailings alone, respectively. With special attention to management, in decreasing order barley can revegetate certain copper mine waste and soil material and produce an acceptable quality grain and straw for livestock feed.[101]

In similar experiments with mine tailings, overburden soil material, mined areas, tailing ponds, and dumps, the reclamation and revegetation was readily accomplished and the waste returned to valuable productive soil.[98]

5. Saline and Alkali Waste Lands

About half of the earth's land is in arid or semiarid climatic zones. The accumulation of soluble salts and sodium in excessive amounts in these areas constitutes a serious problem. In the United States, one fifth of the lands in 14 of our western states is affected by excessive accumulations of salts. Abandoned irrigation farms are commonplace due to the development of saline and alkali conditions.[102] Economically feasible reclamation practices to return fields so affected to productive capacity are necessary, since excessive salts are wastes, whether generated by nature, agricultural industry, or poor water management, and must be corrected. Land treatment of these waste lands can conceivably bring them into productive uses.

A saline soil by definition has a saturation extract conductivity greater than 4 mmhos/cm and an exchangeable sodium percentage of less than 15. A black alkali soil is a "sodium" soil. It has an exchangeable sodium percentage greater than 15. The present trend is to refer to the latter as "sodic soils". The term, "saline-alkali soils" refers to a saline (salty) as well as a sodic (alkali) soil.

Some of the most prominent methods employed in the reclamation and improvement of saline and alkali waste land are listed for convenience; (a) drainage to lower water table when needed, (b) organic material application with leaching to transport salts from root zone, (c) applying waste chemicals to supply or make available soil calcium to replace sodium; (d) tillage to break up soil stratifications, compact layers, hard pans, pressure pans, (e) soil structure improvement, and (f) vegetation management using organic waste residues.

If subsoil drainage is poor and a high water table exists, the provision for adequate drainage is the first step in reclamation. Soluble salts which are present in the seepage water move to the surface by evaporation where accumulate. Poor drainage and intermittent wetting intensify the upward movement of these salts and their accumulation on and near the soil surface.

A pattern of deep ditch construction often has been successful in draining large areas as well as in intercepting seepage from higher ground. The ditch should be constructed deep enough to drain water below the root zone and large enough to remove, rapidly, the seepage under all intensities of irrigation practices.

An effective drainage system must provide for leaching of soils to remove the accumulated salts from the root zone. Tile drainage has been used successfully in many areas, but the cost of installation must always be balanced against effectiveness in improving land values. Certain soil types lend themselves better to tile drains than others. Often open ditch and tile drainage systems are integrated for a more economical and effective program of drainage. Tile drainage is less economical to maintain than open ditch methods.

Drainage wells have been constructed in the Southwest U.S.A. to lower the water level of certain areas where high water tables cause serious salinity problems. The "draw-down" wells are pumped and emptied into canals or pipes to carry the water away.

Leaching is the second step in reclamation following drainage. A favorable balance between the amount of salt brought into an area by irrigation waste and the amount removed from the root zone by drainage and crops must exist in order to maintain permanent agriculture. The leaching requirement of a soil may be defined as, "the fraction of the irrigation water that must be leached through the root zone to control soil salinity at any specific level." The leaching requirement may be calculated by formula. The final value will depend upon the salt concentration permissible in the soil solution for a given crop.

Some saline soils can be improved for crop production by leaching with water only. Other saline soils lose their favorable permeability upon extensive leaching with good water. These soils and those that are inherently poorly permeable may be improved by use of gypsum or some other chemical that will provide soluble calcium to maintain flocculation.

Waste chemicals: alkali "sodic" soils must be supplied with soluble calcium either directly or indirectly in the irrigation water. Calcium is necessary for replacing sodium before the soil can be improved in physical condition sufficiently to permit adequate permeability of water for reclamation. Calcium provides good tilth whereas sodium causes the soil to disperse or puddle if cultivated or compacted when wet and thus to form clods upon drying.

Waste amendments and wastes like sulfuric acid, ferrous sulfate, and plant or crop residues are most commonly used in reclaiming land. They are preferred because they are abundant and not as costly as other chemicals, and also partly because they are not toxic to plants and do not form toxic compounds with the soil.

Calcium is present on most arid soils as lime (calcium carbonate), gypsum (calcium sulfate), calcium phosphate and other salts of lesser importance. The solubility of these native sources is so low that they are not sufficiently effective alone to achieve reclamation. Hence, sulfuric acid, sulfur, iron sulfate, and calcium polysulfide may be applied to soils to react with the native sources to render the calcium more soluble.

Gypsum, calcium sulfate ($CaSO_4 \cdot 2H_2O$), is used more than any other chemical in soil reclamation. It is applied to the soil prior to leaching. The amount to apply depends upon the soil. Laboratory analyses should be used to determine the gypsum requirement for each soil. It is seldom that less than 4.5 t/ha are applied where reclamation is needed.

Table 17
EFFECT OF SURFACE APPLIED SULFURIC ACID ON THE
DEPTH OF WATER PENETRATION DURING 5 HOURS
UNDER A 1 CM DEPTH OF CONTINUOUS PONDING,
LABORATORY TESTS

Soil Samples	Sampling Locations	Exchangeable Sodium	H_2SO_4, tons/ha[a]			
			0	1	5	10
		%		cm		
Playa clay loam	Willcox	100	3	7	16	25
Stewart silt loam	Willcox	100	10	16	33	57
Gila clay loam	Ft. Thomas	85	4	11	14	19
Pima clay loam	Safford	25	13	16	19	21
Gothard clay loam	Willcox	21	23	31	37	40
Gila clay loam	Yuma	11	12	15	21	12
Elfrida loam	Kansas Stl.	2	36	41	42	28
Cave gravelly loam	Tucson	0.2	39	42	44	35

[a]   To convert to U.S. tons/acre, multiply by 0.45.

From Miyamoto, S. and Stroehlein, J. L., *Prog. Agric. Ariz.*, 27, 13, 1975.

Because sodium salts in irrigation water will reduce the permeability of water through soil, irrigation water is often treated with finely ground gypsum. The calcium of gypsum helps to counteract this effect.

Sulfuric acid (H_2SO_4) reacts immediately with the soil lime to make calcium soluble and available for replacing sodium. It may be placed effectively in irrigation water.

Sulfuric acid is an industrial waste of great proportions (estimated at about 4.4 million tons) and the amounts will more than double if air restrictions are enforced. Direct application of sulfuric acid to soils to reclaim sodic and saline soils has been practiced in the Western U.S.A. for over 75 years. The hydrogen ion or acidity is responsible for the benefit in soil reclamation.[103,104] Sulfuric acid reacts with the lime (caliche) to form soluble calcium to replace sodium ions and other monovalent ions, making the soil more permeable and better structured. Improvement in depth of water penetration and in water infiltration rate by application of sulfuric acid, ferrous sulfate, and waste sulfur to soils has been well proved both in field and laboratory, Table 17.[105]

Sulfur oxidizes readily in the soil to form sulfuric acid. Certain soil microorganisms, always present in soils, are responsible for this process. The sulfuric acid in turn reacts with the calcium compounds in the soil forming gypsum. Agricultural sulfur should be finely ground since its rate of reaction in the soil is directly related to fineness of particle size.

Iron sulfate, a mine waste, as well as aluminum sulfate have been applied to soils to make soil calcium more soluble. The process is achieved by their oxidation when placed in contact with the soil to form sulfuric acid. Iron and aluminum sulfates are usually broadcast on the soil as is gypsum. Both are soluble in water. Ferrous sulfate is a mining industry waste.

Calcium polysulfide (lime sulfur) is a brown solution, low in calcium content. Sulfur is its most important constituent for reclamation purposes. Upon contact with water the sulfur changes from the liquid to the solid state and appears as very fine colloidal particles. The sulfur readily oxidizes to sulfuric acid in the soil which in turn forms soluble calcium from soil lime.

Special tillage practices have not been used extensively in the reclamation of saline and alkali soils. Where shallow compact layers of impermeable soil occur, deep plowing has improved water penetration by breaking up the layers. Where a shallow layer of clay overlies sand, deep plowing has been effective in reclamation of "hard spots" and clay "slicks".

Horizontal layers or deposits of soils of different textures in irrigated valleys have been successfully broken up by deep plowing to improve water penetration and percolation so necessary for reclamation.

Chiseling and ripping have also been used effectively to relieve soil compaction and to allow improved water penetration as well as root penetration. The working of straw or crop residue trash into the trenches is advantageous for better water penetration. Exploratory experiments show that vertical mulching offers more promise for improving infiltration rate of water over a longer period of time than do chiseling and ripping.

Soil structure improvement is important to reclamation since structure influences the rate of water infiltration into the soil and consequently leaching of salts from the root zone. Alternate wetting and drying to the wilting point of plants improves soil structure for greater water improvement.

Organic matter additions, if sufficiently large, can also improve soil structure. Organic residues provide a source of energy for the microorganisms that produce slimes and gums, which promote the formation of stable soil aggregates. Manure alone, however, has not proved as effective in reclamation of alkali soils as gypsum or sulfur unless added in very great quantities. Manure plus sulfur is more effective than either alone. Likewise, manure plus gypsum is more effective than either alone. Chopped straw has been shown to improve the infiltration rate of a sodium irrigation water into fine sandy loam as much as gypsum does.

The failure to recognize that saline and alkali soils require special management treatments, even after fairly complete reclamation, has resulted in low revegetative production and often in complete plant failure.

6. Urban Lands

City parks have used solid sewage sludge top-dressing for soil amendment and fertilization programs for many years-some 50 years or more. The older city parks in Tucson and Phoenix, for example, have received sludge as fertilizer for over 65 years. Sun dried Phoenix-sewage sludge, also, was used as filler base for mixed commercial (N-P-K) fertilizer for over 40 years with no problems being reported. The well known commercial organic fertilizer Milorganite® has been sold to home gardners and landscape fanciers all over the U.S.A. Some users were not aware of its Milwaukee city sludge origin. A 5% nitrogen content was guaranteed on the label. Highway embankments and urban street medians accept sewage sludge, forest residues, and agricultural straw residue levels ranging from scatterings to blankets. Some are applied as organic fertilizer, others as mulch to start revegetation, and still others to control soil, water, and wind erosion.

C. Wastewater Loading Rates

The composition of wastewaters varies greatly from one source to another. Soils and landscapes also vary. Rather than attempt to provide detailed loading rates for specific wastewaters, those properties of the soil and wastewater influencing loading rate will be discussed as related to a particular landscape, and geologic, and climatic condition. Attention will center primarily on wastewater quality with respect to its polluting constituents; nitrogen (N), phosphorus (P), organic matter (OM) as evaluated by biological oxygen demand (BOD), human pathogens, organic toxicants, and heavy metals. Wastewater loading rates should not exceed the rate and capacity of the soil for retention and confinement. Thus, the ability of the soil to renovate wastewater for vegetative production and groundwater recharge is a function of the combined effects between the soil biological, chemical, and physical properties and those of the wastewater pollutants as listed above. Methods used to apply the wastewater provide another dimension to loading rates that must be given serious consideration.

1. Nitrogen

The transformations of nitrogen in the soil are controlled almost wholly by biological action. The soil environment must be made favorable for the microorganisms involved in the desired transformations. Aerobic conditions favor mineralization to nitrate (NO_3^-) and anaerobic conditions favor denitrification mostly to elemental nitrogen (N_2), and some NO and NO_2, which escape to the atmosphere as gases. Nitrogen in the organic form is associated mostly with protein and amino acids. In the organic form, ammonium (NH_4^+) and nitrate (NO_3^-) predominate, although traces of nitrite and cyanide sometimes occur in the surface organic layer of soils. In anaerobic soil, nitrite forms predominate with very little-to-none of other, inorganic forms. The behavior of ammonium and nitrate ions differs markedly in soils. Ammonium ions enter into cation exchange with the soil particulate surfaces and, therefore, move through the soil considerably less rapidly than nitrate that is very weakly held as an anion. In fact, nitrates move wherever water moves and at about the same rate unless it is intercepted or denitrified. Because of the high degree of mobility, nitrate pollution of groundwater always is a threat when renovated waters contribute seen from recharge to groundwater. A number of factors influence N accumulation during renovation as seen from Table 18 that relates percentage N renovation with waste loading rate, dose-schedule, soils, vegetation cover, and method of application.

Loading rates of nitrogen-containing wastewaters, usually are controlled more by the nitrogen concentration than any other single pollutant. The concentration of nitrogen reaching the groundwater is primarily altered by (a) soil type (as related to soil characteristics explained earlier), (b) depth of the soil, (c) method and technique of application, and (d) vegetative cover. A relationship between loading rates, method of application, and nitrogen removal rates for municipal wastewaters is reported in Table 18.[106] Nitrogen removal is influenced by all four factors listed above, but most markedly by soil type as reflected in particle size distribution or surface area per unit weight. The more clay, the greater the loss of nitrogen, due most likely to biological denitrification in the microenvironment of soil aggregates during periods of intermittent flooding.[107] Denitrification also requires sources of electrons and active microbial populations. The organic matter already present in the soil and that added in the wastewater represent viable sources. Deep, fine textured soils (minimum 0.7 m [2 ft] depth) maximize denitrification and loss to the atmosphere of nitrogen gases. Unfortunately, soil anaerobic conditions most desirable for denitrification support plant growth only to a limited extent. A balance between the two objectives must be made and a compromise reached between nitrogen and carbon removal and longevity of the soil to function as a permanent nitrogen sink. Moreover, increasing the wastewater applied (hydraulic loading), dosage, and schedule of application results in a larger mass removal of N but a lower removal percentage. Further increases in application result in the buildup of organic matter ''thatch'' on the soil surface slowing the rate of N removal. Such clogging and slow-down of soil effectiveness was controlled at Flushing Meadows by the ''off-on'' application technique and drying of the surface.[60] Further improvement in penetration and infiltration can be effected by surface scarification where the site will permit use of such equipment.[60] The general consensus of opinion as to the best system for nitrogen removal from wastewater is the use of spray irrigation on a silt loam soil, with a moderate-to-high infiltration capacity, and a continuous vegetative cover.[4,6,63,81,106] Removal of nitrogen by crops and other equally dense vegetation usually accounts for about 50% of that applied in wastewaters and sludges (Figure 5).[4] Various methods have been suggested[108] to reduce nitrification rates or to absorb nitrates; such as (a) denitrification, (b) nitrification inhibition, and (c) vegetative cover. Adequate loading rate is achieved when the nitrate concentrations in renovated water meet the U.S. Public Health potable water standards in leachate recharging groundwater.

2. Phosphorus

Movement of inorganic phosphorus through soils is extremely limited in almost any soil.

Table 18
THE EFFECT OF LOADING RATES, DOSE-SCHEDULE, SOIL TYPE AND VEGETATION COVER AND METHOD OF APPLICATION ON EFFICIENCY OF NITROGEN AND PHOSPHORUS REMOVAL

Amount Sprayed cm/h	Dose-Schedule	Duration	Renovation, %[a] N	P	Soil Type	Soil Cover
		Rapid Infiltration Ponds				
36	2 day infiltration, 2 day rest	year round	0—10	62	silty sand	bare—grass
36	14 day infiltration, 14 day rest	year round	30	62	silty sand	grass
55	6 month infiltration, 6 month rest	183 days	92	93	sand	grass
6	7 day infiltration, 7 day rest	year round	11	100	sandy loam	bare
6	14 day infiltration, 7 day rest	year round	42	94	sandy loam	bare
6	continuous 120 days, few months rest		86	55	sandy loam	bare
12	1 dose/day 10 hours	year round	0—58	0—34	sand	bare
		Overland Runoff				
0.9	2 doses of 6 hrs, 5 days/week	150 days	90	65—84	clay loam	grass
0.63	1 dose of 7 hrs, 5 days/week	year round	84	56	sandy loam	grass
0.63	1 dose of 7 hrs, 3 days/week	year round	84	83	clay loam	grass
		Spray Irrigation				
2	1 day/week	Apr.—Nov.	80	99	loam	forest
1.4—2	2—3 days/week	Apr.—Nov.	91	99	loam	forest
2	1 day/week	52 weeks	0	91	loam	forest
4.5	3 day sprayed, 4 day rest	year round	56	91	silty sand	forest
2	1 day/week	year round	97	99	loam	reed canarygrass

[a] Renovation is expressed as percentage of the original effluent element content that is retained in the soil utilized by vegetation or volatilized. Concentrations of N and P in the effluents used in the experiments reported in the table were around 20 and 10 mg/ℓ, N and P, respectively.

Modified from Spyridakis, D. E. and Welch, E. B., 1976.

In acid soils iron and aluminum phosphate complexes form readily and in calcareous (arid and semiarid region) soils, calcium phosphates form in addition. Removal of phosphorus from municipal wastewaters and sludges and retention by the soil is not a difficult problem. Crop and vegetation removal accounts for about 30% of the P applied in most organic wastes and wastewaters.[109] Renovation of phosphorus from municipal wastewater by soil is quite efficient. Soil texture and soil depth influence retention or renovation of P more than any other factors, Table 18. The method of application exercises an indirect effect since the rapid-infiltration techniques of recharge necessarily are associated with coarse-textured soils and high loading rates. Unless overland flow (or runoff) techniques are carefully controlled,

P attached to soil particle surfaces, can be lost to surface waters through erosion. Should this occur, problems of eutrophication can be aggravated in surface waters. Spray irrigation most efficiently removes P (>95%) from wastes, Table 18. The long-term loadings of P require special attention, however, because the capacity of the soil to absorb P can be overreached, particularly if the soil is coarse textured. Movement of P through soils in the organic matter and fine colloidal particulates has been shown to be more pronounced than the movement of inorganic forms of P through soils.[109-111]

3. Organic Matter (BOD, COD, TOC)

Preventing soluble organic matter (as measured by BOD, COD, and TOC) from reaching the groundwater is controlled almost wholly by biodegradation processes. Providing the soil with the most favorable environment and controlling the hydraulic loading rate in a compatible manner to the degradation rate is the best way to manage the soluble organic matter of wastes. The two most limiting elements in organic matter reduction are nitrogen and oxygen. A continuous supply of soil moisture also is essential to provide a favorable microbial habitat for maximum biodegradation. Furthermore, organic matter decomposition stops when the soil freezes, a process linked to dehydration. Usually soil freezing is limited to the upper few centimeters except in extremely cold climates where the cold season remains prolonged. Excessive loading rates of organic matter create problems. Soils become anaerobic, decomposition slows, odors form, and flies and insects accumulate. According to Spyridakis and Welch,[106] a model of secondary effluent with a BOD concentration of 25 mg/ℓ and an application rate of 5.1 cm (2 in.) per week or 12.8 kg/ha/week (11.5 lb/ha/week) should not result in adverse accumulation of undecomposed organic matter in surface soil. The loading rate appears conservative for permeable soils. Hydraulic loading rates much greater (e.g., 100 kg/ha/d and greater) have been made with no appreciable accumulation of organic matter or anaerobic conditions. For example, sulfate pulp mill waste with a BOD of 1150 mg/ℓ was spray irrigated on reed canarygrass sod on sandy soil at a BOD loading rate of 155 kg/ha/d (138 lb/a/d) with a resultant BOD removal of 99%. Yet at a loading rate of 7.6 cm/d, the soil remained saturated and the grass died, a reminder again that hydraulic loading rate and efficiency of organic matter degradation (or BOD loading rate) should be considered together in any long-term disposal program depending on climatic factors of rainfall and temperature.

4. Heavy Metals

Recent research on noncultivated land receiving municipal sludges (solids) and municipal wastewaters, fortunately, does not reveal excessive migration or movement of significant amounts of heavy metals to groundwaters where reasonable precautions are taken to avoid such occurrences. The most prominent soil properties related to attenuation of heavy metals in soils, have been identified as; (a) particle size distribution, clay, (b) surface area per unit weight, (c) hydrous oxides of Fe, and probably Mn and Al, (d) soil pH, (e) cation exchange capacity (CEC), as directly related to texture, (f) total dissolved solids, soluble salts, and (g) total organic carbon, TOC.[112,113] By regression analysis, 80% or more of the attenuation is accounted for by the pH, clay, hydrous oxides of iron (Fe) of the soil, and total organic carbon (TOC), and soluble salt (TDS) of the vehicle (aqueous) of transport or leachate. The higher the soil pH, CEC, and content of clay and hydrous oxides of Fe, the greater is the heavy metal retention. The lower the soluble salt (TDS) and organic carbon components (TOC) of the vehicle of transport, the greater is the retention. Other soil characteristics such as soil organic matter, structure and physical barriers of compaction (bulk density), lime and gypsum layers, silicated hardpans, clay and ortstein (iron pans) and clay minerals composition, shrinking and swelling, and drying and wetting all exert influence on heavy metal retention as they relate to permeability and infiltration. Oxygenic and anoxygenic

Table 19
RELATIVE SOLUBILITIES OF VARIOUS CATION-ANION COMBINATIONS IN WEAKLY ALKALINE AND REDUCING SOLUTIONS

Toxic cations	Major anions			
	CO_4^-	OH^-	SO_3^-	$SO_4^=$
Be^{2+}	Soluble	Insoluble		Insoluble
Cd^{2+}	Insoluble	Insoluble soluble	Slightly	Soluble
Cr^{2+}	—[a]	Insoluble	—[a]	Insoluble
Cu^+	Insoluble	Insoluble	Slightly soluble	Not formed in water
Hg^+	Insoluble	Insoluble	Insoluble	Slightly soluble
Pb^{2+}	Insoluble	Slightly soluble	Insoluble	Insoluble
Zn^{2+}	Very slightly soluble	Insoluble	Slightly soluble	Soluble

Toxic anions	Major cations	
	Ca^{++}	Mg^{++}
AsO_3^-	Slightly	?
SeO_3^-	Insoluble	Insoluble

[a] uncommon.

conditions also must be considered since reducing conditions in soil encourage heavy metal migration. The relative solubilities of certain cation-anion combinations such as those compiled in   Table 19 should be kept in mind. A conservative estimate of a loam-textured soil depth capable of retaining heavy metals for long periods of time for most municipal and many industrial wastes is 1 to 2 m (3 to 6 ft). Under forest vegetation much shallower depths of soil 0.5 to 1.2 m (1.5 to 4 ft) remove heavy metals satisfactorily for several years.[3] In the Pacific Northwest, dewatered sewage sludge applied at loading rates of 25 cm to cleared forest land and 10 cm to forested land showed very little mobility of heavy metals and phosphorus on sandy soils.[111] In the southern Appalachians on steep, wooded slopes (15 to 30% and 90 to 110 m long) treated sewage wastewater sprayed at 76 mm (3 in.) per one day a week at a rate of 8 mm/h (0.32 in./h) was found to be adequately renovated to be returned to stream surface water.[114] The textures ranged from sandy loam in the surface to 25 to 70 cm permeable clay loam to 120 to 155 cm to loam in the C horizon below.

Sandy, well-drained soils in cold northern climates also have been used for wastewater renovation. Loose sandy soils, as expected, retain polluting constituents of wastewaters poorly, but with high density vegetation and surface organic "mull" and/or "duff", interception of the pollutants may be expected. This condition was tested in Barry County, Michigan.[115] The results of field-sprayed irrigation rates of 0, 30, and 70 mm/wk indicate that the effluent had no direct adverse effect from excess moisture or toxic elements on eight different tree species. Some adverse ecological effects are reported. In a similar experiment in Michigan, loading rates were 0.38, and 76 mm/wk of liquid sewage effluent along with surface applied sludge at rates of 7.5, 15, and 23 t/ha and another at rates of 11.5, 23, and 46 t/ha (dry). There was no indication in either well or stream water that either heavy metals or nitrates had migrated to water sources.

A great number of research investigations, experiments, and studies were undertaken in the United States during the 1970s. Experience with renovation of municipal wastewater and fertilization with municipal solid waste indicate considerable benefit can be expected for vegetative growth and soil fertility improvement without significant deterioration of groundwater or surface streams. When cautiously applied, and within reasonable loading rates, other wastes also have been used without reducing the quality of surrounding waters.[116] Most investigators have kept the heavy metal concentrations well within the U.S. Department of Public Health potable water standards in leaching and recharging groundwaters. The length of time these land treatment/utilization experiments have been underway on a quantitative basis, however, has not been long. Information during short-terms should be viewed with caution since the data may not represent the long-term steady state conditions. Furthermore, great attention should be given to site selection, site premonitoring analysis, and "safe" loading rates for each associated pollutant contained in the waste (resource) before the final decision for a disposal operation and management program is initiated. Monitoring programs during waste discharges should include continuous quality measurements of the associated ecosystem and environment stability.

5. Organic Toxicants

Certain wastewaters and sludges from municipal as well as industrial sources have been known to contain some organic toxicants. The data on municipal sewage effluents are lacking or fragmented at best, although, chlorobiphenols, phthalates, phenolic compounds, and some of the chlorinated hydrocarbon pesticides have been identified.[117] Estimated levels of no less than 1 mg/ℓ have been made.[106] Numerous organic toxicants are applied both unintentionally or in pest control. A great number of reactions occur in soils to absorb, attenuate, degrade, or precipitate them. Organic toxicants may be adsorbed on the soil particle surfaces, biodegraded by microorganisms, chemically degraded, volatilized, leached, or taken up by vegetation. Certain monitoring for toxic organic constituents should be made in all land utilization of wastes involving agronomic crops or other vegetation for public consumption.

6. Human Pathogens

Municipal sewage sludges and wastewaters contain human health hazards even though the hazards may be minimal. The transmission of pathogens from disposals through groundwater pollution and through entrances into the food chain and through aerosols represents only part of the problem. Animals grazing on pastures irrigated with contaminated wastewaters and sludges also may be seriously infected and carry diseases back to human beings. The public health implications of land utilization require a great detail of attention, far more than can be offered in a few paragraphs. However, excellent reviews are available.[118-121] Loading rate recommendations for soil management of human pathogens are just emerging from health services in the various states. The control guidelines of land applications are being finalized in Arizona, California, Colorado, Florida, New York, and Texas. Other states either have begun establishing land application guidelines or are moving in that direction, though cautiously.

IV. EFFECTS ON ECOSYSTEMS

The local ecosystems and environment cannot escape from change or alteration as a result of land treatment and waste utilization, particularly in noncultivated areas. Indeed, the purpose of waste utilization on disturbed soils is to provide for improvement in the environment of the land, of both surface and subsurface soils, and to return the landscape to its original beauty and the soils to usefulness. Utilization of the undisturbed soil as a "living filter" also is considered to be a most ecologically appropriate practice.[76] Recycling of plant nutrients and organic carbon is a natural and necessary process for maximizing the aesthetics and usefulness of the plant and soil ecosystem. In any event, waste disposal and utilization on the land will have some kind of ecological and ecosystem impact. If the impact is favorable for agronomic crop production, more rapid tree growth, or better pasture and grazing, the procedures and intensity of disposal are looked upon with favor. All changes and alterations in ecosystems are evaluated in the light of many and varied interests.

The judgment of the effects on ecosystems of waste utilization on noncultivated lands must be limited to actual cases of recent origin. This is partly because only recently has a workable knowledge of land treatment been sufficient to avoid accidental ecological damage, and due to the lack of knowledge, outcries as to adverse environmental impacts have been loud and clear. There has not been sufficient long-term quantitative experimental research data generated to be able to make reliable longtime predictions as to the effect of land treatment on natural ecosystems and most environments. Nevertheless, there are data on disposal management that can be helpful in avoiding the most obvious environmental or ecosystem deterioration. Much of this information, in the previous section on ''Loading Rates'', is a sample of such data.

REFERENCES

1. **Sopper, W. E. and Kardos, L. T.,** Vegetative response to irrigation and municipal wastewater, in *Recycling Treated Municipal Wastewater and Sludge Through Forest and Cropland,* Sopper, W. E. and Kerr, S. N., Eds., The Pennsylvania State University Press, University Park, 1973, 271.
2. **Leach, L. E., Enfield, C. G., and Harlin, C. C., Jr.,** Summary of Long-term Rapid Infiltration System Studies, EPA-600/2-80-165, U.S. Environmental Protection Agency, Cincinnati, Ohio, 1980, 51.
3. **Sopper, W. E.,** Crop selection and management alternatives - perennials, in *Recycling Municipal Sludges and Effluents on Land,* U.S. Environmental Protection Agency, U.S. Department of Agriculture, and National Association of State Universities and Land Grant Colleges, Washington, D.C., 1973, 143.
4. **Breuer, D. W., Cole, D. W., and Schiess, P.,** Nitrogen transformation and leaching association with wastewater irrigation in Douglas fir, poplar, grass, and unvegetated systems, in *Utilization of Municipal Sewage Effluent and Sludge on Forest and Disturbed Land,* Sopper, W. E. and Kerr, S. N., Eds., The Pennslyvania State University Press, University Park, 1979, 19.
5. **Smith, W. H. and Evans, J. O.,** Special opportunities and problems in using forest soils for organic waste application, in *Soils for Management of Wastes and Wastewaters,* Elliott, L. F. and Stevenson, F. J., Eds., Amer. Soc. Agron., Madison, Wis., 1977, 429.
6. **Urie, D. H.,** Nutrient recycling under forests treated with sewage effluents and sludge in Michigan, in *Utilization of Municipal Sewage Effluent and Sludge on Forest and Disturbed Land,* Sopper, W. E. and Kerr, S. N., Eds., The Pennslyvania State University Press, University Park, 1979, 7.
7. **Miner, J. R. and Hazen, T. E.,** Transportation and application of organic waste to land, in *Soils for Management of Organic Wastes and Waste Waters,* Elliott, L. F. and Stevenson, F. J., Eds., Amer. Soc. Agron., Madison, Wis., 1977, 379.
8. **Walker, J. M.,** Trench incorporation of sewage sludge, Proc. Natl. Conf. Municipal Sludge Management, Pittsburgh Pa., June, 1974. 1.
9. **Reed, C. H.,** Equipment for incorporating animal manures and sewage sludges into the soil, in *Managing Livestock Waste,* Proc. Third Int. Symp. on Livestock Waste, ASAE Pub., PROC-275, Urbana, Ill., 1975, 444.
10. **Powers, W. L., Wallingford, G., Murphy, L. S., Whitney, D. A., Manges, H. C., and Jones, H. E.,** Guidelines for Applying Beef Feedlot Manure to Fields, Coop. Est. Serv. Bull. C-502, Kansas State University, Manhattan, 1974, 11.
11. **Gilmour, J. T. and Gilmour, C. M.,** Simulated model for sludge decomposition in soil, *J. Environ. Qual.,* 9, 194, 1980.
12. **Phung, T., Larker, L., Ross, D., and Bauer, D.,** Land Cultivation and Industrial Wastes and Municipal Solid Wastes: State-of-the-Art Study, Vol. II, Field Investigations and Case Studies, EPA-600/2-78-140b, U.S. Environmental Protection Agency, Cincinnati, Ohio, 1978, 156.

13. **Halvorson, G. A.,** Movement of Elemental Constituents in Sagehill Loamy Sand Treated with Municipal Waste, M.S. thesis, Oregon State University, Corvallis, 1975, 35.

14. **Cottrell, N. M.,** Disposal of Municipal Waste on Sandy Soil: Effects on Plant Nutrient Uptake, M.S. thesis, Oregon State University, Corvallis, 1975, 101.

15. **Phung, T., Barker, L., Ross, D., Bauer, D.,** Land Cultivation of Industrial Wastes and Municipal Solid Wastes: State-of-the-Art Study, Vol. I, Technical Summary and Literature Review, EPA-600/2-78-140a, U.S. Environmental Protection Agency, MERL, Cincinnati, Ohio, 1978, 205.

16. **Harris, A. R.,** Physical and Chemical changes in forested Michigan sand soils with effluent and sludge, in *Utilization of Municipal Sewage Effluent and Sludge on Forested and Disturbed Land,* Sopper, W. E. and Kerr, S. N., Eds., The Pennsylvania State University Press, University Park, 1979, 155.

17. **Brockway, D. G., Schneider, G., and White, D. P.,** Dynamics of Municipal wastewater renovation in a young conifer-hardwood plantation in Michigan, in Utilization of Municipal Sewage Effluent and Sludge on Forested and Disturbed Land, Sopper, W. E. and Kerr, S. N., Eds., The Pennsylvania State University, University Park, 1979, 87.

18. **Sopper, W. E.,** Use of the soil-vegetative biosystem for wastewater recycling, in *Land Treatment and Disposal of Municipal and Industrial Wastewater,* Sanks, R. L. and Asano, T., Eds., Ann Arbor Science Pub. Inc., Ann Arbor, 1979, 17.

19. **Garner, H. V.,** Experiments on the direct, cumulative, and residual effect of town refuse, manures, and sewage sludges at Rothamsted and other centers, *J. Agric. Sci.,* 67, 223, 1966.

20. **Kardos, L. T., Sopper, W. E., Myers, E. A., Parizek, R. R., and Nesbitt, J.,** Renovation of secondary effluent for reuse as a water resource, EPA-600/2-74-016, U.S. Environmental Protection Agency, Cincinnati, Ohio, 1978, 495.

21. **Mays, D. A., Terman, G. L., and Duggan, J. C.,** Municipal Compost: Effects on crop yields and soil properties, *J. Environ. Qual.,* 2, 89, 1973.

22. **Kardos, L. T., Scarsbrook, C. E., and Volk, V. V.,** Recycling elements in wastes through soil-plant systems, in Soils for Management of Organic Wastes and Waste Waters, Elliott, L. F. and Stevenson, F. J., Eds., *Amer. Soc. Agron.,* 1977, 301.

23. **Bengston, C. W. and Cornette, J. J.,** Disposal of composted municipal waste in a plantation of young slash pine: Effect on soil and trees, *J. Environ. Qual.,* 2, 441, 1973.

24. **Esptein, E. and Wilson, G. B.,** Composting Sewage Sludge, in *Proc. Natl. Conf. on Municipal Sludge Mgt.,* Information Transfer, Inc., Rockville, Md, 1974, 123.

25. **Epstein, E. and Parr, J. P.,** The application of sludge to land, in *Waste Management Technology and Resource and Energy Recovery,* Proc. 5th Natl. Cong., Natl. Solid Waste Mgt. Assoc. and U.S. Environ. Protect. Agency, U.S. Environmental Protection Agency, Washington, D.C., SW 22 p., 1977, 321.

26. **Bollen, W. L. and Lu, K. C.,** Effects of Douglas-fir sawdust mulches and incorporation on soil microbial activity and plant growth, *Soil Sci. Soc. Amer. Proc.,* 21, 35, 1957.

27. **Aspitarte, T. R., Rosenfeld, A. S., Samle, C. B., and Amberg, H. R.,** Pulp and paper mill sludge disposal and crop production, *Tappi, (Tech. Assoc. Pulp Pap. Ind.)* 56, 140, 1973.

28. **Kincannon, B. C.,** Oily waste disposal by Soil Cultivation Process, EPA-R2-75-110, Office of Resources and Monitoring, U.S. Environmental Protection Agency, Washington, D.C., 1972, 45.

29. **Loynachan, T. E.,** Low-temperature mineralization of crude oil in soil, *J. Environ., Qual.,* 7, 494, 1978.

30. **Mitchell, W. W., Loynachan, T. E., and McKendrick, J. D.,** Effects of tillage and fertilization on persistence of crude oil contamination in the Alaskan soil, *J. Environ. Qual.,* 8, 525, 1979.

31. **Snyder, H. J., Rice, G. B., and Skujins, J. J.,** Disposal of waste oil re-refining residues by land farming, in *Residual Management by Land Disposal,* Fuller, W. H., Ed., Proc. Hazardous Waste Res. Symp., EPA 600-9-76-015, U.S. Environmental Protection Agency, Cincinnati, Ohio, 1976, 195.

32. **Huddleston, C. B. and Cresswell, L. W.,** Oily waste disposal by soil cultivation, in *Proc. Open Forum on Mgt. of Petrol. Refin. Wastewaters, Manning, Ed.,* U.S. Environmental Protection Agency, RSKERL, Ada, Okla., 1976, 101.

33. U.S. Army Engineer Waterways Experiment Station Environmental Laboratory, Effects of Flue Gas Cleaning Waste on Groundwater Quality and Soil Characteristics, EPA-600/2-79-1964, U.S. Environmental Protection Agency, Cincinnati, Ohio, 1979, 117.

34. **Baker, Michael, Jr., Inc.,** State of the Art of FGD Sludge Fixation, Final Rept. Res. Proj. 786-1, Electric Power Res. Inst., Palo Alto, Calif., 1978, 276.

35. **Fry, Z. B.,** The use of liner materials for selected FGD waste ponds, in *Land Disposal of Hazardous Wastes,* EPA-600/9-016, U.S. Environmental Protection Agency, Cincinnati, Ohio, 1978, 256.

36. **Leo, P. P. and Rossoff, J.,** Control of Waste Water Pollution from Power Plant Flue Gas Cleaning Systems: First Ann. R and D Rept., EPA-600/7-76-018, U.S. Environmental Protection Agency, Cincinnati, Ohio, 1976, 110.

37. **Gregg, R. T.,** Development Documents and Effluent Limitations Guidelines and New Source Performance Standards, Soap and Detergents Manufacturing, Paint Source Category, EPA-440-/1-74-018a, U.S. Environmental Protection Agency, Cincinnati, Ohio, 1974, 146.

38. **Helling, C. S.,** Pesticide mobility in soil: II. Application of soil thin-layer chromatography, *Soil Sci. Soc. Amer. Proc.,* 35, 737, 1971.

39. **Riley, J. E.,** Development Document of Effluent Limitations Guidelines and New Source Performance Standards for the Tire and Synthetic Segment of the Rubber Processing Paint Source Category, EPA-440/1-74-013a, U.S. Environmental Protection Agency, Washington, D.C., 1974, 195.

40. **WAPORA, Inc.,** Assessment of Industrial Hazardous Waste Practices, Paint and Allied Products Industry, Contract Solvent Reclamation Operations and Factory Applications of Coatings, U.S. Environmental Protection Agency, Washington, D.C., 1975, 296.

41. **Cheremisonoff, N. P., Cheremisonoff, P. N., Ellenbusch, F., and Perna, A. J.,** *Industrial and Hazardous Waste Impoundment,* Ann. Arbor Sci. Pub., Inc., Ann Arbor, Mich., 1979, 475.

42. U.S. Environmental Protection Agency (OSWMP), Report to Congress: Disposal of Hazardous Wastes, EPA SW-115, U.S. Environmental Protection Agency, Washington, D.C., 1974, 110.

43. Council for Agricultural Science and Technology (CAST), Application of Sewage Sludge to Cropland: Appraisal of Potential Hazards of the Heavy Metals to Plants and Animals, 430/9-76-013, (MCD-33) OWPO, U.S. Environmental Protection Agency, Washington, D.C., 1976, 77.

44. **Elliott, L. F. and Stevenson, F. J., (Eds.),** *Soils and Management of Organic Wastes and Waste Waters,* Amer. Soc. Agron., Madison, Wis., 1977, 650.

45. **Knezek, B. D. and Miller, R. H., Eds.,** *Application of Sewages and Wastewaters on Agricultural Land: A Planning and Educational Guide,* MCD-35, OWPO, U.S. Environmental Protection Agency, Washington, D.C., 1976, 65.

46. **Sanks, R. L. and Asano, T., Eds.,** *Land Treatment and Disposal of Municipal and Industrial Wastewater,* Ann Arbor Sci. Pub., Inc., Ann Arbor, Mich., 1976, 310.

47. **Sopper, W. E. and Kerr, S. N., Eds.,** *Utilization of Municipal Sewage Sludge on Forest and Disturbed Land,* Proc. Symp., School of Forest Resources and Inst. for Research on Land and Water Resources, The Pennsylvania State University Press, University Park, 1979, 61.

48. U.S. Environmental Protection Agency, Use of Water Balance Method for Predicting Leachate Generated from Solid Waste Disposal Sites, EPA SW-168, U.S. Environmental Protection Agency, Washington, D.C., 1975, 40.

49. **Steffger, F. W.,** Energy from agricultural products, in *A New Look at Energy Sources,* McCloud, D. E., Ed., ASA Special Publication No. 22, Amer. Soc. Agron., Madison, Wis., 1974, 23.

50. **Thomas, P. R.,** Ohio Guide for Land Application of Sewage Sludge: Report of the Task Force on Land Application of Sewage Sludge, *Ohio Coop. Ext. Serv. Bull.,* 598, 1, 1975.

51. **Ewal, K. C. and Odum, H. T.,** Cyprus domes: Natures tertiary treatment filter, in *Utilization of Municipal Sewage Effluent and Sludge on Forested and Disturbed Land,* The Pennsylvania State University Press, University Park, 1979, 103.

52. **Reed, S. C.,** Waste Water Management by Disposal on Land Special Rept. 171, Cold Region Research and Engineering Laboratory, Hanover, N.H., 1972, 5.

53. **Pound, C. E. and Crites, R. W.,** Wastewater Treatment and Reuse by Land Application, Vol. I., Summary, EPA-660/2-73-006a, U.S. Environmental Protection Agency, ORD, Washington, D.C., 1973, Sec. VI.

54. **Larson, W. E. and Schuman, G. E.,** Problems and needs of high utilization rates of organic wastes, in *Soils for Management of Organic Wastes and Waste Waters,* Elliott, L. F. and Stevenson, F. J., Eds., Amer. Soc. Agron., Madison, Wis., 1977, 589.

55. **Gilde, L. C., Kister, A. S., Law, J. P., Neely, C. H., and Parmelee, D. M.,** A spray irrigation system for treatment of cannery waste, *J. W.P.C.F.,* 43, 2011, 1971.

56. **Driver, C. H.,** Assessment of the Effectiveness and Effects of Land Disposal Methodologies of Waste Management, Army Corp. Engin., Wastewater Mgt. Rept., 72, 45, 1972.

57. **Ryan, J. R. and Loehr, R. C.,** *Site Selection Methodology for the Land Treatment of Wastewater,* Special Rept., 81-28, U.S. Army Corps of Engin., Cold Reg. Res. and Engn. Lab., Hanover, NH, 1981, 74.

58. **Bouwer, H., Rice, R. C., and Escarcega, E. D.,** Renovating Secondary Sewage by Groundwater Recharge with Infiltration Basins, U.S. Water Cons. Lab., Office of Res. and Monitoring Proj. No. 16060DRV, U.S. Environmental Protection Agency, Washington, D.C., 1972, 182.

59. **Boen, D. F.,** Study of Reutilization and Wastewater Recycled through Groundwater, Vol. 1., Eastern Municipal Water Dist., Off. Res. Monit., Proj. No. 1606DDZ, U.S. Environmental Protection Agency, Washington, D.C., 1971, 35.

60. **McMichael, F. C. and McKee, J. E.,** Wastewater Reclamation at Whittier Narrows, California Statewater Qual. Control Board, Pub. No. 33, 166, 38.

61. **Merrell, J. C., Kato, A., and Pantler, H. E.,** The Santee Recreation Project, Santee, Calif., Summary Rept. 1962-1964, Pub. Health Serv., Pub. No. 999-WP-27, 1965, Washington, D.C., 45, 1967.

62. **Loehr, R. C., Jewell, W. J., Novak, J. D., Clarkson, W. W., and Friedman, G. S.,** *Land Application of Wastes,* Vol. I., Van Nostrand Reinhold, Co., New York, 1979, 89.

63. **Bouwer, H.,** Ground Water Recharge Design for Renovating Waste Water, ASCE, Sanitary Engin. Div., 96, No. SA 1, 59, 1970.

64. U.S. Environmental Protection Agency, U.S. Army Corps of Engineers, and U.S. Department of Agriculture, Process Design Manual for Land Treatment of Municipal Wastewaters, EPA/1-77-008, U.S. Environmental Protection Agency, Washington, D.C., 1977, 102.

65. **Pound, C. E., Crites, R. W., Olson, J. V., and Metcalf and Eddie, Inc.,** Long-term Effects of Land Application of Domestic Wastewaters, Hollister, Calif., Rapid Infiltration Site, EPA-600/2-78-084, U.S. Environmental Protection Agency, R. S., Kerr Environ. Res. Lab., Ada, Okla., 1978, 150.

66. **Bouwer, H.,** Renovating secondary effluent by groundwater recharge with infiltration basins, in *Symposia on Recycling Treated Municipal Wastewater and Sludge Through Forest and Cropland,* The Pennsylvania State University Press, University Park, 1972, 1.

67. **Crites, R. W.,** Land treatment of wastewater by infiltration-percolation, in *Land Treatment and Disposal of Municipal and Industrial Wastewater,* Sanks, R. L. and Asano, T., Eds., Ann Arbor Sci. Pub. Inc., 1976, 193.

68. **Cunningham, R., Tluczek, L., and Urie, D. H.,** Soil Incorporation Shows Promise for Low Cost Treatment of Sanitary Vault Wastes, *U.S. Forest Serv. Res. Note, No.,* 181, 3, 1974.

69. **Spangler, F. L., Sloey, W. E., and Fetter, C. W.,** Wastewater Treatment by Natural and Artificial Marshes, EPA-600/2-76-207, U.S. Environmental Protection Agency, R. S. Kerr Environ. Res. Lab., Ada, Okla., 1976, 171.

70. **Tilton, D. L. and Kadlec, R. H.,** The utilization of a fresh-water wetland for nutrient removal from a secondarily treated wastewater effluent, *J. Environ. Qual.,* 8, 328, 1979.

71. U.S. Environmental Protection Agency, Municipal Sludge Management: Environmental Factors, EPA 430/9-77-004, MCO-28 (OWPA-Constr. Grant Prog.) U.S. EPA, Washington, D.C., 1977, 31.

72. **Lakshman, G.,** An ecosystem analysis approach to the treatment of waste waters, *J. Environ. Qual.,* 8, 353, 1979.

73. **Parrott, H. A. and Boelter,** The use of peat filter beds for wastewater renovation at forest recreation areas, Utilization of Municipal Sewage Effluent and Sludge on Forest and Disturbed Land, The Pennsylvania State University Press, University Park, 1979, 115.

74. **Farnham, R. S. and Brown, J. L.,** Advanced Wastewater Treatment Using Organic and Inorganic Materials: I. Use of Peat and Peat-sand Filtration Media, Proc. 4th Int. Peat Congr., Helsinki, Vol. 4, 271, 1972.

75. **Odum, H. T., Ewel, K. C., Mitsch, W. J., and Ordway, J. W.,** Recycling Treated Sewage Through Cypress Wetlands in Florida Center for Wetland, Occas. Pub. 1., University of Florida, Gainesville, 1975, 5.

76. **Sopper, W. E. and Kerr, S. N.,** Renovation of Municipal Wastewater in Easter Forest Ecosystems, in *Utilization of Municipal Sewage Effluent and Sludge on Forest and Disturbed Land,* Sopper, W. E. and Kerr, S. N., Eds., The Pennsylvania State University Press, University Park, 1979, 61.

77. **Sutton, P.,** Establishment of Vegetation of Toxic Mine Spoil, Proc. Res. Appl. Tech. Symp., Mine-Land Reclamation, Bituminous Coal Research, Pittsburgh, Pa, 81, 1973.

78. **White, R. K.,** Methods and techniques for sludge application on the land, in *Utilization of Municipal Sewage Effluent and Sludge on Forest and Disturbed Land,* Sopper, W. E. and Kerr, S. N., Eds., The Pennsylvania State University Press, University Park, 1979, 285.

79. **Griebel, G. E., Arimiger, W. H., Park, J. F., Steck, D. W., and Adams, J. A.,** Use of composted sewage sludge in renovation of surface-mined areas, in *Utilization of Municipal Sewage Effluent and Sludge on Forest and Disturbed Land,* Sopper, W. E. and Kerr, S. N., Eds., The Pennsylvania State University Press, University Park, 1979, 293.

80. **Kardos, L. T., Sopper, W. E., Egerton, B. R., and Di Lissio, L. E.,** Sewage effluent and liquid digested sludge as aids to revegetation of strip mine spoil and anthracite coal refuse banks, in *Utilization of Municipal Sewage Effluent and Sludge on Forested and Disturbed Land,* Sopper, W. E. and Kerr, S. N., Eds., The Pennsylvania State University Press, University Park, 1979, 315.

81. **Kerr, S. N., Sopper, W. E., and Edgerton, B. R.,** Reclaiming refuse banks with heat-dried sewage sludge, in *Utilization of Municipal Sewage Effluent and Sludge on Forested and Disturbed Land,* Sopper, W. E. and Kerr, S. N., Eds., The Pennsylvania State University Press, University Park, 1979, 333.

82. **Kardos, L. T., Sopper, W. E., Dickerson, J. A., and Hunt, C. F.,** Use of sewage wastewater and sludge in revegetating strip-mining spoil banks, in Chemurgy-For Better Environment and Profits, 32nd Ann. Conf., Chemurgic Council, Washington, D.C., 32, 29, 1970.

83. **Sopper, W. E.,** Revegetation of strip-mine spoil banks through irrigation with municipal sewage effluent and sludge, *Compost Sci.,* 11, 6, 1970.

84. **Sopper, W. E. and Kardos, L. T.,** Municipal wastewater aids revegetation of strip-mine spoils banks, *J. Forestry,* 70, 612, 1972.

85. **Dickerson, J. A.,** 1975, Reclamation of Anthracite Coal Refuse Using Treated Municipal Wastewater and Sludge, M.S. thesis, The Pennsylvania State University, University Park, 1975, 20.

86. **Berg, W. A. and Vogel, W. G.,** Toxicity of acid coal-mine spoil to plants, in *Ecology and Reclamation of Devastated Land, Vol. 1,* Hutnik, R. J. and Davis, D., Eds., Gordon and Breach, 1973, 57.

87. **Jones, M. and Cunningham, R. S.,** Sludge used for strip mine restoration at Palzo: Project development and compliance water quality monitoring, in *Utilization of Municipal Sewage Effluent and Sludge on Forest and Disturbed Land,* Sopper, W. E. and Kerr, S. N., Eds., The Pennsylvania State University Press, University Park, 1979, 369.

88. **Stucky, D. J. and Bauer, J.,** Establishment, yield, and ion accumulation of several forage species on sludge-treated spoils of the Palzo mine, in *Utilization of Municipal Sewage Effluent and Sludge on Forest and Disturbed Land,* Sopper, W. E. and Kerr, S. N., Eds., The Pennsylvania State University Press, University Park, 1979, 379.

89. **Cunningham, J. D., Keeney, D. R., and Ryan, J. A.,** Yield and metal composition of corn and rye grown on sewage sludge-amended soil, *J. Environ. Qual.,* 4, 448, 1975.

90. **Roth, P. L., Jayko, B. D., Weaver, G. T.,** Initial survival and performance of woody plant species on sludge-treated spoils of the Palzo mine, in *Utilization of Municipal Sewage Effluent and Sludge on Forest and Disturbed Land,* Sopper, W. E. and Kerr, S. N., Eds., The Pennsylvania State University Press, University Park, 1979, 389.

91. **Soboda, D., Smout, G., Weaver, G. T., and Roth, P. L.,** Accumulation of heavy metals in selected woody plant species on sludge-treated strip-mine spoil at the Palzo site, Shawnee National Forest, in *Utilization of Municipal Sewage Effluent and Sludge on Forest and Disturbed Land,* Sopper, W. E. and Kerr, S. N., Eds., The Pennsylvania State University Press, University Park, 1979, 395.

92. **Jastrow, J. D., Zinimerman, C. A., Dvorak, A. J., and Hinchman, R. R.,** Comparison of lime and fly ash as amendments to acidic coal mine refuse: Growth responses and trace-element uptake of two-grasses, Argonne Natl. Lab., (UC-88), *Natl. Tech. Info. Serv.,* U.S.D.C., Springfield, Va, 22161, 36, 1979.

93. **Wallace, A. T.,** Land disposal of liquid industrial wastes, in *Land Treatment and Disposal of Municipal and Industrial Wastewater,* Sanks, R. L. and Asano, T., Eds., Ann Arbor Science Pub. Inc., Ann Arbor, Mich., 1976, 147.

94. **Hill, R. D., Hinkle, K. D., and Klingensmith, R. S.,** Reclamation of orphan mined lands with municipal sludges — case studies, in *Utilization of Municipal Sewage Effluent and Sludge on Forest and Disturbed Land,* Sopper, W. E. and Kerr, S. N., Eds., The Pennsylvania State University Press, University Park, 1979, 423.

95. **Williams, B. D. and Packer, P. E.,** Sewage sludge and other organic materials as amendments for revegetation of spent oil shale, in *Utilization of Municipal Sewage Effluent and Sludge on Forested and Disturbed Land,* Sopper, W. E. and Kerr, S. N., Eds., The Pennsylvania State University Press, University Park, 1979, 353.

96. **Day, A. D., Ludeke, K. L., and Tucker, T. C.,** Influence of soil materials in copper mine wastes on growth and quality of barley grain, *J. Environ. Qual.,* 6, 179, 1977.

97. **Day, A. D. and Ludeke, K. L.,** Stabilizing copper mine tailings disposal berms with giant bermuda grass, *J. Environ. Qual.,* 2, 314, 1973.

98. **LeRoy, J. C. and Keller, H.,** How to reclaim mined areas, tailing ponds, and mined dumps into valuable lands, *World Mining,* 8, 34, 1972.

99. **Brooks, K. N., Barovsky, J. P., and Arnet, C. M., Jr.,** Wastewater applications to iron-ore overburden material in northeastern Minnesota: Prospects for renovation and reclamation, in *Utilization of Municipal Sewage Effluent and Sludge on Forest and Disturbed Land,* Sopper, W. E. and Kerr, S. N., Eds., The Pennsylvania State University Press, University Park, 1979, 407.

100. **Liesman, L. A.,** A vegetational and soil chronosequense on the Mesabi from range spoil banks, Minnesota, *Ecol, Monogr.,* 27, 222, 1957.

101. **Ludeke, K. L., Day, A. D., Stith, L. S., and Stroehlein, J. L.,** Pima studies tailings soil makeup as prelude to successful revegetation, *Engin. Mining J.,* 175, 72, 1974.

102. **Fuller, W. H.,** Reclamation of saline and alkali soils, *Plant Food Rev.,* Fall, 7, 1962.

103. **Fuller, W. H. and Ray, H. E.,** Gypsum and sulfur-bearing amendments for Arizona soils, Coop. Ext. Serv. and Agric. Expt. Sta., University of Arizona, Tucson, Bull. 1-27, 8, 1963.

104. **Stroehlein, J. L., Miyamoto, S., and Ryan, J.,** Sulfuric Acid for Improving Irrigation Waters and Reclamating Sodic Soils, University Arizona Col. Agric. Bull. 75-5, 49, 1978.

105. **Miyamoto, S. and Stroehlein, J. L.,** Sulfuric Acid for Increasing Water Penetration into Some Arizona Soils, *Prog. Agric., Ariz.,* 27, 13, 1975.

106. **Spyridakis, D. E. and Welch, E. B.,** Treatment process and waste effluent disposal on land, in *Land Treatment and Disposal of Municipal and Industrial Wastewater,* Sanks, R. L. and Asano, T., Eds., Ann Arbor Sci. Pub. Inc., Ann Arbor, Mich., 45, 1976.

107. **Lance, J. C. and Wistler, F. D.,** Nitrogen balance in soil columns intermittently flooded with secondary sewage effluent, *J. Environ. Qual.,* 1, 180, 1972.

108. **Broadbent, F. E.,** Factors affecting nitrification-denitrification in soils, in *Recycling Treatment Municipal Wastewater and Sludge Through Forest and Cropland,* Sopper, W. E. and Kardos, L. T., Eds., The Pennsylvania State University Press, University Park, 232, 1973.

109. **Hannapel, R. J., Fuller, W. H., Bosma, S., and Bullock, J. S.,** Phosphorus movement in a calcareous soil: I. Predominance of organic forms of phosphorus in phosphorus movement, *Soil Sci.,* 97, 350, 1964.

110. **Hannapel, J. R., Fuller, W. H., and Fox, R. H.,** Phosphorus movement in a calcareous soil: II. Soil microbial activity and organic phosphorus movement, *Soil Sci.,* 97, 421, 1964.

111. **Riekerk, H. and Zosokski, R. J.,** Effects of dewatered sludge applications on a Douglas-fir forest soil on the soil, leachate, and groundwater composition, *in Utilization of Municipal Sewage Effluent and Sludge on Forest and Disturbed Land,* Sopper, W. E. and Kerr, S. N., Eds., The Pennsylvania State University Press, University, Park, 1978, 35.

112. **Fuller, W. H.,** Investigation of Landfill Leachate Pollutant Attenuation by Soils, EPA-600/2-78-158, U.S. Environmental Protection Agency, Cincinnati, Ohio, 1978, 219.

113. **Fuller, W. H.,** Predicting radioactive heavy metal movement through soil, in Proc. of Symp. on Waste Isolation in *U.S., Tech. Programs and Public Education,* Vol. 2, Waste Regulations and Programs: High-Level Waste, Post, R. G. and Wacks, M. E., Eds., University of Arizona, Tucson, 1983, 341.

114. **Nutter, W. L., Schultz, R. C., and Bristler, G. H.,** Renovation of municipal wastewater by spray irrigation on steep forest slopes in the southern Appalachians, in *Utilization of Municipal Sludge Effluent and Sludge on Forest and Disturbed Land,* Sopper, W. E. and Kerr, S. N., Eds., The Pennsylvania State University Press, University Park, 1979, 77.

115. **Cooley, J. J.,** Effects of irrigation with oxidation pond effluent on tree establishment and growth on sand soils, in *Utilization of Municipal Sludge Effluent and Sludge on Forest and Disturbed Land,* Sopper, W. E. and Kerr, S. N., Eds., The Pennsylvania State University Press, University Park, 1979, 145.

116. **Cope, C. B., Fuller, W. H., and Willetts, S. L.,** *The Scientific Management of Hazardous Wastes,* Cambridge University Press, Cambridge, 1983, Chap. 5.

117. **Blakeslee, P. A.,** Monitoring cjonsideration for municipal wastewater effluent and sludge application to the land, in Proceedings of a Joint Conference on Recycling Municipal Sludges and Effluents on Land, University of Illinois, Urbana, IL, Champaign, IL, 1973, 183.

118. **Bernarde, M. A.,** Land disposal of sewage effluents: Appraisal of health effects of pathogenic organisms, *J. Amer. Water Works Assoc.,* 85, 432, 1973.

119. **Krishnaswami, S. K.,** Health aspects of land disposal of municipal wastewater effluent, *Can. J. Pub. Health,* 62, 36, 1971.

120. **Hoodley, A. W. and Goyal, S. M.,** Public health implications of applications of wastewaters to land, in *Land Treatment and Disposal of Municipal and Industrial Wastewater,* Sanks, R. L. and Asano, T., Eds., Ann Arbor Sci. Pub. Inc., Ann Arbor, Mich., 1976, 101.

121. **Stewart, W. S.,** State of the Art Study of Land Impoundments, EPA-600/2-78-196, U.S. Environment Protection Agency, Cincinnati, Ohio, 76, 291, 1978.

Soil as a Waste Treatment System

Chapter 4

WASTE TREATMENT ON LAND PRIMARILY FOR DISPOSAL

I. SCOPE

Land Treatment (LT), for disposal only, relates to a much broader soils and soil modification program than does waste utilization, where wastes are identified as a resource. Although the most desirable approach involves wastes as resources, coping with waste disposal problems must go beyond the present technology for effective utilization. Similarly the economics of benefits and costs and legislative actions disallow the "valuable resource" concept for many cases.

Society's attitudes are changing. People are taking a more active part in waste disposal, and thus all must be willing to pay the price for an uncontaminated or "pollution-free" environment. In the meantime, the generation of wastes is not lessening but, explosively increasing. As once observed by Dr. Cecil Wadleigh,[1] former director of Soil and Water Conservation, Research Division, in the USDA, the public will need to compromise among various objectives. Those who express loud concern over pesticide fear of possible ill effects from accidental or inadvertent ingestion of such chemicals and yet, they are equally adamant against biting into an apple and finding half a worm. The two extremes are easily avoidable despite the sage verse of Samuel Taylor Coleridge in about 1798 referring to the city of Cologne:

> "The river Rhine, it is well known, doth wash your city of Cologne; but tell
> me nymphs! what power divine shall henceforth wash the river Rhine?"

This chapter provides critical information on soil as a treatment system as opposed to a resource system for disposal of a wide variety of wastes: solid, liquid, and gaseous, toxic and nontoxic, hazardous and nonhazardous, and pathogenic and nonpathogenic. Many of the systems and techniques are well known, such as:

- Land treatment, refuse farming, and landspreading
- Sanitary landfills
- Soil encasement and soil liners
- Composting and mulching
- Lagoons, ponds, and marshes
- Soil, water, and wind erosion control
- Soils for gas absorption

The discussion presented in each chapter does not attempt to provide detail on a preferential choice of one disposal system over another. Neither are details of design or operation of a specific system emphasized, but rather the discussion centers on management of soils involved in pollution containment and the way soils function to protect entrance into the food chain and groundwater. Some kinds of wastes cannot be distinctly pigeonholed, whereas other wastes will find no suitable land disposal system that can accommodate their hazardous characteristics. Nevertheless, the full potential of soil, its wise management, and the interrelationships involved with the many different kinds of wastes will be explored.

II. HISTORICAL INTRODUCTION

Land treatment of hazardous wastes is characterized as spreading the waste on the soil surface or incorporating it into the upper few centimeters by mechanical manipulation such as tilling or soil injection. The method of application and extent of soil manipulation depends on the physical, chemical, and toxic nature of the waste and the rate of biodegradation desired. Liquids, for example, lend themselves physically to spray-, flood-, or drip-type application. Because of their fluid nature they penetrate the soil and thus, do not require mechanical soil incorporation unless they carry significant amounts of solids. Solids, on the other hand, require special equipment for spreading and, usually, incorporation. Mixing the waste with soil enables the natural biological, chemical, and physical properties to decompose, destroy, and detoxify both the hazardous and nonhazardous organic constituents of waste more rapidly. Inorganic metals are more readily attenuated when the wastes containing them are mixed with soil.

The single purpose of land treatment as opposed to land utilization is final disposal of the waste with little or no demand of the waste to function as a resource. The major benefit of land treatment is to engage the natural assimilative capacity of the land for disposal. Once the wastes have been destroyed, the land may be used for other beneficial purposes. In some instances, the land is more fertile following land treatment as a result of the accumulation of organic matter or humus as by-products of the decomposition reactions and incidental release of certain plant nutrients. This is in contrast to lands used for landfills and surface impoundments, which are effectively removed from further use.

Modern methods of land treatment, just as the sanitary landfill, developed out of necessity to provide better control of waste disposal than indiscriminant spreading of raw wastes over land surfaces or mixing into surfaces of garden soils. Spreading organic plant residues and animal manures, including "night soil" on croplands and home gardens dates so far back that the origin and development of the practices are lost in prehistory. The gradual increase in green and lush growth of vegetation in driving from rural to densely populated areas in the Orient is still obvious as nutrients in the produce migrate to the cities and find their way into the local home garden soils as human and animal waste. Many countries still practice landspreading sewage as they have for centuries. Animal manures provided the main source of agricultural crop fertilization until about the close of World War I. Today, however, we are confronted not only with the usual waste of animal and plant origin but of industrial origin. Moreover, wastewaters accumulate in population centers unlike during our early history and before the time of indoor plumbing and backyard swimming pools.

Wastewater spreading as well as solid waste disposal was practiced before Christ (B.C.) although somewhat limited as compared with today.[2] In the sixteenth century it is reported that effluents were being used in Bunzlaw, Germany, for beneficial crop production beginning in 1559.[3] From about this time through the nineteenth century, the spreading of sewage effluent on farmland for irrigation was commonly practiced in Western Europe as a practical measure as a consequence of excessive pollution of many rivers. Also, the early sewage farms developed during this period. Only a few were successful. Lack of success was the result of poor management stemming mostly from a lack of basic knowledge of physical, chemical, and microbiological principles of soil behavior. Soil overloading caused anaerobic conditions to develop which, in turn, caused objectionable odors spurring public complaints, and finally ending in crop failure. The lands often were abandoned for crop production, at least until they dried out and aerobic conditions were restored.

Although some of the successful refuse farms produced fruits and vegetables, most were planted to grains, grasses, and root crops. Some historic successful sewage farms have been compiled by Pound and Crites[4] and are reproduced in Table 1.[4]   "Sewage farming" was recognized during the 1870s in England as a soil treatment.[5]

In the United States sewage farming became prominent at about the same time, due to concern over the expanding pollution of natural streams and rivers. These early methods referred to as sewage farming, land farming, garbage farming, refuse farming, landspreading, land application, soil farming, and soil incorporation, rarely represented the best technology. Management of wastes soon yielded to better method of land treatment, sometimes identified as land cultivation. As defined by Phung et al., land cultivation is a process whereby waste is mixed with or incorporated into the surface soil at a land disposal site.[6] It differs from

Table 1
HISTORICAL DATA ON SEWAGE FARMING

Date	Location	Description	Wetted area, acres[e]	Flow, mgd[f]	Average	Reference
Non-United States						
1559	Bunzlaw, Germany	Sewage farm	—	—	—	25
1861	Croydon-Beddington, England	Sewage farm	420	4.5	2.8	81
1864	South Norwood, England	Sewage farm	152	0.7	1.2	81
1860	Berlin, Germany	Sewage farm	27,250[a]	150[a]	1.4	81
1875	Leamington Springs, England	Sewage farm	400	0.8	0.5	96
1880	Birmingham, England	Sewage farm	1,200	22	4.7	96
1893	Melbourne, Australia	Irrigation	10,376[b]	50[b]	1.2	76
	Melbourne, Australia	Overland flow	3,472[b]	70[b]	5.2	76
1902	Mexico City, Mexico	Irrigation	112,000[b]	570[b]	1.3	—
1923	Paris, France	Irrigation	12,600	120	2.5	81
1928	Cape Town, South Africa	Irrigation	—	—	—	21
United States						
1872	Augusta, Maine[c]	Irrigation	3	0.007	0.6	97
1880	Pullman, Illinois[c]	Irrigation	40	1.85	12.0	97
1881	Cheyenne, Wyoming	Irrigation	1,330[d]	7.0[d]	1.3	—
1887	Pasadena, California	Irrigation	300	—	—	81
1895	San Antonio, Texas	Irrigation	4,000[a]	20[a]	1.3	81
1896	Salt Lake City, Utah	Irrigation	180	4	5.7	97
1912	Bakersfield, California	Irrigation	2,400[d]	11.3[d]	1.2	—
1928	Vineland, New Jersey	Irrigation	14	0.8	14.7	83

[a] Data for 1926.
[b] Data for 1971.
[c] Abandoned around 1900.
[d] Data for 1972.
[e] Multiply by 0.4047 to convert to ha.
[f] Multiply by 3785 to convert to m³/d.

Pound, C. E. and Crites, R. W., *Wastewater Treatment and Reuse by Land Application*, U.S. Environmental Protection Agency, Washington, D.C. 1973, 249.

earlier solid waste disposal methods since it is designed as a means of ultimate disposal. It differs from the present concept of land treatment, as perceived by the U.S. EPA and noted in the Resource Conservation and Recovery Act (RCRA, Public Law 94:580).[7] The proposed regulations issued by the Environmental Protection Agency's Office of Solid Waste recognize land treatment as one of the technologies for management of waste. It currently is being utilized for disposal of hazardous industrial waste. Technically, the land or soil is taken as a medium to biodegrade hazardous wastes. A land treatment site or facility is that portion of a facility where hazardous waste is applied onto or incorporated into the surface soil. Thus, land treatment as defined above is broader than land cultivation, but includes land cultivation.

Destruction of the soil for vegetative growth is not a part of land treatment. Land treatment must provide sound, environmentally safe disposal of waste residuals through biological, chemical, and physical interactions occurring in soils. The organic constituents are expected to biodegrade through the activity of the indigenous soil microorganisms. The inorganic metal components are expected to attenuate (or immobilize) primarily through physical-chemical interactions with the soil. The natural processes in all of the soilwaste complexities and interactions must be appreciated, although their specific reactions cannot always be

Table 2
SELECTED INDUSTRIAL WASTES SUITABLE FOR LAND TREATMENT

Industry	Waste Type	Specific Potential Hazards	Recommended Precautions
Food and kindred products	Wastewater, sludge	High sodium and TDS content and resulting detrimental effects on soil properties and plant growth	Gypsum addition; segregation of high sodium and TDS waste streams
Textile finishing	Secondary waste-water treatment sludge	Heavy metal content	Plant and water monitoring; appropriate loading rate
Wood preserving	Wastewater	Pentachlorophenol creosote and possible contamination of waste supplies	Appropriate loading rate about 28 to 37 m³/ha (3,000 to 4,000 gal/ac)
Paper and allied products	Primary waste-water treatment sludge	Contamination with toxic materials may occur at some plants reprocessing secondary materials	Sludge analysis and subsequent appropriate site design and operating precautions
Organic fibers, noncellulosic	Secondary waste-water treatment sludge	High zinc and nitrate content	Appropriate application rate and cover crop
Pharmaceuticals	Waste mycelium	High zinc and TDS content	Appropriate application rate and cover crop
Soap and other detergents	Wastewater	Possible water supply degradation from excess nutrients	Use cover crop with good nutrient uptake characteristics
Organic chemicals	Wastewater treatment sludges	Potential hazards are dependent on the specific chemicals produced	Chemical analysis of sludge to detect potentially hazardous constituents
Petroleum refining	Nonleaded tank bottoms	High nickel, copper, vanadium, and lead content	Monitoring of soil and groundwater concentrations to determine when disposal site life is expended

Phung, T., U.S. Environmental Protection Agency, Cincinnati, Ohio, 1978, 205.

followed, if land treatment is to succeed as a viable disposal option and develop as a realistic waste management program.

An estimate places the number of operational industrial waste land treatment sites at about 200, located in 32 states.[8] Unfortunately, only a limited amount of data are available since most were developed under a trial and error procedure.

III. LAND TREATMENT PRACTICES

A. Constraints

Selected industrial wastes suitable for land treatment have been compiled in Table 2 according to present available information from industry.[6] The list is more limited than may be expected as suitable for the broader concept of land treatment. The list, however, provides a good point of departure for this discussion on the newly emerging disposal practice of land treatment (LT). The selection is based on certain assumptions, namely:

- The waste is biodegradable, in whole or at least in part.
- The soil microorganisms, indigenous to the soil, will survive and function at reasonable and practicable application rates of the waste.
- The long-term toxic effects of accumulated residues and possible ion adsorption on the soil exchange complex can either be prevented or mitigated.
- Reasonable and practical loading rates will not cause pollution of the groundwater by hazardous constituents nor allow toxic substance to enter the food chain. The land treatment site will remain environmentally safe.
- The cost-effectiveness of land treatment vs. other treatment disposal alternatives is within reasonable limits.
- The land treatment will leave the soil in virtually the same (or an even higher) productive condition as originally.

Successful and safe treatment of wastes can be achieved only if biological, chemical, and physical reactions flourish in the soil. The number of land treatment facilities now being built has shown a significant increase in recent years. The wide diversities of industries that have either on-going or pilot-plant systems include industrial solid waste, refractory metal processing, petroleum refining, food processing, textiles, paper and pulp, metal plating, leather tanning, printing ink, munitions, foundries, inorganic chemicals, organic chemicals, pharmaceuticals, soaps and detergents.

Considerable technical information is available for the disposal of solid wastes related to agriculture and forestry.[1] More recently sewage sludge solids have received intensive experimentation as a resource for agricultural production on cultivated lands.[9] Despite the present conscientious efforts to relate plant and animal residues as assets, the economic aspects often turn them back into waste disposal problems.

Accumulations of animal manures, sewage sludge, cannery residues, and other plant materials are now being recommended by some consultants for landfilling, trenching, and sewage farming on dedicated land areas too small to accommodate the volume of waste applied. This trend can be stopped by developing new technologies, new techniques, and above all new attitudes towards willingness to pay for maintaining a quality environment. Disposal of wastes through intelligent land treatment must become an active program, if only for the sake of disposal alone.

Land treatment in this chapter is confined to the notion of a technology for management of hazardous waste as defined in the Resource Conservation and Recovery Act (RCRA, Public Law 94:580)[7] and the proposed regulations issued by the Environmental Protection Agency's Office of Solid Waste. Although land treatment is currently being utilized for disposal of hazardous industrial waste, its present practice is limited and there are large gaps in the knowledge of the best management processes. In fact, the approximately 200 industrial land-treatment waste sites in 32 states constitute the extent of interest. Some of the lack of interest is due to inadequate (a) dissemination of the importance of issues related to this technology, (b) information necessary to establish a high degree of confidence in soil treatment systems, and (c) information on the ways to utilize LT technology to its fullest extent.

The kind of soil modifications necessary for most effective pollution control will depend on the waste being treated. Waste and site-specific characteristics are interdependent. Pollutant attenuation and mobility are highly dependent on the characteristics of the waste in relation to the natural properties of the soil. Adequate design of land treatment sites requires reliable estimates of soil and waste characteristics. This information is needed for establishing site suitability, especially from a quantitative standpoint, and for providing background data for future monitoring.

B. Solid Wastes

1. Municipal Solid Wastes

A form of land treatment was experienced by those in rural western Washington during the 20s and 30s. In one scenario, a farmer supplemented his income by collecting kitchen garbage from the local logging camps and spreading it over idle, stony, gravelly, and sandy areas of his farm strictly for disposal purposes. Biological activity coupled with the 150 cm (60 in.) rainfall rapidly degraded the organic residues. Although organic matter accumulated to an observable extent, the residual effect was short lived, except for dinnerware accidentally thrown away with the food. Along with the intensive development of the packaging industry at about this time, came the more challenging municipal solid waste (MSW) of glass bottles, tin cans, paper, wax cartons, rubber, etc. This same farmer, ingeniously invented many land treatment technologies for disposal purposes. Nearly all failed to permanently improve the quality of the land and its soil.

Today, with the shredder and hammer mill, pretreatment of municipal solid waste can reduce particle size distribution and land disposal can be more effective. For example, in Oregon, shredded municipal solid waste and sewage materials were added to field plots of Sagehill loamy fine sand.[10] Rates of 0, 224, 448, and 846 t/ha were used along with 0, 51.4, 103, 206 m³/ha of sewage. The experiment was terminated with limited success. Grass production for grazing purposes was not seriously curtailed. A similar experiment in Oregon resulted in rapid decomposition of solid waste, except for plastic, rubber, and metal by the end of the first growing season.[11] The soil bulk density was found to decrease with the disposal of solid waste. The moisture holding capacity and organic matter content increased. The LT of solid waste also appeared to effectively control wind erosion. No significant change in N, P, S, Ca, Mg, K, Fe, and Cu of the plants was found, but some slight increases, in Mn, Zn, Mo, Co, and Cr were recorded in the plant tops grown on the treated lands the first year. In conclusion, land treatment of shredded MSW at rates less than 900 t/ha was found to be a suitable method of disposal in Oregon. The most favorable rate of application was about 440 t/ha. The management of the 900 t/ha is enhanced by compacting on the soil surface.

Other experimentations with unsorted, shredded municipal solid waste include those of Stanford[12] of Houston, Texas, King et al.[13] of Ontario, Canada, Tri Service Project at Hueneme, California,[6] and of the City of Odessa, Texas, as reported by Phung.[6] The disposals were shown to be successful, although it was agreed there were some limitations such as:

- The method is not cost-effective mainly because of the expense of sorting and shredding
- The large land requirement close to cities is expensive, if not altogether prohibitive
- Guidelines for application of loading rates have not been developed by regulatory agencies

2. Refractory Metal Processing Waste

Refractory metal processing waste was applied to a silty clay loam soil in Oregon at rates of 0, 11, 22, 56, 112 t/ha along with N-P-K fertilizer and lime. The waste was added as either a slurry or in a dried condition. Refractory waste used contained Ca, F, Fe, Al, S, and Zr metals in excess of one percent. These elements Mg, Na, Pb, P, $NH_4 - N$, and K varied between 0.1 and 1.0% and Cd, Ci, Co, Cr, Ni, Mn, and Mo were less than 1000 µg/g. Leaching of the native soil occurred under the climatic condition of this site, lowering the total dissolved solids in the soil due to the land treatment. Data collected indicated that Ca, Mg, K, P, Fe, Al, Mn, Cr, Zn, Cu, Ni, Co, Mo, F, Zr, Hf, and Pb did not preferentially accumulate, over that of the untreated soil, in rye grass growing on the land. It was concluded that a one time application up to 112 t/ha to the soil would not pose a threat of heavy metal pollution to the groundwater. Fluoride (F) pollution to groundwater, however, was a distinct possibility.

Incorporation of all wastes into the soil is a highly desirable practice when site conditions permit, even though the objective is for disposal only. Biodegradation is accelerated, metal attenuation is more favorable, and overall retention and confinement is better than surface spreading. Wastes are less subjected to forces of wind and water erosion when covered or even partly covered.

3. Petroleum Refining Waste

Land treatment of hazardous wastes owes much of its early successes to the petroleum industries who developed and used LT techniques since the early fifties in the disposal of oily waste, most of which represent tank bottom residues. Much of our present information is based on these successes, as well as on some failures. Estimations place the use of LT for refinery waste disposal, in one form or another, at 40 to 60%.[14] New waste-treatment program will emerge from these early successes and more quantitatively oriented research now underway.

Soil treatment modification procedures of LT in the future will be concerned with the biodegradation and final disposal of the following sludges without pretreatment:

- Crude tank sludge
- Dissolved air flotation float
- API separator sludge (from refinery wastewater)
- Nonleaded tank bottoms
- Soil oil emulsion solids
- Waste biosludge

Water and a number of heavy metals are present in various proportions in refinery waste, depending on the particular source. Land treatment of such sludges, however, is practiced mainly for disposal of heavy metals. The organic oily materials decompose and dissipate by virtue of the soil microbial activity. It is this high organic content of the petroleum sludges that makes them excellent candidates for LT.

Despite the successful responses of petroleum refinery waste, a number of problems can develop as a result of mismanagement or lack of knowledge of management under the diverse site conditions that are often present. The rate of application and loading schedule must be compatible with the site characteristics, climate, topography, hydrogeology, and most importantly, the soil. Excessive rates of sludge application, particularly if coupled with high frequency of application, can change an effective aerobic soil into an ineffective anaerobic medium. Oxygen deficiencies also develop readily when soils become waterlogged and remain saturated for even short periods of time. Intermediate breakdown products and reduced compounds form with accompanying foul odors, delayed biodegradation, and possible solubilization of heavy metals if proper soil management is neglected. Land treatment not only is highly sensitive to management, but to long-term accumulation of residual inorganic constituents (e.g., heavy metal) and biodegradable resistant organics. The maximum loading rate for the site soil should be determined and nonbiodegradable constituents monitored through the LT period.

Many factors must enter into a decision as to how much of a specific metal or combination of metal and organics may be added to a soil at a selected site without adverse impact. Consequently, there is no easy or sure method that can be prescribed.[15-17] A particularly difficult factor to evaluate is the state of soil microbial activity or biotoxicity occurring in the soil during land treatment. Methods and equations for prediction of selected heavy metal movement through soils material from municipal solid waste leachates have some solutions, however. Some suggestions are discussed in Chapter 5, Volume II.

The metal content of the petroleum refinery waste, to the extent it has been supplied from the industry, is reported by Brown and Associates.[17] The organic constituents of petroleum

Table 3
KINDS AND CONCENTRATIONS OF HEAVY METALS IN UNTREATED AND TREATED SOILS AT PETROLEUM REFINERY LT FACILITIES[18]

Analytical Data in mg/kg (dry weight)[a]

Metals	Background Soil			Treatment Area Soil		
	Samples	Range of Values	Average	Samples	Range of Values	Average
Arsenic	36	0.0—66.0	6.8	26	0.0—60.0	6.5
Cadmium	37	0.0—15.0	2.1	26	0.0—11.0	2.2
Total Chromium	39	0.06—208.0	35.3	27	0.4—56,500	453.0
Copper	39	0.21—419.0	25.2	26	0.8—527.0	86.9
Lead	39	0.2—597.0	37.4	26	0.0—1,900.0	213.0
Mercury	36	0.0—50.0	2.7	26	0.002—26.3	2.2
Nickel	39	0.0—155.0	24.1	27	1.0—228.0	38.5
Selenium	36	0.0—173.0	12.8	25	0.0—10.0	2.2
Vanadium	38	0.001—90.0	25.8	27	0.0—77.0	19.6
Zinc	38	0.4—310.0	45.8	27	1.3—3,130.0	470.0

[a]   In solid phase (solids were dried to constant weight at 105°C).

wastes are so extensive that no one literature source is complete. In addition to the heavy metal and hydrocarbon content of petroleum refractory waste, soluble salts often appear along with inorganic sulfur compounds. Table 3 provides some concept of the kinds and concentrations of heavy metals that might appear in soils treated with petroleum refinery waste.[18]

4. Leather Tanning and Finishing

The wastes from leather tanning and finishing vary widely in composition and toxicity of hazardous pollutants. Industrial subcategories that contribute to the waste streams may be described as:

- Hair pulp/chrome tan/retan-wet finish — describes facilities that primarily process cattle-like raw or cured hides into finished leather by chemically dissolving the hair, tanning with chrome, and retanning and wet finishing.
- Hair save/chrome tan/retan-wet finish — describes facilities that primarily process cattle-like raw or cured hides into finished leather by chemically loosening and mechanically removing hair, tanning with chrome, and retanning and wet finishing.
- Hair save/nonchrome tan/retan-wet finish — describes facilities that primarily process cattle-like raw or cured hides into finished leather by chemically loosening and mechanically removing hair, tanning with vegetable tannins, alum, syntans, oils, or other chemicals, and retanning and wet finishing.
- Retan-wet finish — describes facilities that primarily process unhaired and tanned hides into finished leather by retanning and wet finishing.
- No beamhouse — describes facilities that primarily process unhaired and pickled cattlehides, sheepskins, or pigskins into finished leather by tanning with chromium or other agents, followed by retanning and wet finishing.
- Through-the-blue — describes facilities that primarily process cattle-like raw or cured hides into the blue tanned state by chemically dissolving or chemically loosening and mechanically removing hair and tanning with chrome, with no retanning or wet finishing.

- Shearling — describes facilities that primarily process raw or cured sheep or sheep-like skins into finished leather by retaining the hair on the skin; tanning with chrome or other agents, and retanning and wet finishing.

The hazardous waste streams generated by the leather tanning and finishing industry may be broadly categorized as being about 47% wastewater treatment sludge, 45% chrome trimmings and shavings, 4.5% sewer screenings, and 3.5% buffing dust.[19] The metals that occur most abundantly and in the highest concentration as inorganic pollutants are lead and chromium. Significant concentrations of copper, nickel, zinc, and cyanides, sulfides and sulfates are discharged to the waste streams. Chromium may be found in all of the categories but is concentrated mostly in first, second, and fourth categories. Waste streams of chrome blue trimmings and shavings, buffering dust, sewer screenings, and waste treatment sludge concentrate in the fifth and seventh categories. Lead also occurs in all waste streams except the chrome and blue trimming and shavings. The level of concentration varies for each waste, therefore, loading rates and capacities for each soil and site condition should be based on an analysis of the specfiic waste to be land treated. Because of the lack of knowledge of LT of leather tanning and finishing, pilot studies should precede final unrestricted field disposal. Moreover, only a few of the toxic organic pollutants in leather tanning and finishing are known.[19] Some of the volatile organics identified most often are: benzene, methylene chloride, ethylbenzene, and toluene. Whereas, the basic and neutral organics detected most often are; 1,2-dichlorobenzene, 1,4-dichlorobenzene, anthracene, phenanthrene, and naphthalene; and acidic organics most frequently encountered are phenol, 2,4,6-trichloro-phenol and pentachlorophenol. Very little is known about their rate of movement through soil, rate of biodegradation, or soil-water interactions. Not all of the leather tanning and finishing wastes lend themselves to land treatment. Flammable organic waste should preferably be eliminated by better controlled and safer options than land treatment. If wastes high in Cr and Pb are land treated, unusually large acreages of land may be required to avoid point-source problems and excessively long-term soil pollution.

In summary, the suitability of the various leather tanning and finishing wastes for LT has not been decided. The possibility of some wastes being land treated varies with the industrial process and each of the numerous wastes individually. Not only is the soil attenuation of heavy metals and organic solvents not well known but the salt used to preserve the hides before tanning offers an obstacle for disposal, although the sludge from the secondary wastewater treatment at vegetable tanning plants is usually suitable[6] for LT method.

5. Pulp and Paper Industry Waste

Sludges from the pulp and paper industry are generally amenable to LT, partly due to the plant origin and partly due to the comparatively low potential toxicity of the chemicals used during the manufacturing processes. Seven subclasses of the paper industry are often listed:[6,17]

- Sulfite
- Kraft sulfate
- Semi-chemical
- Straw-board
- Groundwood board
- Cardboard
- Boxed and paper board

The primary sludge waste streams to note under the heading of solids are: lime- and cellulosic- fiber sludge. Both of these can be favorably manipulated in the soil like municipal sewage sludges. The lime sludges are valued on acid soils, particularly where they can be substituted for the agricultural ground limestone. Behavior of pulp and paper industry waste in soils has recently received considerable detailed attention in the literature and offers few serious disposal problems other than overloading.

6. Textile Industry Waste

The textile industry generates nonhazardous wastes (approx. 60%) and hazardous wastes.[20] The most abundant hazardous wastes originate from the dying and finishing operations and in wool scouring. The weaving and knitting operations usually do not generate hazardous wastes. The waste falls into 6 classes depending primarily on the nature of the chemical(s) they contain: (a) fiber, (b) reactive dyes, (c) dispersed dyes and pigments, (d) flame retardants, (e) water repellent finishes, and (f) pigments. Textile industry wastewaters receive secondary biological treatment. Sludges contain waste cellular substances and organic materials pass through the treatment system. These wastes are amenable for disposal by land treatment systems. Land treatment research is seriously needed. Phung[6] reported the presence of at least three sites under construction for disposal of textile finishing wastewater sludges by land cultivation.[21] Only limited information is available.

7. Specific Acids

Specific acids considered here include the inorganic or mineral acids: sulfuric, nitric, and phosphoric; and organic lower fatty acids, formic, acetic, butyric, propionic, etc. Inasmuch as pure species of the acids are too valuable to discard, acid wastes may be expected to contain some polluting constituent(s) that adds to their own polluting properties.

Waste sulfuric acid increased phenomenally during the 70s and 80s as U.S. Environmental Protection Agency air pollution standards forced numerous industries to reduce emissions of the sulfur oxides. The major potential sources of sulfuric acid are from the nonferrous smelting industries and coal burning power plants. Although agriculture has absorbed some of the acid for soil reclamation and improvement in water movement in arid and semiarid climatic regions, there still is an overabundance of waste sulfur oxides and hydrous oxides requiring disposal. Furthermore, the estimated 5 million tons of H_2SO_4 produced in 1980 are likely to double by 1988. The amount of sulfuric acid generated exceeds the demand by industry and agriculture, and therefore, the excess sulfur oxides generated must be disposed. Land treatment is a viable option.

The disposal of sulfuric acid waste and associated oxides and hydrous oxides by land treatment requires an understanding of soils and soil management practices. Treatment of soils in arid and semiarid climatic regions offers many more practical options than of soils in humid regions. The agricultural use of sulfuric acid is well established and discussed in Chapter 2. Land treatment strictly for disposal purposes is less well known. Because of the corrosive nature of the acid and its capacity to cause soil failure with respect to pollutant attenuation or stabilization, management procedures must be associated with the capacity of the soil to neutralize the highly active hydrogen in concentration. Land treatment in humid climates where soils naturally are acid requires the acids to be neutralized with lime or some other low cost material prior to disposal. The waste disposal then becomes gypsum ($CaSO_4 \cdot 2H_2O$) in place of the acid. Any tendency of the acid disposal to lower the pH of the soil in a humid region can lead only to unfavorable residual soils of very low productivity and leakage only to unfavorable residual soils of very low productivity and leakage of solubilized soil metals into groundwaters (Table 4).[22] Low application rates of 448 to 560 kg/ha (400 to 500 lb/a) of sulfuric acid made trace-metal plant nutrients of the soil more soluble in the soil solution and, therefore, more available to plant absorption without even altering the soil paste pH values.[23,24] Land treatment of sulfuric acid for disposal only should be limited to arid and semiarid climatic regions where soils are slightly alkaline to alkaline (just as suggested with lime-desulfurization slurries of Chapter 3), before extensive disposals are extended to acid soils of higher rainfall (humid) climates.

Table 4
MAXIMUM METAL CONCENTRATION (μg/g) IN THE SOLUTION DISPLACEMENT FROM REPRESENTATIVE SOILS RECEIVING D-WATER, DILUTE H₂SO₄, DILUTE ACID PLUS 0.025 *M* AlCl₃, + 0.025 *M* FeCl₂ AND MSW LEACHATE[a]

Soil	Al	Cd	Co	Cr	Cu	Fe	Mn	Ni	Pb	Zn
Deionized Water (pH 7.0)										
Davidson c	<.5	<.005	<.05	<.1	<.05	0.10	0.20	<.05	<.5	<.005
Chalmers si c l	<.5	<.005	<.05	<.1	<.05	0.10	0.10	<.05	<.5	<.005
Mohave s l	<.5	<.005	<.05	<.1	<.05	0.20	2.00	<0.5	<.5	<.005
MSW Landfill Leachate (pH 5.4)										
Davidson c	<.5	0.010	<.05	<.1	<.05	0.10	0.80	<.05	<.5	0.5
Chalmers si c l	<.5	<.005	<.05	<.1	<.05	0.10	0.20	<.05	<.5	0.9
Mohave s l	<.5	<.005	<.05	<.1	<.05	0.20	3.00	<.05	<.5	0.1
Dilute Acid (H₂SO₄ at pH 3.0)										
Davidson c	8.0	0.07	<.05	<.1	22	0.38	0.46	<.05	<.5	0.14
Chalmers si c l	4.0	0.120	<.05	<.1	0.11	0.44	0.25	0.15	<.5	0.25
Mohave s l	<.5	<.005	<.05	<.1	0.33	1.80	1.80	<.05	<.5	0.06
Dilute Acid (pH 3.0) plus 0.025 M AlCl₃, + FeCl₂										
Davidson c	860	94	11	2.8	0.9	1,080	950	0.7	<.5	0.6
Chalmers si c l	530	0.1	3.8	0.1	0.3	1,900	365	6.0	0.3	0.9
Mohave s l	610	0.04	2.2	<.1	<.05	1,550	225	1.3	0.7	0.2

bdl in μg/ml are Cd = 0.005; Cr = 0.1; Co = 0.05; Cu = 0.05; Pb = 0.5; Ni = 0.05; Mn = 0.05; Zn = 0.005; Al = 0.5; and Fe = 0.1.

[a] During 20 pore-volume displacements.

Phosphorus in acid wastes unites readily with soil constituents. Under arid and semiarid climatic soil conditions, slowly soluble carbonates form most extensively. The initial combination is dicalcium phosphate which slowly reverts to the less soluble tricalcium phosphate with time. Under humid climates in acid soils the predominate phosphate formations are associated with iron and aluminum, again of low solubility. Phosphorus (P) in dilute aqueous solution thus is fixed quite firmly and prevented from moving through the soil. As a consequence, pollution of dilute phosphoric acid waste is more closely associated with the heavy metals and toxic elements it carries than the acid itself. However, as a result of the low pH value caused by the acid or the high hydrogen ion activity, heavy metal pollutants and the natural soil metals sustain an accelerated mobility. The most serious hazards associated with land treatment of phosphoric acid itself is (a) the chance that soil and water erosion will carry phosphorus overland to streams and lakes, thus resulting in pollution associated with eutrophication and (b) excess phosphorus will accumulate and remain in the soil for long periods of time causing nutritional imbalance in vegetation growing on the land and also resulting in less than maximum growth responses. The damage to land due to excess P can persist for many years, depending on the quantities of P deposited during land treatment.

Concentrated nitric acid waste lends itself to land treatment very poorly at our present state of technology. The high mobility of nitrate ion in soils makes it a prime candidate for polluting underground waters. If soil conditions are developed such as to cause reduction of nitrates without lowering the Eh or red/ox to a point of solubilizing heavy metals, nitrate may be more amenable to land treatment. If large quantities of waste nitrate are to be eliminated, options more reliable than land treatment are available.

A more thorough discussion of land disposal of strong inorganic acids is available in Volume II, Chapter 6.

The organic nature of the lower fatty acids makes them prime material for disposal by LT. However, two major constraints must be considered: (a) the most effective loading rate to maintain maximum biodegradation and (b) the kind of pollutant accompanying the acid. Research is seriously required for the land treatment of the fatty acid wastes.

Soil column research of acetic acid and its effects on soils is presented in Volume II, Chapter 6.

C. Liquid Wastes

Almost, if not all, wastes from industrial production appear as waste streams that are liquid or that flow readily. Transport throughout the industrial production process is most readily conducted by a fluid system. Water and organic solvents, whether by necessary design or convenience, act as the dominant transport media. In LT the separation of waste streams into solid and liquid phases is convenient because it dictates different methods of soil management, but for industry itself such a differential classification is not always convenient or possible.

Waste streams from the industries producing pesticides, pharmaceuticals, cosmetics, soaps and detergents generate unusually large volumes of liquid waste streams, many of which, contain hazardous constituents in appreciable quantities. Only a few examples of land disposal of hazardous waste liquids will be presented because of the lack of experience and knowledge in this area of soil treatment. The great demand for research in the LT area must be undertaken soon and conducted on a case-by-case basis since these waste streams are highly diverse and involve a great number of hazardous components. Many of these components have not been characterized in soil systems and basic chemical, physical, and biological interactions are not known. Just to list all the possible hazardous components without offering some information on soil interactions would serve the objectives of this text poorly.

1. Pesticides

Considerable literature has accumulated on the biodegradation in soils of specific pesticides (Table 5). For the well being of food-production efficiency and the health of consumers, there is a need in the agricultural industry for the acquisition and development of knowledge concerning the persistence and distribution of pesticide chemicals in the food chain environment. The real problem, however, is not so much with the use of the pesticide but with the disposal of the myriad of hazardous chemical wastes related to pesticide production. Associated with this problem is the relatively rapid migration rate of some pesticides through soil (Table 6). Subcategories of wastes from pesticide production have been suggested.[17] as:

- Halogenated organic products
- Organo-phosphorus production
- Organo-nitrogen production
- Metallo-organo production
- Miscellaneous

Waste stream characterization of each class or subgroup is well described.[17]

Table 5
DECOMPOSITION AND PERIOD OF PERSISTENCE
OF SEVERAL HERBICIDES

Name of Compound	Abbreviation	Persistence in Soil	Active Organisms
3-(*p*-Chlorophenyl)-1, 1-dimethylurea	Monuron	4—12 months	*Pseudomonas*
4-Chlorophenoxyacetic acid	4-CPA		*Achromobacter*
			Flavobacterium
2,4-Dichlorophenoxyacetic acid	2,4-D	2—8 weeks	*Achromobacter*
			Corynebacterium
			Flavobacterium
2,4,5-Trichlorophenoxyacetic acid	2,4,5-T	5—11 months	
2-Methyl-4-chlorophenoxyacetic acid	MCPA	3—12 weeks	*Achromobacter*
			Mycoplana
2,2-Dichloropropionic acid	Dalapon	2—4 weeks	*Agrobacterium*
			Pseudomonas
Dinitro-*o*-*sec*-butylphenol	DNBP	2—6 months	*Corynebacterium*
			Pseudomonas
			Corynebacterium
4,6-dinitro-*o*-cresol	DNOC		
Isopropyl *N*-phenylcarbamate	IPC	2—4 weeks	
Isopropyl *N*-(3-chlorophenyl) carbamate	CIPC	2—8 weeks	
Trichloroacetic acid	TCA	2—9 weeks	*Pseudomonas*
2,3,6-Trichlorobenzoic acid	2,3,6-TBA	>2 years	

Alexander, M., *Soil Microbiology*, John Wiley & Sons, New York, 1961, 241. With permission.

Land disposals of pesticides are common practice.[25] The most used occupy the categories of:

- Burial (landfill type)
- Encapsulation
- Well injection
- Infiltration/evaporation basins
- Soil incorporation

The most common method is burial in landfill systems. Soil incorporation as identified with LT is used least often. It includes a variety of surface and soil disposal techniques most of which have already been discussed as spraying, filling, or flooding the waste over the soil surface. Little discrete quantitative recommendations on loading rates, frequency of application, rates of biodegradation and residual effects are available in the literature at this time.

2. Pharmaceuticals, Cosmetics, Soaps, and Detergents

The pharmaceutical industry, also, produces a great number of varied waste streams. In the production of organic acids and antibiotics, fermented residues accumulate as wastes in large quantities. The practice of adding Zn to some of these fermentations to control the microbial growth is a major deterrent to LT for the residues. Lime is sometimes added to fermentations to raise the pH to the neutral to slightly alkaline level. The mycelial residues from pharmaceutical ferments possess a favorable C/N ratio, contain nutrient elements, and can decompose rapidly and quite completely. Fermentation processes create the largest source of wastes in the pharmaceutical industry and the chemical syntheses create the wastes that are the most difficult to treat.

Table 6
THE MOBILITY OF SELECTED PESTICIDES IN HAGERSTOWN SILTY CLAY LOAM

Pesticide	Mobility Class[a]
Amiben	5
Atrazine	3
Azinphosmethyl	2
Bromacil	4
Chloroxuron	1
Dalapon	5
Dicamba	5
Dichlobenil	2
DDT	1
Dieldrin	1
Diquat	1
Diuron	2
Endrin	1
Fluomenturon	3
MCPA	4
Paraquat	1
Propachlor	3
Propanil	2
Propazine	3
Picloram	4
TCA	5
Trifluralin	1
2,4,5-T	3
2,4-D	4

[a]   Based on soil thin-layer chromatography R_f values in increasing order of mobility: Class 1, 0-0.09; Class 2, 0.10-0.34; Class 3, 0.35-0.64; Class 4, 0.65-0.89; Class 5, 0.90-1.00. Mobility is measured as "Rf" relative to the wetting front, so that an R_f of 0.5 means that over the measured distance the pesticide only moved half as far as the water. An R_f approaching 0 indicates the pesticide was immobile.

Hellings, C. S., *Soil Sci. Soc. Amer. Proc.*, 35, 737, 1971. By permission of the Soil Science Society of America, Madison, Wisconsin.

Potentially hazardous waste streams generated by the pharmaceutical industry are given in Table 7 to show the nature and variability of the waste and to provide an insight in to the land treatment potential.

Soap and detergent waste streams contain either inorganic or organic pollutants, usually in low levels. Phosphorus frequently contaminates the wastes. Disposal by LT is not difficult usually, except when salts and nitrates are abundantly present. Both have the potential for contaminating groundwater at high land-loading rates. Soil manipulation of pharmaceutical, cosmetic, soap, and detergent wastes depends considerably on the specific waste involved.

Table 7
POTENTIALLY HAZARDOUS WASTE STREAMS GENERATED BY THE PHARMACEUTICALS INDUSTRY

Product	Waste Stream	Land Treatment Potential[a] (Rate (R) and Capacity (C) Limiting Constituents)
Veterinary pharmaceuticals from arsenic and organoarsenic compounds.	Wastewater treatment sludge	Arsenic salts (C); arsenic (complexed) (C)
Veterinary pharmaceuticals from arsenic and organoarsenic compounds	Tar residues from the distillation of aniline based compds.	Arsenic salts (C); arsenic (complexed) (C)
Veterinary pharmaceuticals from arsenic and organoarsenic compounds	Residue from the use of activated carbon for decolorization	Arsenic salts (C); arsenic (complexed) (C)
Antibiotics and steroids from fermentation	Spent fermentation beers	
Antibiotics and steroids from fermentation	Spent diatomaceous earth	
Antibiotics and steroids from fermentation	Spills & equipment cleanup wastewaters	Depends on material spilled, spoiled, or cleaned
Antibiotics and steroids from fermentations	Barometric condenser wastewater	Chromium (C); zinc (C)
Antibiotics and steroids from fermentations	Spent solvents from extraction processes	Depends on solvent used
Chemical synthesis (variable batch processes)	Spent solvents	
Chemical synthesis (variable batch processes)	Spent filter cakes filter paper and activated charcoal	
Chemical synthesis (variable batch processes)	Spills, spoiled batches & equipment cleanup waters	Depends on material spilled, spoiled, or cleaned
Pharmaceutical formulation	Spills, spoiled batches & equipment cleanup waters	Depends on material spilled, spoiled, or cleaned
Vaccines, serums & plasma derivatives	Spent media broth & egg wastes	Depends on extractant used
Vaccines, serums & plasma derivatives	Spoiled batches, spills & equipment or cage cleanup waters	Depends on material spilled, spoiled, or cleaned
Botanical drugs	Spent plant material	Depends on extractant used
Botanical drugs	Spills, spoiled batches, & equipment cleanup waters	Depends on material spilled, spoiled, or cleaned

[a] Values for waste constituents may vary; hence, loading rates and capacities should be based on the analysis of the specific waste to be land treated and on the results of the pilot studies performed. Modified from Reference 17.

Because of the great diversity of wastes, no one method can be recommended unconditionally or demonstrated to be superior over another.

3. Wastes Related to Agricultural and Forestry Industry

In agriculture and forestry industry, we are accustomed to regarding wastes as nonhazardous resources. It might seem, therefore, that they should not appear in the content of LT. We must be aware that waste disposal problems proliferate and that which was a previous resource may become a waste. Good examples of this are the large point-source accumulations of animal manures, such as that of urban riding stables, feedlots of beef cattle and swine production, lambing yards, and poultry and turkey pens. Nine major categories of entities contaminate the air, water, and land of the environment associated with the industries of agriculture and forestry according to Wadleigh, former Director of the USDA ARS Soil, Water and Plant Sciences, Beltsville, Md:[1]

1.   Radioactive substances
2.   Chemical air pollution
3.   Airborne dust
4.   Infectious agents and allergens
5.   Excessive salts and sodiums
6.   Organic wastes
7.   Sediments
8.   Agricultural and industrial chemicals
9.   Plant nutrients

These nine may be found associated with both rural and urban areas. Since they represent wastes reaching the land in a variety of ways from airborne, waterborne transport, or applied to improve quality of production, or are naturally present as excesses as salts and minerals. They either do not lend themselves to LT or have received attention in other chapters. For example, the first four categories are reviewed in Volume II, Chapter 1, category 5 appears at the end of Volume I, Chapter 3. Organic wastes and plant nutrients (e.g., fertilizers and nitrates), agricultural and industrial chemicals (e.g., pesticides) are scattered in appropriate discussions throughout the book.

Sediments are considered along with soil erosion and conservation. Not the least of natural pollutants associated with soil erosion and conservation is the salt burden carried by rivers. The Colorado River is a case in point. During the flow from the origin to Yuma, Arizona, the river accumulates over 1.25 million tons of salt. Salinity is a hazard on a large proportion of uncultivated acreage in the western states.[1] Soil reclamation practices are fairly well known and understood, yet salts continue to accumulate.[26,27]

In summary, land treatment for disposal purposes differs only in a small way from land treatment for resource recovery. The main differences are:

●   Land for disposal only, need not be limited to agriculturally producing land nor expected to be food producing land. Thus, a wider choice of land disposal sites are available for disposal only.
●   Since crops do not necessarily need to grow on LT areas for disposal, the land treatment can be more intensive.
●   Costs per unit of waste treated will be less for LT for disposal only.
●   A broader selection of waste sources are associated with LT for disposal only. Wastes with a greater potentially hazardous biodegradable pollutant fit into the design better when disposal is the only consideration. More slowly biodegradable substances can be treated more effectively.

On the other hand:

●   Land may be expected to be taken out of production longer with the consequent loss of revenue.[26,27]

IV. WASTE CHARACTERISTICS AND LOADING RATES

A. Waste Factors Affecting Loading Rates

The amount of waste that is applied per unit area of land, often called the waste application rate, is a critical factor of successful land application. In order to avoid problems associated with the overloading of soils, waste application rates should be carefully determined prior to initiating operations. Waste application rates should be specified in terms of the monthly application rate and the total cumulative application rate over the expected lifetime of the facility. In determining waste application rates the objective is to match waste applications with the capacity of the soil-biological system to assimilate the waste. Treatability studies may be necessary for this determination.

Most wastes are complex mixtures that vary in composition. Thus, it is necessary to base loading rates on individual constituents of the waste rather than on bulk waste characteristics. The waste constituents usually considered are nitrogen, salts, oil and grease, toxic organics, metals, water, and anionic toxicants. The quantity of each constituent in the waste must be determined from qualitative and quantitative chemical analyses.

Loading rates for disposal purposes only are not expected to be identical to those suggested for waste for resource recovery purposes. However, the waste characteristics and composition and the required soil properties may be expected to be similar. The design of land application systems for disposal of wastewaters or liquid industrial wastes has primarily been controlled by the hydraulic loading (Table 8). Others[4,6,28-30] have suggested that BOD be considered in soil application rates. The importance of BOD (or TOC) on rate of biodegradation and its effects at different loading rates is well illustrated in Figure 2. Excessive overapplication can lead to serious reduction in the biodegradation rate as aerobic conditions change to anaerobic. There is some evidence[6] that with wastewaters having BOD concentrations of 1000 mg/ℓ or less, the hydraulic loading alone may be considered as the guide for LT loading rates. As we become more knowledgeable in liquid industrial waste land treatment, other characteristics may be limiting and loading rates may be established for each major parameter to be the most compatible with soil properties or other site characteristics. In addition to the chemical composition, the physical composition of consistency: liquid, semi-liquid, low moisture solids, and bulky wastes has been suggested[17] (Table 9). Obviously the mechanics of application of the different consistencies to the land will not be the same and consequently loading rates may differ. For example, injection of liquid beneath the soil surface reaches a limit at about 8% solids. Most applicators prefer to inject liquids with no more than 6.5% solids. Loading rates, therefore, may be less for injection than for solids spread by manure guns, manure spreaders, or graders and dozers after it is dumped.

Two basic constraints emerge as guidelines for establishing realistic loading rates:

- The rate of waste application that does not exceed the capacity of the soil to degrade, immobilize, or attenuate the waste pollutant should be selected.
- The waste must be distributed uniformly and at such a rate as not to create nuisance.

The characteristics of the waste that influencing loading rates most often are

- Water (volume)
- Nitrogen
- Phosphorus
- Human disease hazard
- Phytotoxicity
- BOD or TOC
- Biodegradability
- Heavy metal(s)
- Toxic organic compounds
- Salts and sodium

Table 8
SUMMARY OF HYDRAULIC AND ORGANIC LOADING
RATES USED IN EXISTING LAND DISPOSAL SYSTEMS
FOR INDUSTRIAL WASTES

Type of Waste	Hydraulic Load (gal/ac day[a])	Organic Load (lb BOD/ac day[b])
Biological chemicals	1,500	370
Fermentation beers	1,350	170
Vegetable tanning		
Summer	54,000	360
Winter	8,100	54
Wood distillation	6,850	310
Nylon	1,700	287
Yeast water	15,100	—
Insulation board	14,800	138
Hardboard	6,000	85
Boardmill whitewater	15,100	38
Kraft mill effluent	14,000	26
RI[c]	350,000	120
Semichemical effluent	72,000	90—120
Paperboard	7,600	13—30
Deinking	32,400	108
Poultry	40,000	100
Peas and corn		
57 day pack	49,000	238
35 day pack	34,400	2,020
Dairy		
Low value	2,500	10
High value	30,000	1,000
Soup	6,750	48
Steam peel potato	19,000	80
Instant coffee and tea	5,800	92
Citrus	3,100	51—346
Cooling water — aluminum casting (RI)	95,000	35

[a]   Multiply by 9.35×10^{-3} to convert to m³/ha day.
[b]   Multiply by 0.89 to convert to kg/ha day.
[c]   RI — Rapid Infiltration.

Wallace, A. T., *Land Treatment and Disposal of Municipal and Industrial Wastewaters*, Ann Arbor Science Publishers, Ann Arbor, Mich., 1976, 147. With permission.

Table 9
SUGGESTED WASTE CONSISTENCY
CLASSIFICATION

Consistency	Characteristics
Liquid	Less than 8% solids and particle diameter less than 2.5 cm
Semiliquid	3—15% solids or particle diameters over 2.5 cm
Low moisture solids	Greater than 15% solids
Bulky wastes	Solid materials consisting of contaminated lumber, construction materials, plastic, etc.

B. Loading Rate Limitations

The use of the land treatment option as a means of disposing of wastes is largely an issue of economics, since for many, the permissible rates of application may require occupation of excessive areas of land. One of the most convenient methods of determining loading rates is by calculating the acceptable rates for each constituent and using the most restrictive figures. The term rate limiting constituent (RLC) has been proposed representing the waste fraction which controls yearly loading rates.[17] The land area requirements are calculated by dividing yearly waste receipts (kg/year) by the acceptable loading rate (kg/ha/year).

Two other restrictive terms need to be understood, namely ALC, application limiting constituent, and CLC, capacity limiting constituent. ALC refers to the one constituent limiting the amount of waste which may be applied in a single disposal, but is rapidly biodegraded, lost from the system, or immobilized. CLC refers to some limiting constituent of waste that is a conservative, accumulating species, like low level Cd or As, which can set the upper limit for the total amount of waste that may be applied at a single site (kg waste/ha). The determination of these parameters is justified by their usefulness in several ways. For example, maximum design life may be calculated by dividing CLC controlled waste treatment maximum (kg/ha) by the design loading rate (kg/ha/year), i.e., the design life is capacity limiting constituent (CLC/Design Loading Rate).

There are a number of constituents useful for determining the loading rates of wastes associated with a hazardous waste land treatment disposal system. Since water is practically ubiquitous in hazardous waste streams and may contribute a significant input to the soil, a complete hydrological balance at the site must be calculated.

1. Hydrological Balance

The hydrological balance is most important to humid climatic regions to evaluate waste storage requirements, waste application rates, and runoff retention, and treatment needs. In arid and semiarid climatic regions, the hydrologic balance may be used to calculate the waste application or loading rate directly, provided the soils are highly permeable and runoff is not a problem. In humid climates where rainfall generates runoff at times during the year, the hydrological balance may be used to calculate the soil storage needed to contain the rainfall. In either case, the water balance is written by:

$$P + W = ET + R + L + \Delta S$$

where:

P is precipitation
W is water applied to the waste
ET is the evapotranspiration
R is the runoff which will need to be collected
L is the leachate
ΔS is the change in water stored

Examples of hydrological balance calculations may be found in the literature such as by Brown and Associates.[17] The computer programs have improved on the modeling of components of the hydrological balance for soils hydrology.[31-34]

2. Plant Nutrients

Loading rates for agricultural land, for example, are closely related to the nutritional requirements of the crop being grown on the land. Nitrogen mineralization of municipal sludge determines the loading rate for the specific crop.[35] There are additional considerations in loading rates, such as climate in relation to biodegradation rate, total dissolved solid (salt level) concentration, and method of application. With respect to the latter, greater loading

Table 10
ANNUAL NITROGEN, PHOSPHORUS, AND
POTASSIUM UTILIZATION (lb/acre) BY
SELECTED CROPS[a]

Crop	Yield	Nitrogen	Phosphorus	Potassium
Cotton	250 lb (lint)[c]	250	35	175
Corn	150 bu[d]	185	35	178
	180 bu	240	44	199
Corn silage	32 tons[e]	200	35	203
Soybeans	50 bu[f]	257[b]	21	100
	60 bu	336[b]	29	120
Grain sorghum	8000 lb[c]	250	40	166
Wheat	60 bu[f]	125	22	91
	80 bu	186	24	134
Oats	100 bu[g]	150	24	125
Barley	100 bu[h]	150	24	125
Alfalfa	8 tons[e]	450	35	398
Orchard grass	6 tons	300	44	311
Brome grass	5 tons	166	29	211
Tall fescue	3.5 tons	135	29	154
Bluegrass	3 tons	200	24	149

[a]   Values reported above are from reports by the Potash Institute of America and are for the total above-ground portion of the plants except for cotton. Where only grain is removed from the field, a significant proportion of the nutrients is left in the residues. However, since most of these nutrients are temporarily tied up in the residues, they are not readily available for crop use. Therefore, for the purpose of estimating nutrient requirements for any particular crop year, complete crop removal can be assumed.

[b]   Legumes get most of their nitrogen from the air, so additional nitrogen sources are not normally needed.

[c]   Multiply by 1.12 to convert to kg/ha.

[d]   Multiply by 62.77 to convert to kg/ha.

[e]   Multiply by 2.24 to convert to t/ha.

[f]   Multiply by 67.25 to convert to kg/ha.

[g]   Multiply by 35.87 to convert to kg/ha.

[h]   Multiply by 53.80 to convert to kg/ha.

Sommers, L. E. and Nelson, D. W., *Application of Sludges and Wastewaters on Agricultural Land: Planning and Educational Guide*, Ohio Agricultural Research Division Center, Wooster, 1976, 3. With permission.

rates may be used for surface applications since about half of the $NH_4^+ - N$ is lost by volatilization. If incorporated, the available N added in the sludge should equal the N fertilizer recommendation or vegetation use. Some calculations are available as a guide for municipal sludge annual application rates to the land:[35]

Step 1. Obtain N requirement for the crop grown from Table 10 or obtain N fertilizer recommendation from Cooperative Extension Service or soil analysis laboratory.

Step 2. Calculate tons of sludge needed to meet crop's N requirement.

    a. Available N in sludge

$$\% \text{ Inorganic N } (N_i) = (\% \ NH_4\text{-N}) + (T \ NO_3\text{-N}) \qquad (2)$$

$$\% \text{ Organic N } (N_o) = (\% \text{ Total N}) - (\% \text{ inorganic N}) \qquad (3)$$

Table 11
RELEASE OF RESIDUAL NITROGEN DURING
SLUDGE DECOMPOSITION IN SOIL

Years After Sludge Application	Organic N Content of Sludge, %						
	2.0	2.5	3.0	3.5	4.0	4.5	5.0
	Lb N Released per Ton Sludge Added[a]						
1	1.0	1.2	1.4	1.7	1.9	2.2	2.4
2	0.9	1.2	1.4	1.6	1.8	2.1	2.3
3	0.9	1.1	1.3	1.5	1.7	2.0	2.2

[a]   Multiply by 0.5 to convert to kg/t.

Sommers, L. E. and Nelson, D. W., *Application of Sludges and Wastewaters on Agricultural Land: Planning and Educational Guide*, Ohio Agricultural Research Division Center, Wooster, 1976, 3. With permission.

i. Surface applied sludge
Pounds available N/ton sludge = $(\%NH_4\text{-}N \times 10) +$
$(\% NO_3\text{-}N \times 20) + (\% N_o \times 4)$ (4)

ii. Incorporated sludge
Pounds available N/ton sludge = $(\% NH_4\text{-}N \times 20) +$
$(\% NO_3\text{-}N \times 20) + (\% N_o \times 4)$ (5)

b. Residual sludge N in soil
If the soil has received sludge in the past 3 years, calculate residual N from Table 11.

c. Annual application rate

i.
$$\text{Tons sludge/acre} = \frac{\text{crop N requirement} - \text{residual N}}{\text{lb available N/ton sludge}}$$ (6)

ii.
$$\text{Tons sludge/acre} = \frac{2 \text{ lb Cd/acre}}{\text{ppm Cd} \times .002}$$ (7)

iii. The lower of the two amounts is applied.

Step 3. Calculate total amount of sludge allowable.

a. Obtain maximum amounts of Pb, Zn, Cu, Ni, and Cd allowed for CEC of the soil from Table 12 in lb/a.

b. Calculate amount of sludge needed to exceed Pb, Zn, Cu, Ni, and Cd limits, using sludge analysis data.

Metal

Pb: $\text{Tons sludge/acre} = \dfrac{\text{lb Pb/acre}}{\text{ppm Pb} \times .002}$ (8)

Zn: $\text{Tons sludge/acre} = \dfrac{\text{lb Zn/acre}}{\text{ppm Zn} \times .002}$ (9)

Cu: $\text{Tons sludge/acre} = \dfrac{\text{lb Cu/acre}}{\text{ppm Cu} \times .002}$ (10)

Ni: $\text{Tons sludge/acre} = \dfrac{\text{lb Ni/acre}}{\text{ppm Ni} \times .002}$ (11)

Cd: $\text{Tons sludge/acre} = \dfrac{\text{lb Cd/acre}}{\text{ppm Cd} \times .002}$ (12)

(Note: sludge metals should be expressed on a dry-weight ppm (mg/kg) basis.)

The lowest value is chosen from the above five calculations as the maximum tons of sludge per acre which can be applied.

Step 4. Calculate amount of P and K added in sludge

Tons of sludge $\times$ % P in sludge $\times$ 20 = lb. of P added (13)

Tons of sludge $\times$ % K in sludge $\times$ 20 = lb. of K added (14)

Step 5. Calculate amount of P and K fertilizer needed

(lb. P recommended for crop)* - (lb. P in sludge) = lb. P fertilizer needed (15)

(lb. K recommended for crop)* - (lb. K in sludge) = lb. K fertilizer needed (16)

Generally, phosphorus will not be deficient if nitrogen application rates meet the demand of the crop. In fact, phosphorus fertilization may be excessive for some crops when sludge is added at rates equivalent to the nitrogen utilization rate (i.e., 250 lb N/a).

3. Salt and Sodium

Almost all wastes contain soluble salts. Soils and waters are no exception. Soluble salts accumulate more abundantly in arid and semiarid climatic regions than in humid. Surplus precipitation in regions east of the Mississippi River, in particular, remove salts by leaching. Because of the high potential evapotranspiration imposed on hot desert soils and the generally higher concentration of soluble salts in the MSW leachaes, feedlot manures and sewage sludge, for example, from this region, salt accumulation must be given prime attention. Salt injury to sensitive crops can occur if sewage sludge of wastewater contains more than about 1250 ppm dissolved solids (EC of about 2.0 mmhos/cm) and is applied without sufficient irrigation water to keep the salts moving down out of or below, the root zone. Fortunately, there are crops such as alfalfa, cotton, barley, sorghum, and other small grain crops that have a relatively high salt tolerance.

For example, high rates of sludge application (112 to 224 t/ha dry matter) were found to support less than maximum yields of small grain and soybeans on Pima silt loams in Arizona.[36] These levels, however, are many times greater than proposed for soil injection at 475 to 750 m^3/ha. A 3.5% solids slurry will provide only about 16 to 26 t/ha dry matter. Also, fortunately, the sodium absorption ratio (SAR) of the Tucson sludge is favorable. Using data from this publication the SAR is calculated:[36]

$$\text{SAR} = \frac{Na^+}{\sqrt{Ca^{++} + Mg^{++}/2}} \quad \text{or} \quad \frac{3.91}{1.65} = 2.37 \quad (17)$$

Since SAR values from 5 to 15 can lead to poor water penetration and deterioration of soil structure in soils. The value of 2.37, however, is lower than 5, therefore, the sludge should be satisfactory for long-time use.

Fortunately, boron as a soluble salt is usually favorably low in municipal sewage sludges where industry is minimally represented.

4. Metals

The total amount of sludge metal permitted on agricultural land is reported in Table 12.[37] The cation exchange capacity (CEC) determines the level allowed. Fine textured soils (silt and clay) have higher CEC than coarse textured soils (sands). For example, silt loams, silty clay loams, clay loams, and sandy clay loams all should have CEC above 15. Sands and gravelly sands represent the lower end of 0.5 to 5 meq/100 g CEC. Approximately four times as much sludge may be applied to upland loams than in the sand and gravelly soils of the flood terraces and river floodplains.

The CEC employment in Table 12 does not imply that metals added to soils become attached to soil particles as exchangeable cations. The CEC is closely correlated with other soil characteristics such as (a) surface area of the soil particles, (b) hydrous oxides of Fe,

Table 12
TOTAL AMOUNT OF SLUDGE
METALS ALLOWED ON
AGRICULTURAL LAND

Metal	Soil Cation Exchange Capacity (meq/100 g)[a]		
	0—5	5—15	15
	Maximum Amount of Metal (lb/ac)[b]		
Pb	500	1000	2000
Zn	250	500	1000
Cu	125	250	500
Ni	125	250	500
Cd	5	10	20

[a] Determined by the pH 7 ammonium acetate procedure.
[b] Multiply by 1.12 to convert to kg/ha. The values in Table 4.12 are the total amounts of metals which can be added to soils. With metal contaminated sludges, one of the above criteria may be met with a single application, whereas 5, 10 or 20 applications may be needed for "clean" domestic sludges. Furthermore, when the metal limits are reached, sludge application must be terminated. A soil pH $\geq$ 6.5 must be maintained in all sites after sludge is applied to reduce the solubility and plant uptake of potentially toxic heavy metals.

Knezek, B. D. and Miller, R. H., Ohio Agriculture Research Center, Res. Bull. 1090, Wooster, 1976, 1. With permission.

Mn, and Al, and (c) organic matter, all of which react with heavy metals to attenuate. The CEC determination is fairly easily accomplished on most soils (except calcareous) and is reasonably proportional to the ability of a soil to minimize the metal effects on crops, if any.

5. BOD and TOC

Organic constituents appear in significant amounts in wastes and waste streams usually considered acceptable for land treatment since biodegradation is the primary advantage attributed to the LT option for disposal, although losses of volatilization, leaching, and perhaps runoff accompany biodegradation. Waste loading based on BOD concentration (i.e., above 1000 mg/ℓ) receives considerable attention by several investigators. Wallace,[28] considers the usefulness of hydraulic loading and, in a few cases, cation loading, especially sodium. Through the use of data generated by the lysimeter technique (Figure 1), he has shown that highly different rates of biodegradation can appear when loading rates of waste sulfite liquor alter aerobic and anaerobic conditions in the soil (Figure 2). Aerobic degradation rates may be seen to be much higher than anaerobic rates in the waterlogged soils.

All natural plant carbonaceous constituents decompose at differing rates.[38-41] Lignin, for example, decomposes more slowly than cellulose, and hemicellulose decomposes more rapidly than cellulose (Figure 3). The presence of lignin and its concentration in native plant tissues also influences decomposition rates (Table 13). Thus BOD and TOC both are im-

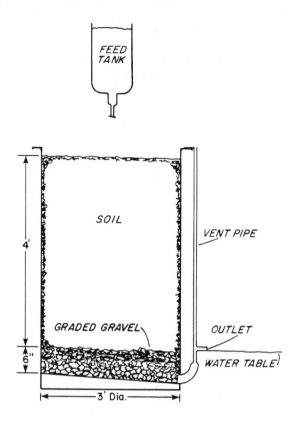

FIGURE 1.   Cross section of lysimeter. (From Wallace, A. T., Land disposal of liquid industrial wastes, in *Land Treatment and Disposal of Municipal and Industrial Wastewater*, Sanks, R. L. and Asano, T., Eds., Ann Arbor Science Publishers, Ann Arbor, Mich., 1976, 153. With permission.)

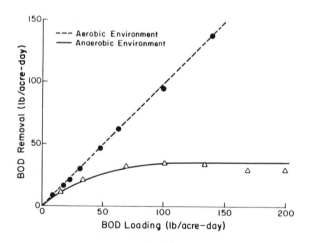

FIGURE 2.   Comparative BOD removal rates in aerobic and anaerobic land treatment of wood pulping wastes. (From Wallace, A. T., Land disposal of liquid industrial wastes, in *Land Treatment and Disposal of Municipal and Industrial Wastewater*, Sanks, R. L. and Asano, T., Eds., Ann Arbor Science Publishers, Ann Arbor, Mich., 1976, 153. With permission.)

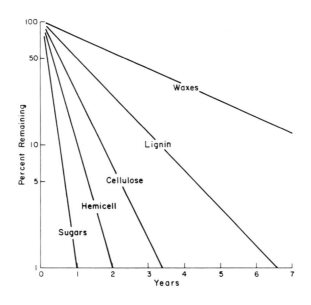

FIGURE 3. Decomposition curves for various organic constituents in forest litter. (From Minderman, G., Addition, decomposition, and accumulation of organic matter in forests, *J. Ecol.*, 56, 355, 168. With permission.)

Table 13
DECOMPOSITION BY *PSEUDOMONAS EPHEMEROCYANEA* OF JUTE PREPARATIONS CONTAINING DIFFERENT QUANTITIES OF CELLULOSE AND LIGNIN

Preparation %	Cellulose Content (%) %	Lignin Content (%)	% of Cellulose Decomposed
A	99.2	0.0	100.0
B	95.5	3.3	95.6
C	89.2	6.3	83.1
D	82.7	11.9	37.9
E	75.6	12.6	17.7

Incubation period of 21 days.

Fuller, W. H. and Norman, A. G., *J. Bact.*, 46, 273, 1943. With permission.

portant parameters to know when comparing organic waste biodegradation rates as means of establishing long time loading rates.

The importance of environmental influence on biodegradation of organics in soils, also, must be considered.[42] The effect of some environmental conditions on cellulose decomposition rate is well demonstrated in Figure 4.[28] The environmental effects may be further demonstrated by comparing the earlier Figure 2 with that of Figure 5 intended to include much higher loadings, 10, 752 kg/ha/d (9600 lb/a/d).[28] As interpreted by Wallace (1976), at the highest BOD loading the wood sugars were 100% decomposed, however, the BOD reduction was only 10%, due probably to a shift from aerobic to anaerobic metabolism.[28]

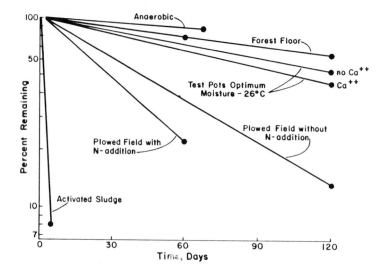

FIGURE 4.   Rates of cellulose decomposition in soil and water under various environmental conditions. (From Wallace, A. T., Land disposal of liquid industrial wastes, in *Land Treatment and Disposal of Municipal and Industrial Wastewater*, Sanks, R. L. and Asano, T., Eds., Ann Arbor Science Publishers, Ann Arbor, Mich., 1976, 155. With permission.)

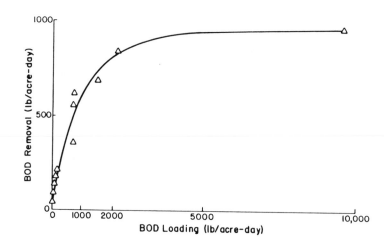

FIGURE 5.   BOD loading and removal relationship for waste sulfite liquor applications to 10-foot deep soil columns. (From Wallace, A. T., Land disposal of liquid industrial wastes, in *Land Treatment and Disposal of Municipal and Industrial Wastewater*, Sanks, R. L. and Asano, T., Eds., Ann Arbor Science Publishers, Ann Arbor, Mich., 1976, 156. With permission.)

The shift starts near 1000 kg/ha/d. Also to be noted is the great difference between maximum BOD removal between and anaerobic curve of Figure 2 and the aerobic curve in Figure 5. The anerobic condition of Figure 2 was caused by waterlogging and not by organic over-loading. This presentation is made in an attempt to illustrate certain problems in formulating a rational design equation for LT disposal processes and to suggest that with more data and better knowledge of the lysimeter or soil column parameters, rational models are possible.

Among the several mathematical models proposed, the one by Pirt[43] appears to give reasonable results but requires further critical testing. Both aerobic and anaerobic soil columns were used and mathematical equations were developed for both. The expression and functional relationships relating to the aerobic columns was

$$S_r - S_e = nKe/PF \qquad (18)$$

where

S_r = feed substrate concentration
S_e = effluent substrate concentration
n = number of theoretical compartments in the column
c = dissolved oxygen concentration in the liquid medium
P = oxygen demand constant (g oxygen consumed/g substrate oxidized)
F = liquid flow rate
K = $A(2D\rho q)$
    A = cross-sectional area of column
    D = diffusion coefficient for limiting substrate through layer of biomass
    ρ = density of biomass
    q = specific rate of substrate consumption by biomass (gram substrate/gram dry biomass-hour)

The assumption in this equation is that oxygen is the rate-limiting factor. With some evaluation and exclusion of variables due to deviation from piston-flow, carbon/nitrogen relationships, temperature and possible nonbiological removal, the above approach has realistic possibilities for land treatment systems.

6. Toxic Organics

Toxic organics (solids, liquids, and solvents) have been discharged onto or into soil as long as industry has been operating. Today some of those disposals appear in groundwater (e.g., TCE, PCB, etc.), others remain undecomposed to contaminate the soil (e.g., dioxane) and still others may have entered the food chain (e.g., PCB). Toxic organics have not been researched sufficiently, except for a few isolated cases, to form guidelines for land treatment. The diversity of this group allow for very few generalities. Land treatment of toxic organics must be managed on a case-by-case basis under strict supervision.

Evaluation-screening methodologies and assay techniques for biodegradation rates and residual toxicity are required before land treatment can be recommended. Where information is not available for a particular type of hazardous waste, laboratory and/or field tests must be conducted to demonstrate this capability. The Hazardous Waste Land Treatment (HWLT) Regulations issued by EPA's Office of Solid Waste (OSW) requires a permit to operate a hazardous waste land treatment site. The screening method(s) should be rapid and simple and effective enough to assist in demonstrating land treatability of a particular waste. Operators and applicators need a method which will assist in determining a rate of waste application at which the site soil is capable of achieving waste attenuation in an environmentally acceptable way, and also in determining when attenuation of a waste has reached the time for reapplication. Table 3, as presented earlier, shows the kind of information needed for biodegradation assessment.

V. SOIL MANAGEMENT

The success of land treatment depends almost wholly on soil management once the site, design, and layout have been selected. Soil management must insure adequate treatment and retention of the potential pollutants contained in the waste at or as near to the place of

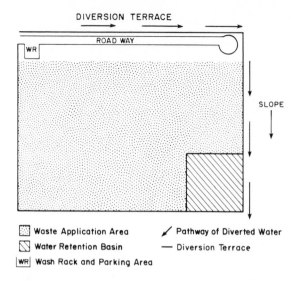

FIGURE   6. Possible layout of a land treatment facility in a gently sloping uniform terrain when only one cell is to be used. (From Brown, K. W. and Associates, Inc., Hazardous Waste Land Treatment, SW-874, OWWM, U.S. Environmental Protection Agency, Washington, D.C., 1983, 769.)

deposit as possible. Transport in any direction, across the surface, downward into the groundwater sources or up into the atmosphere by wind action must not be allowed to occur. Water and air transport vehicles may be controlled by management of the soil. Experts with many years of experience in this field are centered in the USDA Soil Conservation Service and the State Agricultural Experiment Stations. Although the dominant physical as well as chemical nature of the waste often determines the design and layout, soil management will differ little whether single, rotating, or progressive cell land configurations are installed (Figures   6,   7 and   8). One of the most important factors in soil management is the amount of runoff which will be allowed to collect and the options available for disposal. The most ideal option, but also the most difficult to maintain, is on-site disposal of runoff by evaporation, transpiration, percolation, and reapplication. Several possible layouts of a land treatment facility have been suggested.[17] They are reproduced in Figures   6,   7, and   8   to provide a background for a better understanding of the discussion on soil management.

A. Predisposal Land Preparation

Land treatment for disposal only, utilizes both land that is capable of being tilled and that which is too shallow, rough, or wooded to permit tilling. Some land surfaces require only clearing of brush or scattered trees to allow tilling. Still other surfaces can be made more acceptable for surface application by removing trees and bushes even though tilling is not practical. Surface grading is expensive, but a limited amount may be sufficiently advantageous to call for slope modification or construction of diversion ditches, terraces, levees, and other grade changes for water control. Distrubance of the soil, its natural sod and plant cover should be kept to a minimum. Breaking of the natural soil surface for whatever reason will encourage accelerated soil and water erosion and runoff. Traffic across the land, also, must be kept to a minimum with the necessary traffic scheduled in the most conservative patterns. The same principles that apply to cleared and range land apply to forested land also.

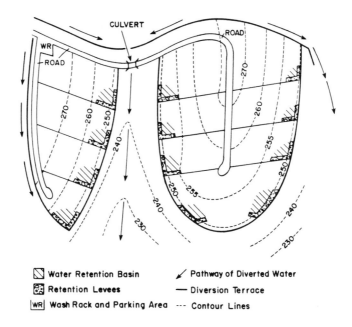

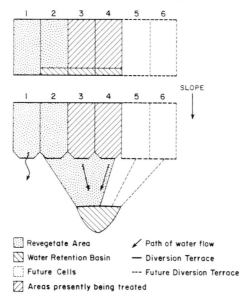

FIGURE 7. Possible layout of a land treatment facility in rolling terrain showing 12 cells, and associated runoff retention basins. (From Brown, K. W. and Associates, Inc., Hazardous Waste Land Treatment, SW-874, OWWM, U.S. Environmental Protection Agency, Washington, D.C., 1983, 770.)

FIGURE 8. Possible layouts of a land treatment facility in a gently sloping uniform terrain when a progressive cell configuration is to be used. (From Brown, K. W. and Associates, Inc., Hazardous Waste Land Treatment, SW-874, OWWM, U.S. Environmental Protection Agency, Washington, D.C., 1983, 772.)

B. Soil Erosion Control

"Natural" or "geological" erosion of the surface of the earth is a constant and continual process. The mountain valleys and other physical patterns of the earth's surface cannot be attributed to any one single event or even a few rare events, but primarily to a long and

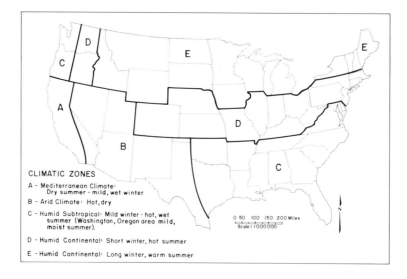

FIGURE 9.   Generalized climatic zones for land application. (Courtesy of the USDA Soil Conservation Service.)

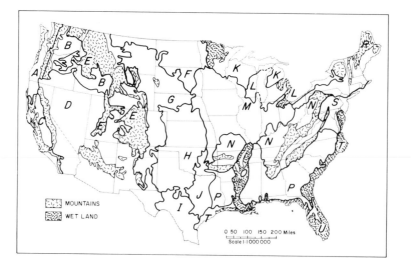

FIGURE 10.   Major land resource regions of the United States. (From Schwab, G. O., Frevert, R. K., Edminster, T. W., and Barnes, K. K., *Soil and Water Conservation Engineering*, 3rd ed., John Wiley & Sons, 1981, 4. With permission.)

continuing series of erosion processes in association with the more dramatic sculpturing geological changes. The erosion that concerns us is not "geological" erosion but "accelerated" erosion caused by activities of mankind. The objective is to predict the probable severity of man-induced erosion and select those land and soil management procedures that will control erosion at an acceptable level. There is a vast literature on soil, water, and wind erosion control, including many popular books and technical college level texts.[44-48]

Erosion by the activity of all agents, water, wind, temperature, and biological, must be kept under certain control if land treatment is to be utilized as a viable option in waste disposal. The acceptable limits of erosion vary according to the soil type, climatic region, or country (Figures 9 and 10). In the United States the limits of soil loss range from 2.5 to

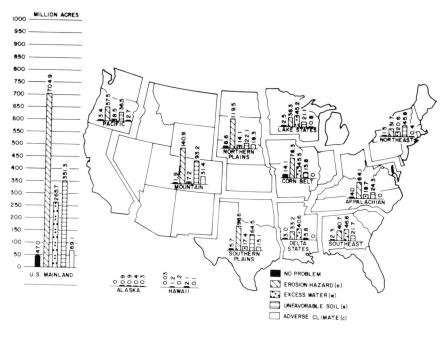

FIGURE 11.   Dominant conservation problems on non-Federal land in the United States. (Numbers shown are in million acres. From U.S. Conservation Needs Inventory, USDA, Washington, D.C., 1967, Preliminary data.)

12.5 t/ha/yr (1 to 5T/a/yr) (Figure 11).   The maximum soil loss figure which erosion should not exceed has been set most often at 12.5 t/ha/yr, and that is the figure recommended for farming practices by the Universal Soil Loss Equation (USLE).

Water erodes the soil surface in a number of ways. Classifications correspond to a progressive concentration of water starting with raindrop (a) "splash" erosion, continuing to (b) rill, (c) gully, and (d) streambank. Wind erosion is classified into (a) detrusion, (b) abrasion, (c) efflation, (d) extrusion, and (e) effusion. The several processes of water and wind erosion may occur at the same time. The two major erosion phases of detachment and transport characterize both water and wind erosion.

Not only is erosion a transport system for chemical pollutants, but also the soil sediments themselves are a problem. Sediments accumulate in streams from soil erosion due to unconcerned poor soil management practices. Plugging of city storm sewers may also occur as a result of poor landscaping and urban carelessness in soil protection of home property. About 4 billion tons of sediment waste wash into tributary streams each year in the U.S.[1] Valuable plant nutrient losses from soil surfaces accompany the sediment movement. Accelerated eutrophication often takes place as the nitrates and phosphates enrich streams and lakes. This type of pollution, from soil surface erosion (e.g., sediment), is not new. Only what is put on the surface, e.g., fertilizers, pesticides, soil conditioners, and wastes of all nature is new. If erosion cannot be controlled, then the other pollutants will not be controlled either. The USDA Soil Conservation Service has been occupied with the control of erosion resulting in transport of soil sediments, plant nutrients, and pesticides since the beginning of the Soil Conservation Service Act in the early 1930s. Now, waste disposal programs, including land treatment, have placed an added burden on erosion control associated with soil, water, and wind. For the most part, those techniques and methods in management of the land for erosion control already developed may be readily and easily applied to land treatment or any other system involving land disposal of wastes. Because of the widespread literature on conventional soil conservation management practices, only a very brief discussion will be presented. The ramifications of pollution and soil conservation and soil

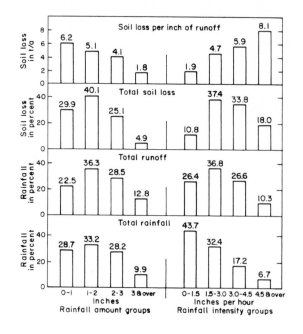

FIGURE 12.   Effect of rainfall and runoff on soil loss from a
sandy clay loam soil in North Carolina. (From Schwab, G. O.,
Frevert, R. K., Barnes, K. K., and Edminster, T. W., *Elementary
Soil and Water Engineering*, 2nd ed., John Wiley & Sons, New
York, 1971, 83. With permission.)

erosion control are endless, therefore, only pollution on land where it is directly and significantly linked with soil erosion will be discussed.

1. Erosion by Water

Water is the primary agent responsible for erosion in humid regions and strangely enough
can be a a dominant force in arid regions also. Those conditions influencing runoff are the
same as those affecting erosion. The effects of rainfall and runoff on soil loss from a sandy
loam soil in North Carolina are illustrated in Figure 12.[49]   The soil loss and runoff are
compared on a basis of different amounts and intensities of rainfall. Also, according to
"rainfall-amount-groups", storms of less than 7.6 cm (3 in.) account for 90% of the total
rainfall and 95% of all soil loss. The total loss differes with the total runoff for the different
rainfall amount groups but, the soil loss per centimeter (in.) of runoff decreases as the size
of the storm increases. On the other hand, when storms are grouped by rainfall intensities,
rains of less than 3.8 cm/h (1.5 iph) intensity represent 44% of the total rainfall but cause
only 11% of the soil loss. Intensities greater than 11.4 cm (4.5 iph) represent only 7% of
the total rainfall but cause 18% of the soil loss. Thus, soil loss per unit of runoff increases
as the intensity increases although the total rainfall may decrease.[49]

Many different variables enter into soil erosion by water and also by wind. To quantify
the erosion caused by soil disturbances, such as tilling, cropping, roadbuilding, overgrazing,
etc., has required the accumulation of a vast fund of knowledge.

The Universal Soil Loss Equation (USDA ARS Agricultural Handbook 282, 1965) is an
attempt to isolate each variable and reduce its effect to a number such that when the numbers
are multiplied together the value represents the amount of soil loss. The equation presented
more recently by Wischmeier estimates the average annual soil loss by the equation:[50-52]

$$A = 2.24RKLSCP \qquad (19)$$

where A =   average annual soil loss in Mg/ha (metric tons/ha)

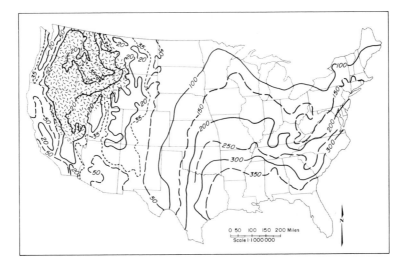

FIGURE 13.   Rainfall and runoff erosivity index R by graphic location. (From U.S. Department Agricultural Research Service, and U.S. Environmental Protection Agency, *Control of Water Pollution from Cropland,* Vol. 1, U.S. GPO, Washington, D.C., 1975, Chap. 1.)

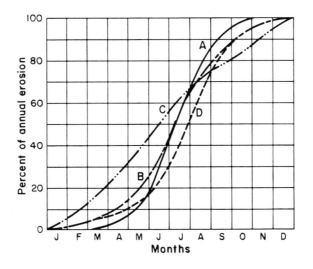

FIGURE 14.   Monthly distribution of the rainfall erosion index. (From Schwab, G. O., Frevert, R. K., Barnes, K. K., and Edminster, T. W., *Elementary Soil and Water Engineering,* 2nd ed., John Wiley & Sons, New York, 1971, 91. With permission.)

R = rainfall and runoff erosivity index by geographic location as given in Figures 13   and 14, or Table 14

K = Soil-erodibility factor (see Table 15),   which is the average soil loss in t/a per unit of erosion index for a particular soil in cultivated continuous fallow with an arbitrarily selected slope length L of 22 m (73 ft) and slope steepness S, of 9% (if K is Mg/ha, change constant 2.24 to 1.0)

LS = topographic factor evaluated in Figure 15

Table 14
FREQUENCY OF ANNUAL AND SINGLE-STORM EROSION INDEX R

Location	Return Period in Years			
	2	5	10	20
ANNUAL EROSION INDEX, R				
Little Rock, Ark.	308	422	510[a]	569
Indianapolis, Ind.	166	225	275[a]	302
Devils Lake, N.D.	56	90	120[a]	142
SINGLE-STORM EROSION INDEX, R				
Little Rock, Ark.	69	115	158	211
Indianapolis, Ind.	41	60	75	90
Devils Lake, N.D.	27	39	49	59

[a]   Interpolated values

Wischmeier, W. H. and Smith, D. D., U.S. Department Agricultural Handbook, 537, 58, 1976.

Table 15
K, SOIL-ERODIBILITY FACTOR BY SOIL TEXTURE IN t/ac[a]

Textural Class	0.5	2	4
Fine sand	0.16	0.14	0.10
Very fine sand	0.42	0.36	0.28
Loamy sand	0.12	0.10	0.08
Loamy very fine sand	0.44	0.38	0.3 –
Sandy loam	0.27	0.24	0.19
Very fine sandy loam	0.47	0.41	0.33
Silt loam	0.48	0.42	0.33
Clay loam	0.28	0.25	0.21
Silty clay loam	0.37	0.32	0.26
Silty clay	0.25	0.23	0.19

[a]   Selected from USDA-EPA, Vol. I (1975) and are estimated averages of specific soil values. For more accurate values by soil types use local recommendations of Soil Conservation Service or state agencies. (1 t/a = 2.24 Mg/ha)

Schwab, G. O., Frevert, R. K., Edminster, T. W., and Barnes, K. K. *Soil and Water Conservation Engineering*, John Wiley & Sons, New York, 1981, 525. With permission.

C = cropping-management factor, which is the ratio of soil loss for given conditions to soil loss from cultivated continuous fallow as given in Table 16

P = conservation practice factor, which is the ratio of soil loss for a given practice to that for up and down the slope farming as given in Table 17

A predicted soil loss by water erosion, therefore, can be made for a given set of conditions. The equation is established for planning farm practices and is designed for cultivated land,

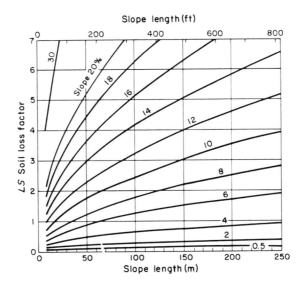

FIGURE 15. Topographic factor, LS, soil loss factor by length and steepness of slope in the Universal Soil Loss Equation. (From Wischmeier, W. H. and Smith, D. D., Predicting Rainfall Erosion Losses- A Guide to Conservation Planning, *U.S. Dept. Agric., Agric. Handbk.,* No. 537, 1976, 58.)

primarily in humid regions. Consequently, it receives the most application where these conditions parallel land treatment/utilization on agricultural land and cultivation is a part of the disposal program. Research to provide an extension of the Universal Soil Loss equation to allow for better adoption to land cultivation and land treatment/utilization is needed. Factors C and P appear to need the most attention and redefining. Even when applied to agriculture the factors in the equation differ with different climatic regions. For example, the equation for arid lands is not the same as for the Midwest.

2. Erosion by Wind

Wind erosion varies only by degree. A number of different types of soil movement occurs:

- Suspension of particles in the air
- Saltation — skipping and bouncing along the surface
- Surface creep — rolling and sliding in almost continuous contact with the surface.

Soil particles that move by suspension usually are less than 0.1 mm in diameter, by saltation 0.1 to 0.5 mm, and by surface creep 0.5 to 1.0 mm. About 50% of soil losses by wind take place as a result of saltation. Soil loss estimations are possible by knowing certain factors.[53] A guide for estimating soil losses by wind has been suggested:[54]

$$E = f(I, C, K, L, V) \qquad (20)$$

where

E = the soil loss by wind
I = the erodibility, i.e., vulnerability to wind erosion
C = a factor representing the local wind conditions
K = the soil surface roughness
L = the length of the field in the direction of the prevailing wind
V = a measure of the vegetative cover

Table 16

RATIO OF SOIL LOSS FROM CROPS TO CORRESPONDING LOSS FROM CONTINUOUS FALLOW[a]

Cover, Sequence, and Management	Crop Yields		Crop-Stage Period[b]				
	Meadow (tons)[e]	Corn (bu)[f]	0 (%)	1 (%)	2 (%)	3 (%)	4 (%)
1st-year corn after meadow, RdL[c]	2	60	15	30	27	15	22
2nd-yr corn after meadow, RdL	3	70	32	51	41	22	26
2nd-yr corn after meadow, RdR[d]	3	70	60	65	51	24	65
3rd- or more yr corn, RdL	—	70	36	63	50	26	30
Small grain w/meadow seeding:							
(1) In disked corn residues							
After 1st corn after meadow	2	60	—	30	18	3	2
After 2nd corn after meadow	2	60	—	40	24	5	3
(2) On disked corn stubble, RdR							
After 1st corn after meadow	2	—	—	50	40	5	3
After 2nd corn after meadow	2	—	—	80	50	7	3
Established grass and legume meadow	3	—	—	—	0.4	—	—

[a] Portion of 100-line published table (Wischmeier, 1960).
[b] Crop-stage periods are defined below:
  0 Turnplowing to seedbed preparation.
  1 Seedbed — first month after seeding.
  2 Establishment — second month after seeding.
  3 Growing cover — from 2 months after seeding to harvest.
  4 Stubble or residue — harvest to plowing or new seedbed.
[c] RdL, crop residues left and incorporated by plowing.
[d] RdR, crop residues removed.
[e] Multiply by 2.24 to convert to t/ha.
[f] Multiply by 62.77 to convert to kg/ha.

Schwab, G. O., Frevert, R. K., Edminster, T. W., and Barnes, K. K., *Soil and Water Conservation Engineering,* John Wiley and Sons, New York, 1981, 525. With permission.

More descriptive identification of these factors is as follows:

- I, soil erodability is determined by the percentage of dry soil particles greater than 0.84 mm in diameter, which is inversely related to soil loss as related to different slopes
- C, climate is based on wind velocity and surface moisture as relative climate percentages (Figure 16)
- K, soil surface roughness is based on the microrelief. Ridges of soil 5 to 12 cm (2 to 5 in.) high are the most effective in reducing soil loss
- L, field length factor is calculated on the median travel distance across the soil surface and includes the magnitude, the prevailing wind direction, and the ratio of the wind erosion forces parallel to the prevailing wind direction to those perpendicular to the prevailing wind direction
- V, vegetation factor depends on plant height, density, and surface area covered, as illustrated in Figure 17

The estimated soil losses by wind equation varies greatly with the climatic region. The above discussion, however, forms a starting point and can be helpful in selecting LT sites.

Table 17
RECOMMENDED CONSERVATION PRACTICE
FACTORS P[a]

	P_c Contouring (maximum slope length in m)	P_{sc} Strip Cropping[b]	P_{tc} Terracing and Contouring[c]
Parallel to Field Boundary	0.8[d]	—	—
1.1—2	0.6(150)	0.30	—
2.1—7	0.5(100)	0.25	0.10
7.1—12	0.6(60)	0.30	0.12
12.1—12	0.8(20)	0.40	0.16
18.1—24	0.9(18)	0.45	—

[a]   Factor up and down slope is 1.0.
[b]   A system using 4-year rotation of corn, small grain, meadow, meadow. Use with terraces for farm planning.
[c]   Recommended only for computing soil loss from the field or loss to the terrace channel with upslope plowing.
[d]   For slopes up to 12% only.

Data from Reference 52; also see References 50 and 51.

Refinements can be included by studying site conditions and contacting the local USDA Soil Conservation Services offices.

C. Water Control

The main transport system for pollutants of wastes located on soil surfaces or below the soil surfaces is water. Waste materials and hazardous substances dissolved or suspended in water are vulnerable to removal to off site lands and waterways or downward percolation into the groundwater. Water control in this discussion relates to pollution control. Water control must be complete and consider:

- The amount passing through the soil containing waste
- The amount passing across the soil surface to land or collecting water bodies beyond the site as runoff
- The amount passing onto the site from off site watersheds or runons
- The treatment of water runoff from the land treatment site prior to release to detoxify or deactivate the contaminating constituents

1. Runoff

Rainfall that makes its way toward water channels, lakes, and/or oceans as surface flow is called runoff. Runoff is a serious factor in establishing the success of a LT site. Soil manipulation or management by design of channels or structures to control natural flow of surface water can determine

- Peak rate of runoff
- Runoff volumes
- Time and distribution of runoff rates

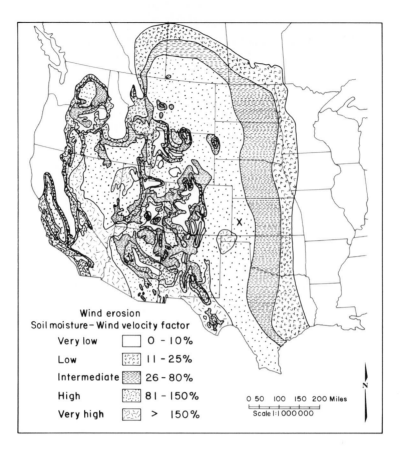

FIGURE 16.   Annual wind erosion climatic factor as a percentage of that in the vicinity of Garden City, Kansas, marked by X. In the Eastern U.S. and Canada the factor is less than 18%. (From Chepil, W. S., Skiddoway, F. H., and Armbrust, D. V., Climatic factor for estimating wind erodability of farm fields, *J. Soil Water Cons.* 17, 162, 1962. With permission of and copyright held by the Soil Conservation Society of America.)

Both precipitation and watershed characteristics influence runoff. The precipitation characteristics include rainfall

- Intensity
- Duration
- Distribution

The watershed characteristics include

- Size and shape
- Topography
- Orientation
- Geology
- Surface culture and vegetation

Several methods have been proposed to express a design peak-rate of runoff, e.g.,

$$Q = CiA \qquad (21)$$

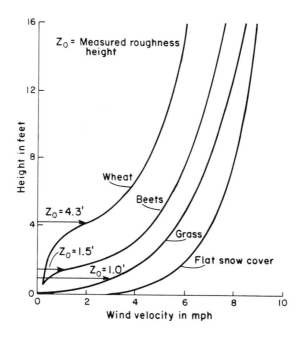

FIGURE 17. Wind-velocity distribution over different types of plant cover and soil surfaces. (From Schwab, G. O., Frevert, R. K., Barnes, K. K., and Edminster, T. W., *Elementary Soil and Water Engineering,* 2nd ed., John Wiley & Sons, New York, 1971, 136. With permission.)

where

$\quad$ Q $\;=\;$ the design peak runoff rate in feet per second (m³/s)
$\quad$ C $\;=\;$ the runoff coefficient which is the ratio of the peak runoff rate to the rainfall intensity and is dimensionless
$\quad$ i $\;=\;$ the rainfall intensity in inches per hour (iph) for the design period and for a duration equal to the "time-of-concentration" of the watershed
$\quad$ A $\;=\;$ the watershed area in acres

Table 4.18 provides data for "i", the "time-of-concentration", and Table 19 gives the rational runoff coefficient, C. The relationship is

$$q = 0.0028 \ CiA \tag{22}$$

where

$\quad$ q $\;=\;$ the design peak runoff rate (m³/s) (Figure 4.18a)
$\quad$ i $\;=\;$ the rainfall intensity in mm/h (Figure 4.18a)
$\quad$ C $\;=\;$ the same dimensionless runoff coefficients
$\quad$ A $\;=\;$ the area in hectares

The rainfall intensity-duration-return period data for any location in the U.S. may be obtained from the U.S. Weather Bureau of that selected location.

There is a lack of small agreement among hydrologists as to the best method for computing "time-of-concentration".[47,55] The data presented in Figure 18 for "i", therefore, may only be used as an example or for the specific location, soil, and hydrologic condition from which the data were obtained. The runoff coefficient "C" in Table 19 also is presented for illustration purposes. The estimates are for hydrologic soil group B derived from small

Table 18
TIME OF CONCENTRATION FOR SMALL WATERSHEDS[a]

Maximum length of flow (ft)	m	Time of Concentration (min) Watershed Gradient (%)					
		0.05	0.1	0.5	1.0	2.0	5.0
500	152	18	13	7	6	4	3
1000	305	30	23	11	9	7	5
2000	610	51	39	20	16	12	9
4000	1220	86	66	33	27	21	15
6000	1830	119	91	46	37	29	20
8000	2440	149	114	57	47	36	25
10,000	3050	175	134	67	55	42	30
20,000	6100	306	234	117	97	74	52

[a]   Computed from $T = 0.0078 L_{0.77} S_{0.385}$, where L is the maximum length of flow in feet, S^c is the watershed gradient in feet per foot, and T^c is the time concentration in minutes.

Schwab, G. O., Frevert, R. K., Edminster, T. W., and Barnes, K. K., *Soil and Water Conservation Engineering*, John Wiley and Sons, New York, 1981, 525. With permission.

Table 19
RUNOFF COEFFICIENT "C" FOR DIFFERENT WATERSHEDS (SOIL GROUP B)

Cover on hydrologic conditions	Coefficient C for rainfall rates of		
	25 mm/hr (1 iph)	100 mm/hr (4 iph)	200 mm/hr (8 iph)
Row crop, poor practice	0.63	0.65	0.66
Row crop, good practice	0.47	0.56	0.62
Small grain, poor practice	0.38	0.38	0.38
Small grain, good practice	0.18	0.21	0.22
Meadow, rotation, good	0.29	0.36	0.39
Pasture, permanent, good	0.02	0.17	0.23
Woodland, mature, good	0.02	0.10	0.15

Horn, D. L. and Schwab, G. O., *Amer. Soc. Agric. Engin. Trans.*, 6, 195, 1963. With permission.

single-crop watersheds at Coshocton, OH.[47] The most important contributing factors here are infiltration rate, surface cover, and rainfall intensity. The runoff coefficients in Table 19   can be converted to other hydrologic soil groups by use of Table 20   whose soil groups are defined in Table 21.   Several assumptions must be made in the rational method:

- The frequency of rainfall and runoff are similar
- Rainfall occurs at uniform intensity for a duration at least equal to the time of concentration of the watershed
- Rainfall occurs at a uniform intensity over the entire watershed area

Despite the oversimplification of the runoff process of the rational method, the method is sufficiently accurate to be a useful and reliable field tool for predictive purposes. As presented here, with certain examples, the method is applicable to areas less than 800 ha (2000 a).

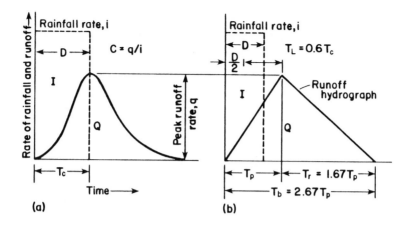

FIGURE 18. Rainfall and runoff with assumptions (a) for the rational equation and (b) for the Soil Conservation Service triangular hydrograph method of runoff estimation. (From Schwab, G. O., Frevert, R. K., Edminster, T. W., and Barnes, K. K., *Soil and Water Conservation Engineering*, 3rd ed., John Wiley & Sons, New York, 1981, 76. With permission.)

Table 20

HYDROLOGIC SOIL GROUP CONVERSION FACTORS

Cover and hydrologic condition	Factors for converting the runoff coefficient C from group B soils to[a]		
	Group A	Group C	Group D
Row crop, poor practice	0.89	1.09	1.12
Row crop, good practice	0.86	1.09	1.14
Small grain, poor practice	0.86	1.11	1.16
Small grain, good practice	0.84	1.11	1.16
Meadow, rotation, good	0.81	1.13	1.18
Pasture, Permanent, good	0.64	1.21	1.31
Woodland, mature, good	0.45	1.27	1.40

[a] Factors were computed from Table 8.19 by dividing the curve number for the desired soil group by the curve number for group B.

Schwab, G. O., Frevert, R. K., Edminster, T. W., and Barnes, K. K., *Soil and Water Conservation Engineering*, John Wiley & Sons, New York, 1981, 525. With permission.

Soil Conservation Service (SCS) Method — this is one of the latest methods for design runoff rates and is described by the U.S. SCS.[48] It uses the assumptions for a triangular hydrograph as illustrated in Figure 18 for runoff estimation.

$$T_p = D/2 + T_L = D/2 + 0.6T_c \tag{23}$$

where

T_p = time to peak, is necessary to develop a design hydrograph for routing runoff

T_L = is time of lag, is an approximation of the mean travel time

T_c = time of concentration is the longest travel time. It does not correspond to time of peak as in the rational equation

D = duration of excess rainfall

Table 21
RUNOFF CURVE NUMBER FOR HYDROLOGIC SOIL-COVER COMPLEXES FOR ANTECEDENT RAINFALL CONDITION II, AND $I_a = 0.25$

Land Use or Cover	Treatment or Practice	Hydrologic Condition	*Hydrologic Soil Group A	B	C	D
Fallow Row crops	Straight row	—	77	86	91	94
	Straight row	Poor	72	81	88	91
	Straight row	Good	67	78	85	89
	Contoured	Poor	70	79	84	88
	Contoured	Good	65	75	82	86
	Terraced	Poor	66	74	80	82
	Terraced	Good	62	71	80	81
Small grain	Straight row	Poor	65	76	84	88
	Straight row	Good	63	75	83	87
	Contoured	Poor	63	74	82	85
	Contoured	Good	61	73	81	84
	Terraced	Poor	61	72	79	82
	Terraced	Good	59	70	78	81
Close-seeded legumes or rotation meadow	Straight row	Poor	66	77	85	89
	Straight row	Good	58	72	81	85
	Contoured	Poor	64	75	83	85
	Contoured	Good	55	69	78	83
	Terraced	Poor	63	73	80	83
	Terraced	Good	51	67	76	80
Pasture or range		Poor	68	79	86	89
		Fair	49	69	79	84
		Good	39	61	74	80
	Contoured	Poor	47	67	81	88
	Contoured	Fair	25	59	75	83
	Contoured	Good	6	35	70	79
Meadow (permanent)		Good	30	58	71	78
Woods (farm wood-lots)		Poor	45	66	77	83
		Fair	36	60	73	79
		Good	25	55	70	77
Farmsteads	—		59	74	82	86
Roads and right-of-way (hard surface)	—		74	84	90	92

*Soil Group	Description	Final infiltration Rate (mm/h)
A	*Lowest Runoff Potential.* Includes deep sands with very little silt and clay, also deep, rapidly permeable loess.	8 — 12
B	*Moderately low Runoff Potential.* Mostly sandy soils less deep than A, and loess less deep or less aggregated than A, but the group as a whole has above-average infiltration after thorough wetting.	4 — 8
C	*Moderately High Runoff Potential.* Comprises shallow soils and soils containing considerable clay and colloids, though less than those of group D. The group has below-average infiltration after presaturation.	1 — 4

Table 21 (continued)
RUNOFF CURVE NUMBER FOR HYDROLOGIC SOIL-COVER COMPLEXES FOR ANTECEDENT RAINFALL CONDITION II, AND $I_a = 0.25$

*Soil Group	Description	Final infiltration Rate (mm/h)
D	*Highest Runoff Potential.* Includes mostly clays of high swelling percent, but the group also includes some shallow soils with nearly impermeable subhorizons near the surface.	0 — 1

Schwab, G. V., Frevert, R. K., Edminster, T. W., and Barnes, K. K., *Soil and Water Conservation Engineering,* John Wiley and Sons, New York, 1981, 525. With permission.

The peak runoff rate as calculated from the triangular hydrograph in Figure 18b:

$$q = 0.0021 \ QA/T_p \tag{24}$$

where

Q = runoff volume in mm depth (area under the hydrograph)
q = runoff rate in m^3/s
A = watershed area in ha
T_p = time of peak in hours

Runoff volume also is of importance for land treatment planning and helpful in design of flood control reservoir. A Soil Conservation Service Method has been developed for this purpose and includes both equations and curves   (Figure 19) showing relationships between rainfall and runoff.[47]

$$Q = \frac{(I - 0.2S)^2}{I + 0.8S} \tag{25}$$

where

Q = the direct surface runoff in depth in mm
I = the storm rainfall in mm
S = the maximum potential difference between rainfall and runoff in mm, starting at the time the storm begins

Where watershed have gauging equipment, I can be plotted against Q and S obtained directly.

2. Rainfall

The rainfall characteristics in general and at the specific LT site must be known in order to develop design procedures for soil-and-water conservation structures. The intensity, duration and frequency of occurrence return period are the most essential features since precipitation occurs randomly for time and amounts. Predictions, therefore, must depend on statistical analyses of data from past records. The oldest datebook dates only to 1890. The intensity of precipitation or average rate of rainfall (mm/h) is not uniform throughout a storm. The average rate for a few minutes is greater than the average rate for the whole storm. The length of time of a storn is the duration. The two characteristics, intensity and duration, relate to each other roughly as illustrated in Figure 20.   Very intense storms of

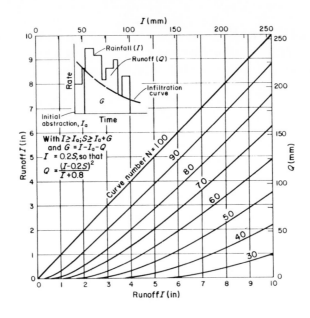

FIGURE 19.   Relationship between rainfall and runoff. (From Schwab, G. O., Frevert, R. K., Edminster, T. W., and Barnes, K. K., *Soil and Water Conservation Engineering,* 3rd ed., John Wiley & Sons, New York, 1981, 79. With permission.)

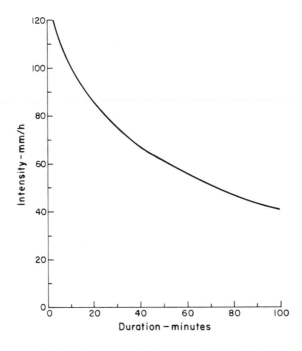

FIGURE 20.   The relationship between rainfall intensity and duration.

high intensity last for a short period and fall on small areas. Those covering large areas seldom are of high intensity but may last for a long period, extending into days. The high intensity storms usually cause more erosion damage and move more soil than those of less

Table 22
RELATIONSHIP OF STORM DURATION AND
RAINFALL INTENSITY AND DEPTH AT
ST. LOUIS FOR A RETURN PERIOD OF 10 YEARS

	Duration of rainfall in minutes					
	5	10	30	60	360	1440
Intensity (in./h)[a]	6.5	5.3	3.5	2.5	0.7	0.23
Depth (in.)[a]	0.54	0.9	1.8	2.5	4.2	5.52

[a] Multiply by 25.4 to convert to mm.

Schwab, G. O., Frevert, R. K., Barnes, K. K., and Edminster, T. W.,
Elementary Soil and Water Engineering, 2nd ed., John Wiley & Sons,
Inc., New York, 1971, 68. With permission.

intensity. The relationship can be expressed mathematically in one of several forms, such as:

$$I = \frac{a}{t + b}$$

where

I = the average intensity of the storm in mm/h
t = the duration of the storm in minutes
a and b are constants: Figure 20 uses values of a = 6000 and b = 50
  which are not suitable for durations of 5 min or less

A second empirical expression between intensity and duration is:[52]

$$I = \frac{I_0}{(1 + BD)^n}$$

where

I   = the average intensity in mm/h
I_0 = the 'instantaneous' intensity or the maximum sustained for very short periods of about 15 seconds. In the UK the values range from 145-170 mm/h
B   = the constant which varies with the mean annual rainfall, and ranges from 15 to 45
D   = duration in hours
n   = the 'continentality' factor, and is low for a rainfall regime with high total but low intensity rain, and high for more intense rainfall patterns. The range is 0.44 to 0.78.

The relationship of storm duration and rainfall intensity and depth at St. Louis[47] is given in Table 22 for a return period of 10 years and in Figure 21 as an example. (The return period is the inverse of the probability an observed event in a given year and is equal or greater than a given event.) The intensity of rainfall increases with the return period, and these intensities vary greatly according to geographic location. Because of the need for voluminous in-depth data to calculate the return period, only some examples are given here. A water-control structure or facility usually is designed for a desired return period that will be compatible with the economy of the structure size.

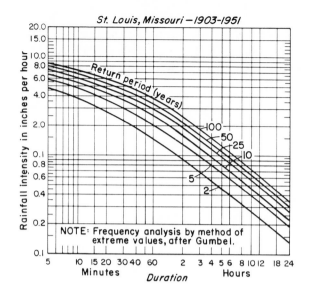

FIGURE 21.   Rainfall intensity-duration-return period data for St. Louis, Mo. (From Schwab, G. O., Frevert, R. K., Barnes, K. K., and Edminster, T. W., *Elementary Soil and Water Engineering*, 2nd ed., John Wiley & Sons, New York, 1971, 32. With permission.)

3. Watershed

The characteristics of the watershed most prominently affecting water runoff behavior include:

- topography
- orientation
- soil
- geology
- size
- shape
- surface culture

The runoff rate and volume per unit of watershed area decreases as the size of the runoff area increases. Size, also, influences the time of the year showing the greatest runoff.[56] The volume of runoff is greater on steep, well-developed slopes than on flat, gentle slopes, or depressed areas. Some watersheds with specific orientations in relation to prevailing storm or wind direction may yield more runoff than those oriented away from the prevailing or incoming storm pattern. The nature of the predominant soil type, texture (sand, silt, and clay), structure, and depth, determines the rate of water infiltration and consequently the runoff volume. The type of vegetation, its height, density, soil cover, and volume, all influence runoff by detaining the rainfall differently, and this relates to the extent of infiltration and evaporation. Finally, surface culture imposed by man, e.g., urbanization, roads, culverts, levees, and dams enter into watershed runoff characteristics.

The shape and size of the watershed can greatly influence the "gathering time" of the runoff in the watershed. The "gathering time" or "time of concentration" is defined as the longest time taken for water to travel by overland surface flow from any point in the watershed to the outlet. The short-wide watershed requires less time of concentration than the long narrow watershed (Figure 22).   Furthermore, in watersheds of nearly identical character-

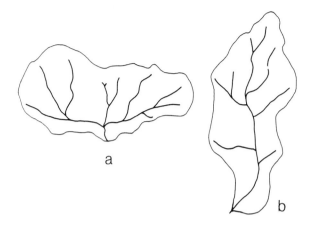

FIGURE 22. The (a) short-wide shaped watershed provides a shorter rainfall collection time than the (b) long-narrow watershed.

istics, except for shape, the long narrow catchment will demonstrate a longer gathering time with a corresponding lower intensity. Should a storm be localized at one end of the watershed, the time of catchment will be different than if it occurred over the whole area or perhaps the opposite end. Thus, the storm orientation pattern can be an important factor in runoff on any watershed.[57,58]

4. Water Modification Practices

The main factors affecting erosion by water are

- climate
- soil
- vegetation
- topography

The reason for emphasizing these factors is to center attention on possible means of modifying them to control erosion as a contributor to nonpoint pollution. The erosion control practices may well be divided into (a) nonspecialized design procedures used on nonpoint sources of tilled, range, pasture, and forest lands and other exposed soil surfaces such as mine spoils and construction sites, and (b) specialized design procedures.

The most efficient and effective control measure for erosion as a contributor to nonpoint pollution is to encourage soil protection by vegetative cover. Several designs that suit large as well as small areas of land both cultivated and noncultivated are well described in the literature, for example:

- Contouring is a field operation of tilling, aerating, plowing, planting, and harvesting (transport) approximately on the topographic contour. Water is impounded behind small ridges and in small depressions until it has an opportunity to infiltrate the soil.
- Strip cropping is the practice of growing, allowing, or encouraging the growth of plants of different types, depending on the need to detain water runoff, in alternating strips on the contour. Strip vegetation for wind modification in dry regions is placed cross-wise to the prevailing wind direction (Figure 23).
- Tillage practices (other than contour or strip methods) consist of the mechanical manipulation of the soil to provide conditions best suited to plant growth, weed control, and maintenance or further development of infiltration and aeration.

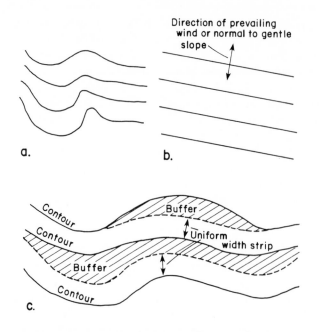

FIGURE 23.   Illustration of three types of strip cropping: (a) contour, (b) field, and (c) buffer. (From Schwab, G. O., Frevert, R. K., Edminster, T. W., and Barnes, K. K., *Soil and Water Conservation Engineering,* 3rd ed., John Wiley & Sons, New York, 1981, 114. With permission.)

These three erosion modification practices were originally developed for agricultural land to counteract the long-term cultivation practices that allow excessive exposure of bare soil with accompanying accelerated erosion and often uncontrolled land deterioration. Tillage practices have a place in land treatment to provide an adequate soil and water environment for pollution stabilization on land, particularly in dry soil and sloping topographic areas. Tillage for subsoiling to correct poor infiltration due to restrictive subsoils of compacted layers, texture stratifications, cemented and indurated accumulations has been found to be very effective in many western soils. Other specialized design procedures for water control that should be mentioned here are vegetated waterways and diversion channels.

5. Water Control Practices

The purpose of developing a discussion on water control is not to provide instruction on the ways to construct or design structures for the land. However, we briefly identify some of the practices that can be helpful in land treatment to prevent redistribution of hazardous wastes and contaminated soil and water beyond the LT site facility.[59,60] In-depth treatment of soil and water conservation and erosion control engineering are referenced frequently throughout this text for those interested in the more technical and engineering aspects of design.[61]

Surface drainage is a land-forming process which relates to the changing of the natural topography of the land to control the movement of water onto or away from the site land surface.[47] Land forming involves a number of land surface practices, such as leveling for uniformity of water movement onto and off of the land, grading and leveling for irrigation, drainage, and moisture conservation. Other purposes are for erosion control, contour benching and terracing, and land smoothing to adjust land elevations. Generally the terms land grading, shaping, and leveling are used interchangeably, depending on the geographic location.

Surface drainage practices necessary to establish field LT sites vary with topography, natural drainage pattern(s), soil characteristics, vegetation, and the planting program. Lands that often require some form of surface drainage and land forming to best establish site facilities are those with rough topography, pot-hole areas, flat fields of poor drainage, areas of soils with subsoil drainage and infiltration properties and unevenly sloping lands. Surface water of undesirable excesses may be eliminated by several processes, alone or together; infiltration, evaporation and transpiration; and/or channels (ditches) either natural or constructed. Random field ditches have been suggested with layout diagrams for drainage as having the possibility of being least expensive if surface drainage problems are not too complicated.[47] Bedding (hump and hollow) consists of narrow-width plow lands in which the dead furrows run parallel to the predominant land slope. Other systems are used under the descriptive terminology of parallel field ditch, parallel lateral ditch, and cross-slope ditch.[47]

Vegetative waterways are constructed to carry excess water around and off of land-treatment sites to isolate the disposal area. Runoff from sloping land must flow to lower lands in a controlled manner to prevent gully development. Vegetation, like grass and sod, can be very effective in reducing the peak flow rate and stabilize the soil in the natural waterway or man-made terraces, contour furrows or diversion channels against gully formation. In general, vegetated waterways are used to carry runoff, however, the best natural growth habit of the vegetation in such waterways is discouraged by the intermittent, or often continuous flow, that keeps the soil more or less permanently saturated over long periods of time.

The vegetated watercourse may best be designed to accommodate the 10 year return period storm. Other factors in the design must include (a) shape and length, (b) recommended and adapted vegetation for the specific area, (c) permissible water velocity to maintain vegetation, (d) roughness coefficient, and (e) adequate channel capacity (Table 23).

Terraces are broad channels across a grade of gently sloping land for control of erosion (Figure 24). Diversion structures are terraces used to intercept and change the flow of water. The use of diversion channels to protect land treatment and other disposal sites, bottom land buildings, and special lands from destructive runon, runoff, washing, and erosion of hillsides is more prominent today than ever before. Land treatment sites, landfills, and other disposal areas may be protected from water by the construction of berms constituted of compacted native soil, some of which extend several feet in height. Diversion and terrace channels, and diversion berms must discharge into suitable grassed waterways to fully control the water and protect the soil from erosion. Diversion terraces and berms also have been used successfully within disposal and LT facilities to stabilize lateral flow and direct internal clear surface flow away from the treatment area. The general function of terraces is to decrease the length of the hillside slope. The major types of terraces are (a) bench terrace designed to reduce land slope and (b) broadbase terrace designed to remove or retain water on sloping land. Terraces are classed into (a) broadbase, (b) conservation, and (c) bench (Figure 24).[47] Many special design grades and spacing expressions occur within the 3 classifications.[61]

The construction of earth embankments may be necessary to conserve water and protect (a LT facility) the land from large water excesses which can then be controlled by the terraces. Earth embankments appear in the form of dikes, levees, and detention dams as protective structures. The success of embankments for water control requires the close adherence to basic principles of soil mechanics, soil physics, and sound engineering design and construction. The stability of the rolled-fill type of earth embankment depends on excellent foundation properties, depth to impervious strata, relative permeability, drainage, and the texture and nature (lime free) of the construction soil material. Only well experienced and qualified engineers should attempt to construct the earth embankment as a water control modification.

Table 23
PERMISSIBLE VELOCITIES FOR CHANNELS LINED
WITH VEGETATION

	Permissible velocity (fps)[b]					
	Erosion resistant soils (per cent slope)			Easily eroded soils (per cent slope)		
Cover	0—5	5—10	Over 10	0—5	5—10	Over 10
Bermuda grass	8	7	6	6	5	4
Buffalo grass	7	6	5	5	4	3
Kentucky bluegrass						
Smooth brome						
Blue grama	3.5	NR[a]	NR	2.5	NR	NR
Tall fescue						
Lespedeza serica						
Weeping lovegrass						
Kudzu						
Alfalfa	5	4	NR	4	3	NR
Crabgrass						
Grass mixture						
Annuals for temporary protection	3.5	NR	NR	2.5	NR	NR

[a] Not recommended.
[b] Multiply by 0.3048 to convert to m/s.

Schwab, G. O., Frevert, R. K., Edminster, T. W., and Barnes, K. K., *Soil and Water Conservation Engineering,* John Wiley and Sons, New York, 1981, 525. With permission.

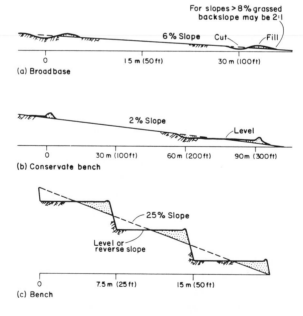

FIGURE 24.   Various general classes of terraces, (a) broadbase, (b) conservation, and (c) bench. (From Schwab, G. O., Frevert, R. K., Edminster, T. W., and Barnes, K. K., *Soil and Water Conservation Engineering,* 3rd ed., John Wiley & Sons, 1981, 166. With permission.)

REFERENCES

1. **Wadleigh, C. H.,** Wastes in Relation to Agriculture and Forestry, Misc. Pub. No. 1965, U.S. Department of Agriculture, Washington, D.C., 1968, 1.
2. **Metcalfe and Eddy,** Inc., *Wastewater Engineering*, McGraw and Hill Book Co., New York, NY, 1972, 1.
3. **DeTurk, E. E.,** Adaptability of sewage sludge as a fertilizer, *Sewage Works J.*, 7, 597, 1935.
4. **Pound, C. E. and Crites, R. W.,** *Wastewater Treatment and Reuse by Land Application*, Vol. II., EPA-660/2-73-0066, U.S. Environmental Protection Agency, Washington, D.C., 1973, 249.
5. **Kirkwood, J. P.,** The Pollution of rivers, in Seventh Ann. Rept., Mass. State Board of Health, Boston, Arno Press, Inc., 1970, 18.
6. **Phung, T.,** Land Cultivation of Industrial Wastes and Municipal Solid Wastes: State of the Art Study, Vol. I. Technical Summary and Literature Review, EPA-600/2-78-140a, U.S. Environmental Protection Agency, MERL, Cincinnati, Ohio, 1978, 205.
7. U.S. Environmental Protection Agency, Hazardous Waste Management System: Permitting Requirement for Land Disposal Facilities, CFR 47 (143), U.S. Environmental Protection Agency, Washington, D.C., 1982, 32324.
8. U.S. Environmental Protection Agency, Land Treatment Research Plan for Hazardous Waste, Myers, L. H., Ed., U.S. Environmental Protection Agency, RSKERL, ORD, Ada, Okla., 1982, 32.
9. **Elliott, L. F. and Stevenson, F. J., Eds.,** *Soils for Management of Organic Wastes and Waste Waters*, Soil Science Society of America, Madison, Wis., 1977, 650.
10. **Kardos, L. T., Scarsbrook, C. E., and Volk, V. V.,** Recycle elements in Waste through soil-plant systems, in *Soils for Management of Organic Wastes and Wastewaters*, Elliott, L. F. and Stevenson, F. J., Eds., Am. Soc. Agron., 1977, 301.
11. **Volk, V. V.,** Application of trash and garbage to agricultural lands, in *Land Application of Waste Materials*, *Soil Conservation Society of America*, USDA SCS, Ankeny, Iowa, 1976, 154.
12. **Stanford, G. B.,** The Houston Landmark Trial, Agron. Abst., Ann. Meeting, Houston, TX, Amer. Soc. Agron., Madison, Wisc., 1976, 34.
13. **King, L. D., Rudgers, L. A., and Webber, L. R.,** Application of municipal refuse and liquid sewage sludge to agricultural land: I. Field Study, *J. Environ. Qual.*, 3, 361, 1974.
14. U.S. Environmental Protection Agency, Oily Waste Disposal by Soil Cultivation Processes, EPA-R-2-72-110, Shell Oil Co., and U.S. Environmental Protection Agency, Washington, D.C., 1972, 110.
15. **Fuller, W. H., Amoozegar, A., Niebla, E., and Boyle, M.,** Behavior of Cd, Ni, and Zn in single and mixed combinations in landfill leachate, in *Land Disposal: Hazardous Wastes*, Proc. 7th Ann. Res. Symp., Shultz, D., Ed., EPA-600/9-81-0026, U.S. Environmental Protection Agency, MERL, Cincinnati, Ohio, 1981, 18.
16. **Fuller, W. H.,** Soil-Waste Interactions, in *Disposal of Industrial and Oily Sludges by Land Cultivation*, *Resource Systems and Management Assoc.*, Northfield, N.J., 1980, 79.
17. **Brown, K. W. and Associates, Inc.,** Hazardous Waste Land Treatment, SW-874, OWWM, U.S. Environmental Protection Agency, Washington, D.C., 1983, 974.
18. Engineering Science Inc., The 1976 API Refinery Solid Waste Survey, Part I: Solid Waste Quantities and Part II: Solids Management Practices, Eng. Sci., Inc., for The Amer. Petroleum Inst., 1978, 1.
19. U.S. Environmental Protection Agency, Subtitle C. Resource and Conservation Recovery Act, RCRA, Section 261.31 and 261.32, U.S. Environmental Protection Agency, OSW, Washington, D.C., 1980a, 2.
20. **Abrams, E. F., Guinan, D. K., Derkics, D.,** Assessment of Industrial Hazardous Waste Practices, Textile Industry, PB-258, 953, 1976, 1.
21. U.S. Environmental Protection Agency, Development Document for Effluent Limitations Guidelines and New Source Performance Standards for the Textile Mills Paint Source Category, U.S. EPA 440/1-74-022a, U.S. Environmental Protection Agency, Washington, D.C., 1974, 1.
22. **Fuller, W. H.,** Predicting radioactive heavy metal movement through soil, in *Waste Management '83*, The University of Arizona Press, Tucson, 1983, 341.
23. **Miyamoto, S. and Stroehlein, J. L.,** Improving water penetration in some Arizona soils — Sulfuric acid., *Prog. Agric. Ariz. Coll. Agric., Bull.* 27, 13, 1975.
24. **Stroehlein, J. L., Miyamoto, S., and Ryan, J.,** Sulfuric acid for Improving Irrigation Water and Reclamating Sodic Soils, *The Univ. of Ariz. Coll. Agric., Bull.* 78, 49, 1978.
25. **Wilkinson, R. R., Kelso, G. L., and Hopkins, F. C.,** State-of-the-Art Report: Pesticide Disposal Research, EPA-600/2-78-183, U.S. Environmental Protection Agency, Cincinnati, Ohio, 1978, 225.
26. **Fuller, W. H.,** Reclamation of saline and alkali soils, *Plant Food Rev.*, Fall, 7, 1962.
27. **Fuller, W. H.,** Water, Soil and Crop Management Principles for the Control of Salts, University Arizona, Agric. Expt. Sta., and Coop. Ext. Serv. Bull., A-42, 21, 1965.

28. **Wallace, A. T.,** Land disposal of liquid industrial wastes, in *Land Treatment and Disposal of Municipal and Industrial Wastewaters,* Sanks, R. L. and Asano, T., Eds., Ann. Arbor Sci. Pub., Ann Arbor, Mich., 1976, 147.

29. **Loehr, R. C., Jewell, W. J., Novak, J. D., Clarkson, W. W., and Friedman, G. S.,** *Land Application of Wastes,* Vol. 1, Van Nostrand Rheinhold, Co., New York, N.Y., 1979, 308.

30. **Thomas, P. R.,** Ohio Guide to Land Application of Sewage Sludge: Report of the Task Force on Land Application of Sewage Sludge, Ohio Coop. Ext. Serv. Bull., 598, 1975, 12.

31. **Linsley, R. K., Kohler, M. A., and Paulhus, J. L. H.,** *Hydrology for Engineers,* McGraw-Hill, Inc., New York, 1975, 260.

32. **Knisel, W. J., Jr., Ed.,** CREAMS. A Field Scale Model From Chemicals, Runoff and Erosion from Agricultural Management Systems, U.S. Dept. Agric. Con. Res. Rept., 26, 640, 1980.

33. **Perrier, E. R. and Gibson, A. C.,** Hydrological Simulation of Solid Waste Disposal Sites (ASSWDS), Prepared for the Municipal Environmental Research Laboratory, ORD, Water Resource Engineering Group, U.S. Army Engn. Waterways Expt. Sta., Vicksburg, Miss., 1980, 1.

34. **Fleming, G. S.,** *Computer Simulator Techniques in Hydrology,* Elsenier Sci. Pub. Amsterdam, 1975, 1.

35. **Sommers, L. E. and Nelson, D. W.,** Analyses and their interpretation for sludge application to agricultural land, in *Application of Sludges and Wastewaters on Agricultural Land: A Planning and Educational Guide,* N. Ctr. Res. Pub. 235, Ohio Agric. Res. Div. Ctr., Wooster, 1976, 3.

36. **Watson, J. E., Pepper, I. L., Unger, M., and Fuller, W. H.,** Yields and leaf elemental composition of cotton grown on sludge-amended soil, (accepted for publication; J. Environ. Qual., 1984.)

37. **Knezek, B. D. and Miller, R. H., Eds.,** *Application of Sludges and Wastewaters on Agricultural Land: A Planning and Educational Guide,* N. Cent. Reg. Res. Pub. 235; Ohio, Agric. Res. and Develop. Ctr., Res. Bull. 1090, Wooster, Ohio, 1976, 1.

38. **Fuller, W. H. and Norman, A. G.,** Cellulose decomposition by aerobic mesaphilic bacteria from soil, I, II, III, *J. Bact.,* 46, 273, 1943.

39. **Fuller, W. H. and Norman, A. G.,** The Retting of Hemp., III. Biochemical changes accompanying Retting of Hemp, *Iowa State Cal. Res. Bull.,* 344, 928, 1946.

40. **Fuller, W. H.,** Soil Organic Matter, *Univ. Ariz. Gen. Bull.,* 240, 19, 1951.

41. **Minderman, G.,** Addition, decomposition, and accumulation of organic matter in forests, *J. Ecol.,* 56, 355, 1968.

42. **Wisiewski, T. K., Wiley, A. J., and Lueck, B. J.,** Ponding and soil infiltration for disposal of spent sulfite liquor in Wisconsin, Proc. 19th Industrial Waste Conf. Purdue, University, 89, 480, 1955.

43. **Pirt, S. J.,** A quantitative theory of the action of microbes attached to a packed column: Relevant to trickling filter effluent purification and to microbial action in soil, *Appl. Chem. Biotechnol.,* 23, 389, 1973.

44. **Bennett, H. H.,** *Elements of Soil Conservation,* 2nd Ed., McGraw-Hill Book Co., Inc., 1955, 358.

45. **Beasley, R. P.,** *Erosion and Sediment Pollution Control,* Iowa State University Press, Ames, 1972, 320.

46. **Hudson, N.,** *Soil Conservation,* Cornell University Press, 2nd Ed., Ithaca, New York, N.Y., 1975.

47. **Schwab, G. O., Frevert, R. K., Edminister, T. W., and Barnes, K. K.,** *Soil and Water Conservation Engineering,* 3rd Ed., John Wiley and Sons, N.Y., 1981, 525.

48. **Stallings, J. H.,** *Soil Conservation,* Prentice-Hall Englewood Cliffs, N.J., 1964, 1, 156.

49. **Schwab, G. O., Frevert, R. K., Barnes, K. K., and Edminster, T. W.,** *Elementary Soil and Water Engineering,* 2nd Ed., John Wiley and Sons, Inc., New York, N.Y., 1971, 68.

50. **Wischmeier, W. H.,** Predicting Rainfall-Erosion Losses from Cropland East of the Rocky Mountains, *USDA-ARS Agric. Handbook,* 282, 1965.

51. **Wischmeier, W. H.,** Cropland erosion and sedimentation in USDA-EPA Control of Water Pollution, Vol. II, U.S. GPO Washington, D.C., 1976, Chap. 3.

52. **Wischmeiser, W. H. and Smith, D. D.,** Predicting Rainfall Erosion Losses — A Guide to Conservation Planning, U.S. Dept. Agric. Handbk. No. 537, 58, 1976.

53. **Skidmore, E. L. and Woodruff, N. P.,** Wind Erosion Forces in the United States and Their Use in Predicting Soil Loss, Agric. Hdbk. No. 346, U.S. Department of Agriculture, ARS, 1968, 26.

54. **Chepil, W. S. and Woodfull, N. P.,** Vegetative and nonvegetative materials to control wind and water erosion, *Soil Sci. Soc. Amer. Proc.,* 27, 86, 1963.

55. **Horn, D. L. and Schwab, G. O.,** Evaluation of rational runoff coefficients for small agricultural watersheds, *Amer. Soc. Agric. Engin. Trans.,* 6, 195, 1963.

56. U.S. Soil Conservation Service, Guide for Rating Limitations of Soils for Disposal of Wastes, Interim Guide, Advisory Soils, U.S. Dept. Agric., Washington, D.C., 1973, 14.

57. Institute of Hydrology, Flood Studies Report, Vol. II., Meterological Studies, Nat'l. Environ. Res. Counc., 1975, 10.

58. **Harrold, L. L.,** Soil Loss as determined by watershed measurements, *Agric. Engin.,* 30, 37, 1949.

59. **Alexander, M.,** *Soil Microbiology,* John Wiley and Sons, New York, 1961, 241.

60. **Hellings, C. S.,** Pesticide mobility in soil, II. Application of soil thin-layer chromatography, *Soil Sci. Soc. Amer. Proc.,* 35, 737, 1971.
61. **Smith, D. D. and Wischmeier, W. H.,** Rainfall Erosion, *Advan. Agron.,* 14, 109, 1962.

Appendixes

APPENDIX A

USDA AND USCS PARTICLE CLASSIFICATIONS

The USDA classification which may be compared most directly with the soil types in the USCS system is soil texture (distribution of grain or particle size) and associated modifiers such as gravelly, mucky, diatomaceous, and micaceous. The size ranges for the USDA and the USCS particle designations (e.g., sand and gravel) are listed in Table 1. The soil texture (USDA — sandy loam, silt loam, etc.) or the soil type (USCS — GC, clayey gravel; SC, clayey sand, etc.) is based on the relative amounts of different-sized particles in a soil. The USDA system for classifying soil texture is compared in Volume I, Chapter 1, Figures 25 and 26; an abbreviated description for the USCS classification is listed in Table 2. Correlation of the USCS and USDA systems on the basis of texture is presented in Tables 2 and 3. These correlations are not precise because texture is a major criterion in the USCS, while texture is a minor criterion in the USDA system. A soil of a given texture can be classified into only a limited number of the 15 USCS soil types, while in the USDA system, soils of the same texture may be found in many of the 10 orders and 43 suborders because of differences in their chemical properties or the climatic areas in which they are located.

Table 1
USDA AND USCS PARTICLE SIZES

USDA		USCA	
Particle	Size range (mm)	Particle	Size range (mm)
Cobbles	76.2—254.0	Cobbles	>76.2
Gravel	2.0—76.2	Gravel	4.76—76.2
Coarse gravel	12.7—76.2	Coarse gravel	19.1—76.2
Fine gravel	2.0—12.7	Fine gravel	4.76—19.1
Sand	0.05—2.0	Sand	0.074—4.76
Very coarse sand	1.0—2.0		
Coarse sand	0.5—1.0	Coarse sand	2.0—4.76
Medium sand	0.25—0.5	Medium sand	0.42—2.0
Fine sand	0.1—0.25	Fine sand	0.074—0.42
Very fine sand	0.05—0.1		
Silt	0.002—0.05	Fines[a]	<0.074
Clay	<0.002	(silt and clay)	

[a] USCS silt and clay designations are determined by response of the soil to manipulation at various water contents rather than by measurement of size.

From Fuller W. H., *CRC Crit. Rev. Environ. Control*, 9, 261, 1980. With permission.

Table 2
MAJOR DIVISIONS, SOIL TYPE SYMBOLS, AND TYPE DESCRIPTIONS FOR THE UNIFIED SOIL CLASSIFICATION SYSTEM (USCS)

Major divisions	Symbol	Description
Coarse-grained soils — More than half of material is larger than No. 200 sieve size		
Gravels — More than half of coarse fraction is larger than No. 4 sieve size.		
Clean gravel — little or no fines	GW	Well graded gravels, gravel-sand mixtures, little or no fines
	GP	Poorly graded gravels or gravel-sand mixtures, little or no fines
Gravels with fines (appreciable fines)	GM	Silty gravels, gravel-sand-silt mixture
	GC	Clayey gravels, gravel-sand-clay mixtures
Sands — More than half of coarse fraction is smaller than No. 4 sieve size.		
Clean sands (little or no fines)	SW	Well-graded sands, gravelly sands, little or no fines
	SP	Poorly graded sands or gravelly sands, little or no fines
Sands with fines (appreciable fines)	SM	Silty sands, sand-silt mixtures
	SC	Clayey sands, sand-clay mixtures
Fine-grained soils — More than half of material is smaller than No. 200 sieve size		
Silts and clays — liquid limit is less than 50	ML	Inorganic silts and very fine sands, silty or clayey fine sands or clayey silts with slight plasticity
	CL	Inorganic clays of low to medium plasticity, gravelly clays, sandy clays, silty clays, lean clays
	OL	Organic silts and organic silty clays of low plasticity
Silts and clays — liquid limit is greater than 50	MH	Inorganic silts, micaceous or diatomaceous fine sandy or silty soils, elastic silts
	CH	Inorganic clays of high plasticity fat clays
	OH	Organic clays of medium to high plasticity, organic silts
Highly organic soils	Pt	Peat and other highly organic soils

Notes: ML includes rock flour. The No. 4 sieve opening is 4.76 mm (0.187 in.); the No. 200 sieve opening is 0.074 mm (0.0029 in.)

From Fuller, W. H., *CRC Crit. Rev. Environ. Control*, 9, 262, 1980. With permission.

Table 3
CORRESPONDING USCS AND USDA SOIL CLASSIFICATIONS

USCS soil types	Corresponding USDA soil textures
1. GW	Same as GP — gradation of gravel sizes not a criterion
2. GP	Gravel, very gravelly[a] sand less than 5% by weight silt and clay
3. GM	Very gravelly[a] sandy loam, very gravelly[a] loamy sand, very gravelly[a] silt loam, and very gravelly[a] loam[b]
4. CG	Very gravelly clay loam, very gravelly sandy clay loam, very gravelly silty clay loam, very gravelly silty clay, very gravelly clay[b]
5. SW	Same — gradation of sand size not a criterion
6. SP	Coarse to fine sand; gravelly snad[c] (less than 20% very fine sand)
7. SM	Loamy sands and sandy loams (with coarse to fine sand), very fine sand; gravelly loam sand[c] and gravelly sandy loam[c]
8. SC	Sandy clay loams and sandy clays (with coarse to fine sands); gravelly sandy clay loams and gravelly sandy clays[c]
9. ML	Silt, silt loam, loam very fine sandy loam[d]
10. CL	Silty clay loam, clay loam, sandy clays with <50% sand[d]
11. OL	Mucky silt loam, mucky loam, mucky silty clay loam, mucky clay loam
12. MH	Highly micaceous or diatomaceous silts, silt loams — highly elastic
13. CH	Silty clay and clay[d]
14. OH	Mucky silty clay
15. PT	Muck and peats

[a] Also includes cobbly, channery, and shaly.
[b] Also includes all of textures with gravelly modifiers where <1/2 of total held on No. 200 sieve is of gravel size.
[c] Gravelly textures included if less than 1/2 of total held on No. 200 sieve is of gravel size.
[d] Also includes all of these textures with gravelly modifiers wither <1/2 of the total soil passes the No. 200 sieve.

From Fuller, W. H., *CRC Crit. Rev. Environ. Control*, 9, 263, 1980. With permission.

APPENDIX B

ENGLISH - SI
CONVERSION CONSTANTS

Length	in.	ft	yd	mi	cm	m	km
1 in.	1	0.083	0.027	—	2.54	—	—
1 ft	12	1	0.333	—	30.48	0.305	—
1 yd	36	3	1	—	91.44	0.914	—
1 mi (statute)	—	5280	1760	1	—	1609	1.61
1 cm	0.394	0.033	0.011	—	1	0.1	—
1 m	39.37	3.281	1.094	—	100	1	0.001
1 km	—	3281	1094	0.621	—	1,000	1

Areas	in.^{2}	ft^2	yd^2	acre	cm^2	m^2	ha
1 in.^{2}	1	0.007	—	—	6.45	0.00064	—
1 ft^2	144	1	0.1111	—	—	0.0929	—
1 yd^2	1,296	9	1	—	—	0.8361	—
1 acre	—	43,560	4,840	1	—	4,047	0.405
1 cm^2	0.155	—	—	—	1	0.0001	—
1 m^2	1550	10.76	1.20	—	10,000	1	0.0001
1 ha	—	107,650	11,961	2.47	—	10,000	1

Volume	in.^{3}	ft^3	Am. gal	ℓ	m^3	ac—ft	ha—m
1 in.^{3}	1	—	0.0043	0.0164	—	—	—
1 ft^3	1,728	1	7.481	28.32	0.0283	—	—
1 Am. gal	231	0.134	1	3.785	0.0038	—	—
1 ℓ	61.02	0.0353	0.2642	1	0.001	—	—
1 m^3	61,022	35.31	264.2	1,000	1	0.00081	0.0001
1 ac—ft	—	43,560	325,872	—	1,233.4	1	0.1233
1 ha—m	—	353,198	—	10×10^6	10,000	8.108	1

Note: 1 yd^3 = 0.765 m^3; 1 m^3 = 1.308 yd^3.

ABBREVIATIONS, SIGNS, AND SYMBOLS

Abbreviations

AASHO	Old Agricultural Department system for particle sizes (pre-1938)
AEC	Positive charge of anion exchange capacity
API	American Petroleum Institute also used as API separator sludge as waste residue of first centrifuge of refinery waste water
ASC	Anion sorption capacity
BOD	Biological oxygen demand
CEC	Total cation exchange capacity
CEC_c	Negative charge of cation exchange capacity
CEC_v	Variable charge of cation exchange capacity
CFU	Colony forming units
CLAY	Percent clay
CLC	Loading capacity limiting constant
COD	Chemical oxygen demand
DL	Design life
DLR	Design loading rate
EC	Electrical conductivity
Eh	Reduction/oxidation factor
EPA	U.S. Environmental Protection Agency
erf (μ)	Error Function of argument μ
erfc (μ)	Complementary error function of argument μ
esu	Electro static unit
exp (μ)	Exponential function of argument μ
FeO	Free iron oxide (only in equations)
ISW	Industrial solid waste
LT	Land treatment
MeV	Million electron volts
MSW	Municipal solid waste
ppb	Parts per billion
pH	A measure of hydrogen ion activity $[H^+]$ (i.e., pH $= -\log [H^+]$)
ppm	Parts per million
PV	Pore volume (units L^3)
PVC	Polyvinyl chloride
PVD	Pore volume displacement
red/ox	Reduction/oxidation ratio
s	Second
SALTS	Soluble salts
SAR	Sodium-absorption ratio
SLF	Sanitary landfill
SSD	Sum of square of differences
SAND	Percent sand
SILT	Percent silt
TC	Total carbon
TIC	Total inorganic carbon
TOC	Total organic carbon
TR	Toxicity reduction
USCS	Unified Soil Classification System
USDA	United States Department of Agriculture
WHC	Water holding capacity

Symbols and Signs
General

A	Delay factor
c	Concentration of the element in soil water (M/L^3)
c_o	Concentration of the element in liquid entering the soil profile (or column)(units M/L^3)
c/c_o	Relative concentration (unitless)
D	Apparent diffusion coefficient (also known as dispersion coefficient)(units L^2/T)
D_e	Modified apparent diffusion coefficient (units L^2/T)
I_0	Modified Bessel function of the first kind of zero order
I_1	Modified Bessel function of the first kind of first order
K_1	Forward reaction term (units $1/T$)
K_2	Backward reaction term (units $1/T$)
n	Amounts of element adsorbed per unit volume of soil (units M/L^3)
R_n	Error of integration by Gauss' formula
r^2	Coefficient of determination
t	Time (units)
V_i	Velocity of relative concentration i (units L/T)
v	Pore water velocity, convective velocity (units L/T)
z	Depth in profile (units L)
α	Fractional pore volume of soil (units L^3/L^3)
ρ_b	Bulk density of soil (units g/cm^3)
ρ_p	Particle density of soil (units g/cm^3)

Texture of Soils

c	clay, < 2 μm size particles
c l	clay loam
s	sand
s l	sandy loam
l	loam
si	silt
si l	silt loam
si c l	silty clay loam

Units

a	acre
cfs	cubic feet per second
d	day
dyne	unit of force, in cgs ($g\ cm/s^2$)
eq	equivalent
meq	milliequivalent
meq/100	milliequivalent per 100 grams
g	gram
kg	kilogram (10^3 grams)
lb	pound
mg	milligram (10^{-3} grams)
μg	microgram (10^{-6} grams)

Symbols and Signs
Units (Continued)

h	hour
ℓ	liter (0.946 quarts)
mℓ	milliliter (1 mℓ = 1 cm^3 = 1 c.c.)
m	meter (39.4 inches)
km	kilometer (0.621 miles)
cm	centimeter (0.394 inches)
mm	millimeter
mhos	reciprocal of ohms
mmhos	millimhos
μmhos	micromhos
μmhos/cm	micromhos/cm (measure of EC)
μ	micron
gal	gallon
gal/min	gallon/minute
lb	pound
min	minute
pvd	pore volume displacement
s	second
t	metric ton
t/ha	metric tons/hectare
T	English tons
T/a	English tons/acre

Elements

Ag	silver
Al	aluminum
As	arsenic
Au	gold
B	boron
Be	beryllium
C	carbon
Ca	calcium
Cd	cadmium
Cl	chloride
Co	cobalt
Cr	chromium
Cu	copper
Fe	iron
Fe^{++} (FeII)	ferrous
Fe^{+++} (FeIII)	ferric
H	hydrogen
Hg	mercury
I	iodine
K	potassium
Mg	magnesium
Mn	manganese
Mo	molybdenum

Symbols and Signs
Elements (Continued)

N	nitrogen
Ni	nickel
O	oxygen
P	phosphorus
Pb	lead
Se	selenium
Sn	tin
Si	silicon
Ti	titanium
Zr	zircon
Zn	zinc

Chemicals

CO_2	carbon dioxide
CO_3	carbonate
HCO_3	bicarbonate
NH_4	ammonium
NO_2	nitrite
NO_3	nitrate
PO_4	phosphate
SO_4	sulfate

Index

INDEX

T